AutoCAD Electrical 2022

A Tutorial Approach

(3rd Edition)

CADCIM Technologies
525 St. Andrews Drive
Schererville, IN 46375, USA
(www.cadcim.com)

Contributing Authors
Sham Tickoo
Professor
Department of Mechanical Engineering Technology
Purdue University Northwest
Hammond, Indiana
USA

Arti Deshpande
CADCIM Technologies
USA

CADCIM Technologies

AutoCAD Electrical 2022: A Tutorial Approach
Sham Tickoo

CADCIM Technologies
525 St Andrews Drive
Schererville, Indiana 46375, USA
www.cadcim.com

ISBN 978-1-64057-133-4

NOTICE TO THE READER

Publisher does not warrant or guarantee any of the products described in the text or perform any independent analysis in connection with any of the product information contained in the text. Publisher does not assume, and expressly disclaims, any obligation to obtain and include information other than that provided to it by the manufacturer.

The reader is expressly warned to consider and adopt all safety precautions that might be indicated by the activities herein and to avoid all potential hazards. By following the instructions contained herein, the reader willingly assumes all risks in connection with such instructions.

The Publisher makes no representation or warranties of any kind, including but not limited to, the warranties of fitness for particular purpose or merchantability, nor are any such representations implied with respect to the material set forth herein, and the publisher takes no responsibility with respect to such material. The publisher shall not be liable for any special, consequential, or exemplary damages resulting, in whole or part, from the reader's use of, or reliance upon, this material.

DEDICATION

To teachers, who make it possible to disseminate knowledge
to enlighten the young and curious minds
of our future generations

To students, who are dedicated to learning new technologies
and making the world a better place to live in

THANKS

To the faculty and students of the MET department of
Purdue University Northwest for their cooperation

To Anju Jethwani for copy-editing

Online Training Program Offered by CADCIM Technologies

CADCIM Technologies provides effective and affordable virtual online training on various software packages including Computer Aided Design and Manufacturing and Engineering (CAD/CAM/CAE), computer programming languages, animation, architecture, and GIS. The training is delivered 'live' via Internet at any time, any place, and at any pace to individuals as well as the students of colleges, universities, and CAD/CAM/CAE training centers. The main features of this program are:

Training for Students and Companies in a Classroom Setting

Highly experienced instructors and qualified Engineers at CADCIM Technologies conduct the classes under the guidance of Prof. Sham Tickoo of Purdue University Northwest, USA. This team has authored several textbooks that are rated "one of the best" in their categories and are used in various colleges, universities, and training centers in North America, Europe, and in other parts of the world.

Training for Individuals

CADCIM Technologies with its cost effective and time saving initiative strives to deliver the training in the comfort of your home or work place, thereby relieving you from the hassles of traveling to training centers.

Training Offered on Software Packages

CADCIM Technologies provides basic and advanced training on the following software packages:

***CAD/CAM/CAE**: CATIA, Pro/ENGINEER Wildfire, Creo Parametric, SOLIDWORKS, Autodesk Inventor, Solid Edge, NX, AutoCAD, AutoCAD LT, Customizing AutoCAD, AutoCAD MEP, EdgeCAM, AutoCAD Electrical, and ANSYS*

***Animation and Styling**: Autodesk 3ds Max, 3ds Max Design, Maya, Alias Design, Adobe Flash, Adobe Premiere, and MAXON CINEMA 4D*

***Civil, Architecture, and GIS**: Autodesk Revit Architecture, AutoCAD Civil 3D, AutoCAD Map 3D, Autodesk Revit MEP, Autodesk Navisworks, Bentley STAAD.Pro, Oracle Primavera P6, RISA 3D, MS Project, GIS, and Robo Structural Analysis*

***Computer Programming Languages**: C++, VB.NET, Oracle, AJAX, and Java*

For more information, please visit the following link:

www.cadcim.com

Table of Contents

Chapter 2: Working with Projects and Drawings

Chapter 3: Working with Wires

Chapter 5: Schematic Components

Chapter 6: Schematic Editing

Chapter 7: Connectors, Point-to-Point Wiring Diagrams, and Circuits

Chapter 8: Panel Layouts

This page is intentionally left blank

Preface

AutoCAD Electrical 2022

AutoCAD Electrical, a product of Autodesk, Inc., is one of the world's leading application designed specifically to create and modify electrical control systems. This software incorporates the functionality of AutoCAD along with a complete set of electrical CAD features. In addition, its comprehensive symbol libraries and tools help you automate electrical engineering tasks and save your time and effort considerably, thereby providing you more time for innovation.

Prior to the introduction of this software, the electrical control designers had to rely on generic software applications requiring manual layout of electrical schematics that were often prone to design errors and user could not share design information using these applications. However, with the introduction of AutoCAD Electrical, the chances of error have reduced considerably, thereby enabling you to design 2D industrial controls faster and accurately. Moreover, AutoCAD Electrical 2022 is used to automate various control engineering tasks such as building circuits, numbering wires, creating bill of materials, and many more.

The **AutoCAD Electrical 2022: A Tutorial Approach** is a tutorial based textbook that introduces the readers to AutoCAD Electrical 2022 software, designed specifically for creating professional electrical control drawings. The book has a wide range of tutorials covering the tools and features of AutoCAD Electrical such as schematic drawings, panel drawings, parametric and nonparametric PLC modules, ladder diagrams, Circuit Builder, point-to-point wiring diagrams, report generation, creation of symbols, and so on. These tutorials will enable the users to create innovative electrical control drawings with ease. The chapters in this textbook are arranged in a pedagogical sequence that makes it very effective in learning the features and capabilities of the software. To enhance the knowledge of users, in this edition, the author has added some new tutorials on concepts such as Customizing the Templates and Title block as well as on tools such as **Show Wire Sequence** and **Insert Wblocked Circuit**.

The salient features of this textbook are as follows:

- **Learn-by-doing Approach**
  The author has adopted the tutorial point-of-view and the learn-by-doing approach in this textbook. This approach guides the users through the process of creating and managing electrical control drawings. In addition, one student project and 27 exercises are added in the textbook for the users to practice and apply the skills learned in the text. Note that all tutorials and exercises in this textbook are based on the NFPA(US) standard.

- **Tips and Notes**
  Additional information related to various topics is provided to the users in the form of tips and notes.

- **Learning Objectives**
  The first page of every chapter summarizes the topics covered in that chapter.

- **Self-Evaluation Test, Review Questions, and Exercises**
  Every chapter ends with a Self-Evaluation Test so that the users can assess their knowledge of each chapter. The answers to Self-Evaluation Test are given at the end of the chapter. Also, Review Questions and Exercises are given at the end of each chapter and they can be used by the instructors as test questions and exercises.

Symbols Used in the Textbook

Note
The author has provided additional information related to various topics in the form of notes.

Tip
The author has provided special information to the users in the form of tips.

New — The author has provided this symbol next to the new topics added in this edition of the textbook.

Enhanced — This symbol indicates that the command or tool being discussed is enhanced in the current release.

Formatting Conventions Used in the Textbook

Please refer to the following list for the formatting conventions used in this textbook.

• Command names are capitalized and bold.	Example: The **MOVE** command
• A key icon appears when you have to respond by pressing the ENTER or the RETURN key.	Enter
• Command sequences are indented. The responses are indicated in boldface. The directions are indicated in italics and the comments are enclosed in parentheses.	Command: **MOVE** Select object: **G** Enter group name: *Enter a group name (the group name is group1)*
• The method of invoking a tool/option from the Command prompt is enclosed in a shaded box.	**Command:** LINE or L

Naming Conventions Used in the Textbook

Tool

If you click on an item in a toolbar or a panel of the Ribbon and a command is invoked to create/ edit an object or perform some action, then that item is termed as Tool.

For example:
To Create: **Line** tool, **Circle** tool, **Extrude** tool
To Edit: **Fillet** tool, **Array** tool, **Stretch** tool
Action: **Zoom** tool, **Move** tool, **Copy** tool

If you click on an item in a toolbar or a panel of the Ribbon and a dialog box is invoked wherein you can set the properties to create/edit an object, then that item is also termed as **tool**, refer to Figure 1.

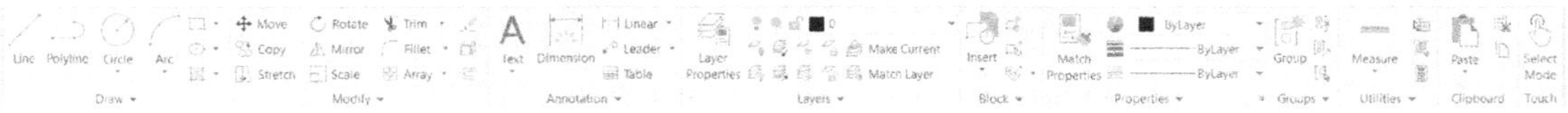

Figure 1 *Various tools in the Ribbon*

For example:
To Create: **Define Attributes** tool, **Create** tool, **Insert** tool
To Edit: **Edit Attributes** tool, **Block Editor** tool

Button

If you click on an item in a Application Status Bar and the display of the corresponding object is toggled on/off, then that item is termed as Button. For example, **Grid** button, **Snap** button, **Ortho** button, **Properties** button, and so on, refer to Figure 2. The item in a dialog box that has a 3D shape like a button is also termed as Button. For example, **OK** button, **Cancel** button, **Apply** button, and so on, refer to Figure 3.

Figure 2 *Various buttons displayed in the Status Bar*

Dialog Box

In this textbook, different terms are used for referring to the components of a dialog box, refer to Figure 3 for the terminology used.

Drop-down

A drop-down is the one in which a set of common tools are grouped together for creating an object. You can identify a drop-down with a down arrow on it. These drop-downs are given a name based on the tools grouped in them. For example, **Edit Components** drop-down, **Modify Wires** drop-down, and so on, refer to Figure 4.

Drop-down List

A drop-down list is the one in which a set of options are grouped together. You can set various parameters using these options. You can identify a drop-down list with a down arrow on it. To know the name of a drop-down list, move the cursor over it; its name will be displayed as a tool tip. For example, **Lineweight** drop-down list, **Linetype** drop-down list, **Object Color** drop-down list, and so on; refer to Figure 5.

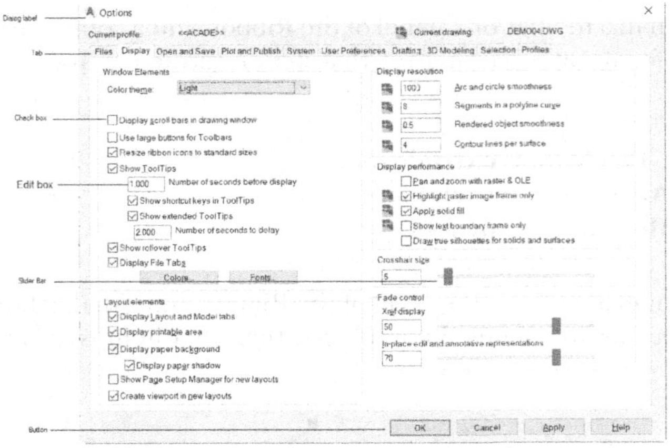

***Figure 3** The components in a dialog box*

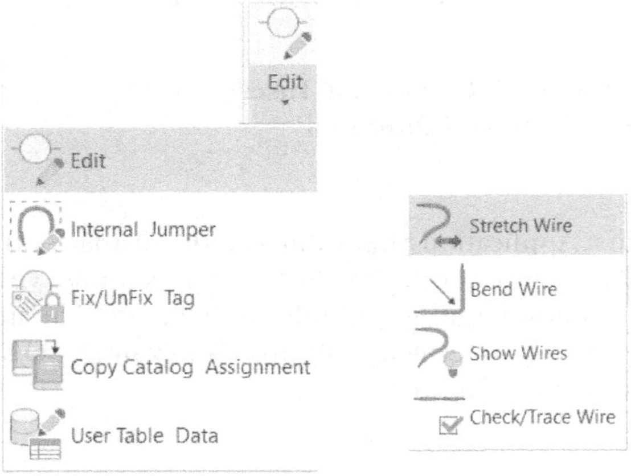

***Figure 4** The **Edit Components** and **Modify Wires** drop-downs*

Options

Options are the items that are available in shortcut menu, drop-down list, Command Prompt, **Properties** panel, and so on. For example, choose the **Properties** option from the shortcut menu displayed on right-clicking on the active project, refer to Figure 6.

Tools and Options in Menu Bar

A menu bar consists of both tools and options. As mentioned earlier, the term **tool** is used to create/edit something or perform some action. For example, in Figure 7, the item **Insert Wire** has been used to create a wire, therefore it will be referred as **Insert Wire** tool.

Similarly, an option in the menu bar is the one that is used to set some parameters. For example, in Figure 8, the item **Zip Project** has been used to zip a project, therefore, it will be referred as an option.

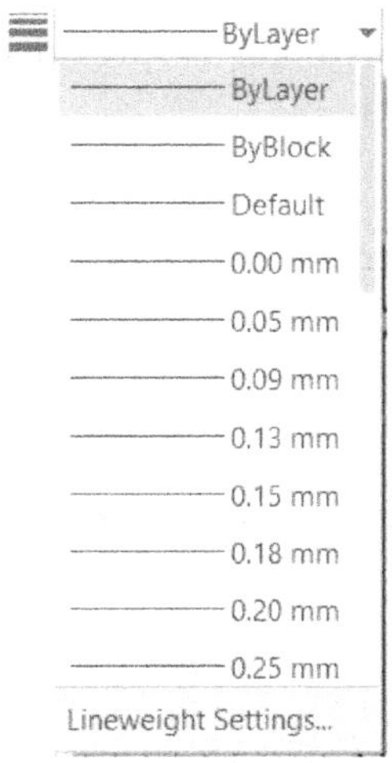

Figure 5 *The **LineWeight** drop-down list*

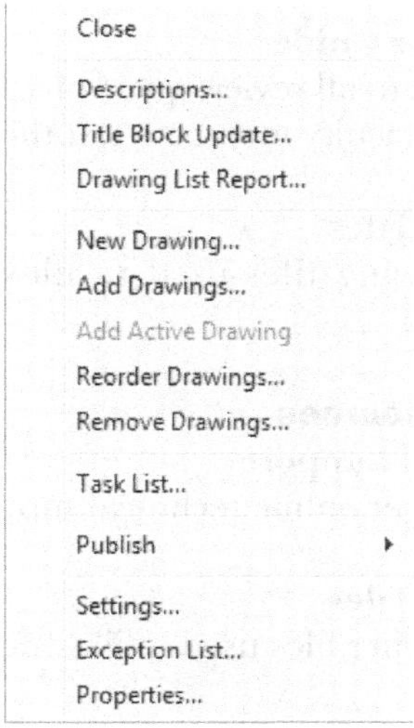

Figure 6 *Options in the shortcut menu*

Figure 7 *Tools in the menu bar*

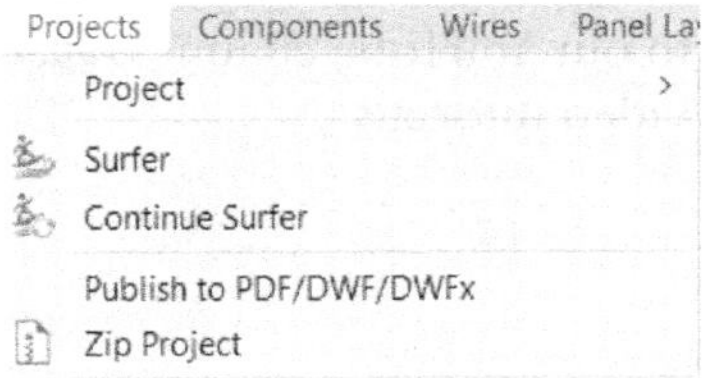

Figure 8 *Options in the menu bar*

Free Companion Website

It has been our constant endeavor to provide you the best textbooks and services at affordable price. In this endeavor, we have come out with a Free Companion website that will facilitate the process of teaching and learning of AutoCAD Electrical 2022. If you purchase this textbook, you will get access to the files on the Companion website.

The resources available for the faculty and students in this website are as follows:

Faculty Resources

- **Technical Support**
  You can get online technical support by contacting ***techsupport@cadcim.com***.

- **Instructor Guide**
  Solutions to all review questions and exercises in the textbook are provided in this guide to help the faculty members test the skills of the students.

- **Drawing Files**
  The drawing files used in illustration, tutorials, and exercises are available for free download.

Student Resources

- **Technical Support**
  You can get online technical support by contacting ***techsupport@cadcim.com***.

- **Drawing Files**
  The drawing files used in illustrations and tutorials are available for free download.

If you face any problem in accessing these files, please contact the publisher at ***sales@cadcim.com*** or the author at ***stickoo@pnw.edu*** or ***tickoo525@gmail.com***.

Video Courses

CADCIM offers video courses in CAD, CAE Simulation, BIM, Civil/GIS, and Animation domains on various e-Learning/Video platforms. To enroll for the video courses, please visit the CADCIM website using the following link: *https://www.cadcim.com/video-courses*

Stay Connected

You can now stay connected with us through Facebook and Twitter to get the latest information about our textbooks, videos, and teaching/learning resources. To stay informed of such updates, follow us on Facebook ***(www.facebook.com/cadcim)*** and Twitter (***@cadcimtech***). You can also subscribe to our YouTube channel ***(www.youtube.com/cadcimtech)*** to get the information about our latest video tutorials.

Chapter 1

Introduction to AutoCAD Electrical 2022

Learning Objectives

After completing this chapter, you will be able to:

- *Install and configure AutoCAD Electrical 2022*
- *Start AutoCAD Electrical*
- *Understand components of the initial AutoCAD Electrical screen*
- *Invoke AutoCAD Electrical commands*
- *Use various commands to save a file*
- *Exit AutoCAD Electrical*
- *Create and manage workspaces*
- *Use various options in AutoCAD Electrical help*

INTRODUCTION

AutoCAD Electrical is a purpose-built controls design software. This software is used to create electrical schematic drawings and panel drawings. AutoCAD Electrical contains various schematic and panel symbols. These symbols, which are mostly the AutoCAD blocks with attributes, carry the intelligence of AutoCAD Electrical drawings. The standard symbol libraries such as JIC, IEC, JIS, and GB, which contain these symbols also get installed with the installation of AutoCAD Electrical. Besides all tools of AutoCAD software, AutoCAD Electrical also contains other electrical tools. You can use these tools for designing control systems speedily, economically, and accurately. As AutoCAD Electrical is compatible with AutoCAD, it is recommended to use the AutoCAD Electrical tools instead of AutoCAD tools when designing electrical circuits. However, AutoCAD Electrical drawings can be edited by using AutoCAD LT or AutoCAD. The all-inclusive symbol libraries and mechanized tasks help in increasing productivity and removing errors, and in providing exact information to the users.

GETTING STARTED WITH AutoCAD Electrical 2022

You can start AutoCAD Electrical by double-clicking on its shortcut icon on the desktop of your computer. You can also load AutoCAD Electrical from the Windows taskbar by using the **Start** button at the bottom left corner of the screen (default position). To do so, choose the **Start** button to display a menu. Next, choose **AutoCAD Electrical 2022 - English > AutoCAD Electrical 2022 - English**, as shown in Figure 1-1; the AutoCAD Electrical 2022 interface is displayed with the **Start** tab chosen by default. In this tab, you need to choose the **New** button to create a new drawing or you can choose the **Open** button to open an existing drawing.

AutoCAD Electrical INTERFACE COMPONENTS

There are various components in the interface of AutoCAD Electrical including the drawing area, **Command window**, **Ribbon**, **Application Menu**, **menu bar**, **Tool Palettes**, **PROJECT MANAGER**, **Model** and **Layout** tabs, Application Status Bar, several toolbars, and so on, refer to Figure 1-2. A title bar containing AutoCAD Electrical symbol and the current drawing name is displayed on top of the screen. Also, the screen has the standard window buttons such as close, minimize, and maximize on the top right corner. These buttons have the same functions as in any other standard window.

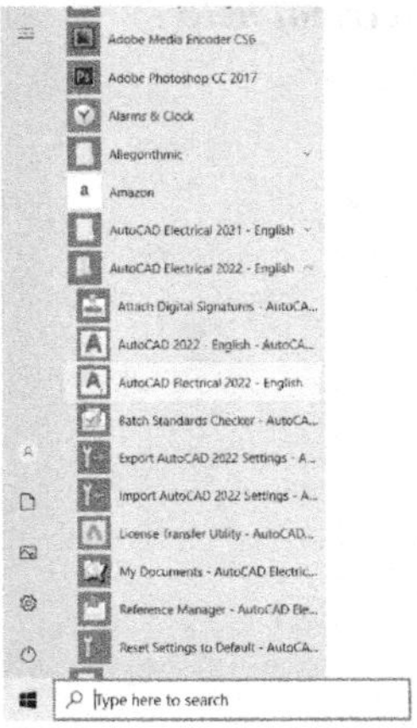

Figure 1-1 *Starting AutoCAD Electrical 2022*

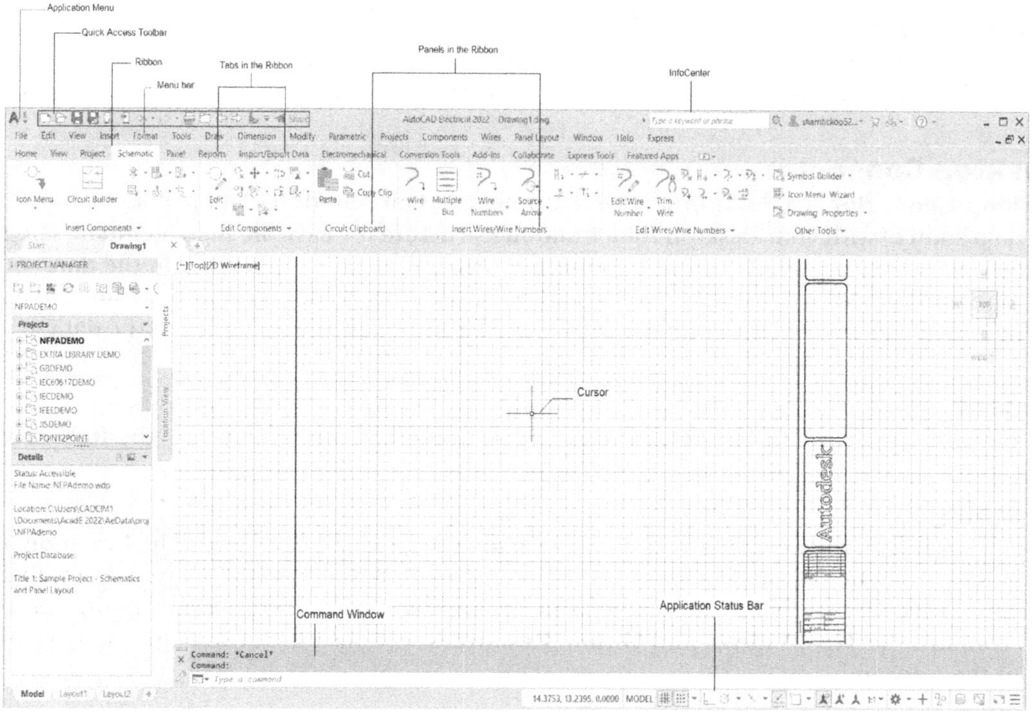

***Figure 1-2** AutoCAD Electrical interface components*

Start Tab

The **Start** tab remains open by default in the initial interface of AutoCAD Electrical 2022. It displays the commonly used options used in AutoCAD Electrical, refer to Figure 1-3. These options are briefly explained next.

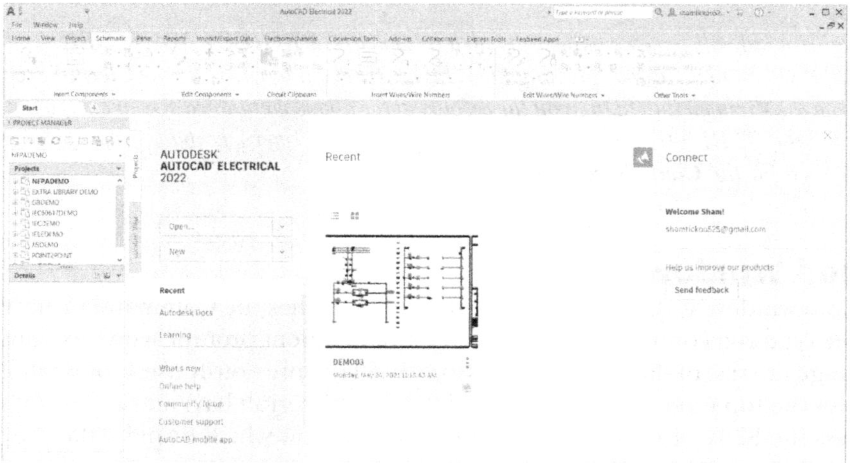

***Figure 1-3** The **Start** tab in AutoCAD Electrical 2022 interface*

Open/New

Choose the **Open** button to open an existing file and choose the **New** button to start a new work from a blank slate or template.

Autodesk Docs

Using **Autodesk Docs,** you can open or save files to the drives connected to your system. Note that you need to have **Desktop Connector** installed to access the connected drives on **Autodesk Docs**.

Desktop Connector

Desktop Connector is a desktop service that integrates an Autodesk data management source (or data source) with your desktop folder and file structure for easy file management. The files in the data source are replicated in a connected drive. You can manage files in the data source through the connected drive just as you would do in any other folder on your machine. Note that changes that you make in the connected drive will be automatically uploaded to the data source.

Recent

When you choose this option, all files you recently worked on are displayed in the **Recent** area.

Learning

When you click on the **Learning** option, the **Learning** page will be displayed. The **Learning** page provides tools to help you learn AutoCAD Electrical, explore the product, learn new or improve existing skills, discover what has changed in the product, or receive relevant notifications. It also contains the **Tips**, **Videos**, and **Online Resources** sections which can help you learn more about the software.

Drawing Area

The drawing area covers the major portion of the screen. In this area, you can draw objects and use commands. To draw objects, you need to define the coordinate points that can be selected by using the pointing device. The position of the pointing device is represented on the screen by the cursor. There is a coordinate system icon at the lower left corner of the drawing area.

> **Tip**
> *You can hide all screen components such as toolbars,* ***PROJECT MANAGER****, and* ***Ribbon*** *displayed on the screen by pressing the CTRL+0 keys or by choosing the* ***Clean Screen*** *option from the* ***View*** *menu. To turn on the display of the above mentioned screen components again, press the CTRL+0 keys. Note that the 0 key on the numeric keypad of the keyboard cannot be used for the* ***Clear Screen*** *option.*

Command Window

The command window at the bottom of the drawing area has the Command prompt where you can enter the required commands. It also displays subsequent prompt sequences and messages. You can change the size of the window and also view all the previously used commands by placing the cursor on the top edge (double line bar known as the grab bar) and then dragging it. You can also press the F2 key to display **AutoCAD Text Window** which displays the previously used commands and prompts.

AutoCorrect the Command Name

If you type a wrong command name at the Command prompt, a suggestion list with most relevant commands will be displayed, refer to Figure 1-4. You can invoke the desired command by selecting the respective option from this list.

AutoComplete the Command Name

When you start typing a command name at the Command prompt, the complete name of the command will be displayed automatically. Also, a list of corresponding commands will be displayed, as shown in Figure 1-5. The commands that have not been used for a long time will be grouped in folders at the bottom of the list.

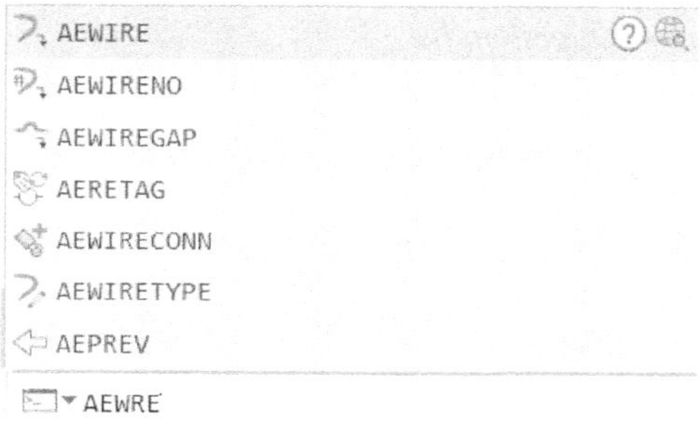

Figure 1-4 Suggestion list with relevant commands

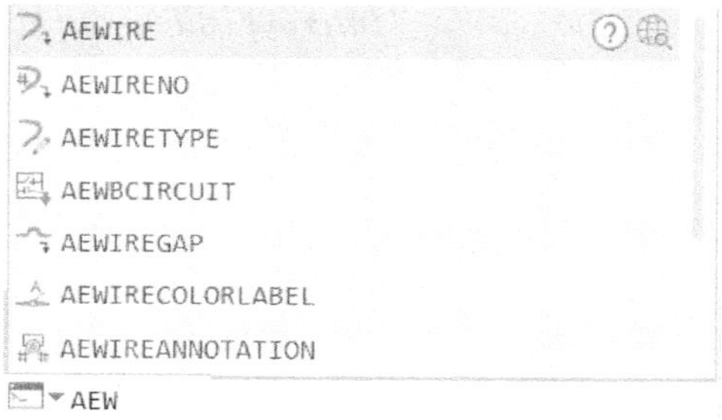

Figure 1-5 Command line displaying complete command name

Internet Search

You can get more information about a command by using the **Search in Help** and **Search on Internet** buttons available adjacent to the command name in the Command line, refer to Figure 1-6. If you choose the **Search in Help** button, the **AutoCAD Electrical 2022 - Help** window will be displayed. In this window, you can find information about the command. By using the **Search on Internet** button, you can find information about the command on the internet.

Input Search Options

In AutoCAD Electrical, you can enable or disable the functions such as AutoComplete and AutoCorrect by using the options available in the **Input Search Options** dialog box. To invoke this dialog box, right-click on the Command prompt; a shortcut menu will be displayed. Next, choose **Input Search Options** from the shortcut menu; the **Input Search Options** dialog box will be displayed, refer to Figure 1-7. Now, you can enable or disable the required functions by using this dialog box.

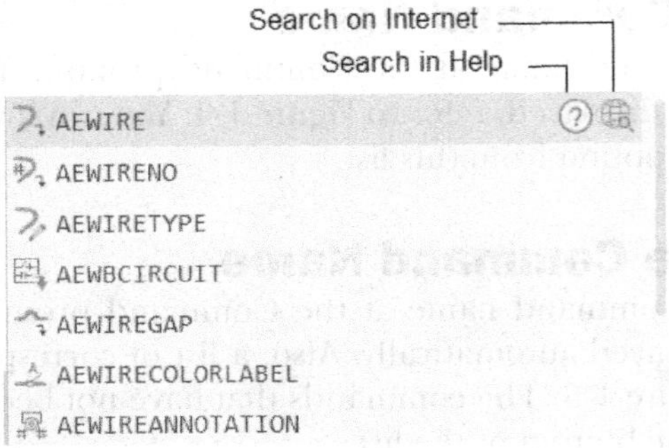

Figure 1-6 *The* ***Search in Help*** *and* ***Search on Internet*** *buttons displayed in the suggestion list*

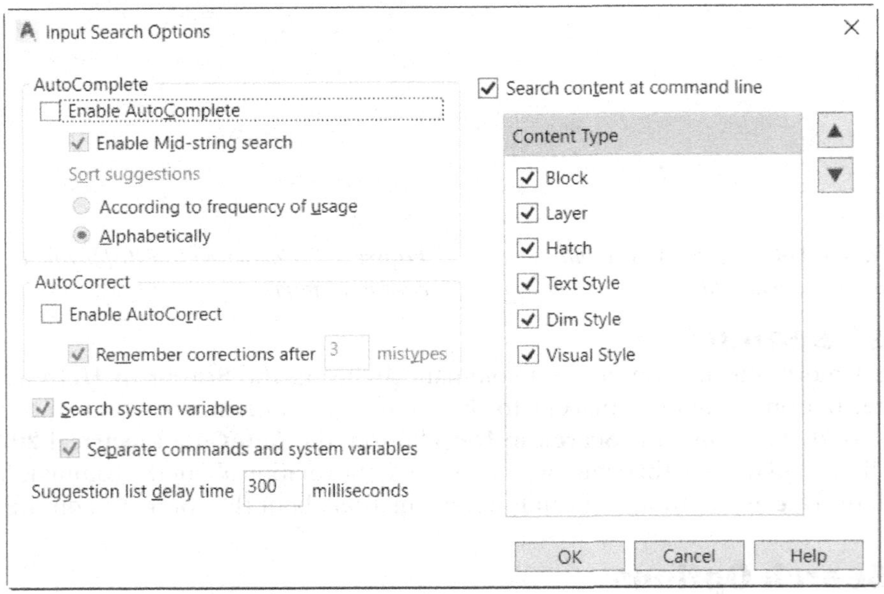

Figure 1-7 *The* ***Input Search Options*** *dialog box*

Application Status Bar

The Application Status Bar is displayed at the bottom of the screen, refer to Figure 1-8. It contains some useful information and buttons that help in changing the status of some AutoCAD and AutoCAD Electrical functions easily. You can toggle between on and off states of most of these buttons by choosing them. Some of these buttons are not available by default. To display/hide buttons in the Application Status Bar, choose the **Customization** button at the extreme bottom right corner of the Application Status Bar; a flyout will be displayed. Next, choose the desired options from this flyout. The most commonly used buttons in the Application Status Bar are discussed next.

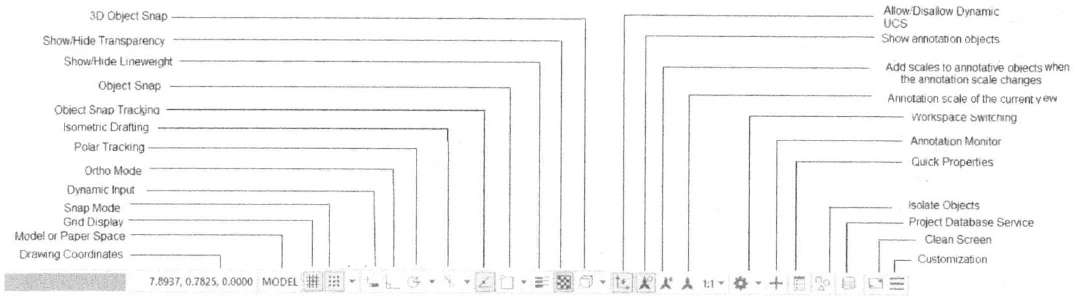

***Figure 1-8** The Application Status Bar*

Drawing Coordinates

The coordinate information is displayed on the left corner of the Application Status Bar. You can choose the **Coordinates** option from the flyout that is displayed on choosing the **Customization** button to turn the coordinate display on and off.

Snap Mode

The snap mode allows you to move the cursor in fixed increments. If the snap mode is on, the **Snap Mode** button will be chosen in the Application Status Bar; otherwise, it will be deactivated. You can also use the F9 function key as a toggle key to turn the snap off or on.

Grid Display

The grid lines are used as reference lines to draw objects in AutoCAD Electrical. Choose the **Grid Display** button to display the grid lines on the screen. The F7 function key can be used to turn the grid display on or off.

Ortho Mode

If the **Ortho Mode** button is chosen in the Application Status Bar, you can draw lines at right angles only. You can use the F8 function key to turn the ortho mode on or off.

Polar Tracking

If you turn the polar tracking on, the movement of the cursor will be restricted along a path based on the angle set in the polar angle settings. Choosing the **Polar Tracking** button in the Application Status Bar turns the polar tracking on or off. You can also use the F10 function key to turn the polar tracking on or off. Note that turning the polar tracking on automatically turns off the ortho mode.

Object Snap

The **Object Snap** button is a toggle button and is used to turn object snap on or off. You can also use the F3 function key to turn the object snap on or off.

3D Object Snap

The **3D Object Snap** button is a toggle button and is used to turn 3D object snap on or off. You can also use the F4 function key to turn the 3D object snap on or off.

Object Snap Tracking

This button is used to turn the object snap tracking on or off. You can also use the F11 function key to turn the object snap tracking on or off.

Dynamic UCS

This button enables you to use the dynamic UCS. You can also use the F6 function key or CTRL + D keys to turn the dynamic UCS on or off.

Dynamic Input

The **Dynamic Input** button is used to turn the **Dynamic Input** mode on or off. You can also use the F12 function key to turn this mode on or off. Turning it on facilitates the heads-up design approach because all commands, prompts, and dimensional inputs will now be displayed in the drawing area and you do not need to look at the Command prompt all the time. This saves the design time and also increases the efficiency of the user. If the **Dynamic Input** mode is turned on, you can enter commands through the **Pointer Input** boxes, and numerical values through the **Dimensional Input** boxes. Also, you can select the command options with the help of **Dynamic Prompt** options in the graphics window.

Show/Hide Lineweight

This button is used to turn the display of lineweights on or off in the drawing. If this button is not chosen, the display of lineweight will be turned off.

Transparency

This button is used to turn on or off the transparency of layers and objects.

Quick Properties

If you select a sketched entity and choose the **Quick Properties** button in the Application Status Bar, the properties of the selected entity will be displayed in a panel.

Model or Paper Space

This button is used to toggle between model space and paper space. The model space is used to work in a drawing area. Paper space is used to prepare your drawing for printing.

Show annotation objects

This button is used to control the visibility of the annotative objects that do not support the current annotation scale in the drawing area.

Add Scales to annotative objects when the annotative scale changes

This button, if chosen, automatically adds all the annotation scales that are set current to all the annotative objects present in the drawing.

Annotation Scale of the current view

This button controls the size and display of the annotative objects in the model space.

Workspace Switching

When you choose the **Workspace Switching** button, a flyout is displayed. This flyout contains predefined workspaces and options such as **Workspace Settings**, **Customize**, and so on. The options in this flyout will be discussed later in detail.

Customization

The **Customization** button is available at the lower right corner of the Application Status Bar and is used to add or remove buttons from the Application Status Bar.

Navigation Bar

In AutoCAD Electrical, the navigation tools are grouped together and are available in the drawing area, as shown in Figure 1-9. The tools in the **Navigation Bar** are discussed next.

Full Navigation Wheel

The **Full Navigation Wheel** has a set of navigation tools that can be used for panning, zooming, and so on.

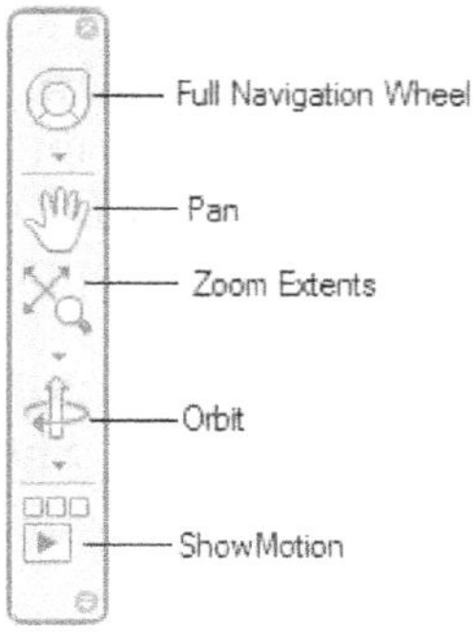

Figure 1-9 *Tools in the* ***Navigation Bar***

Pan

This tool allows you to view the portion of the drawing that is outside the current display area. To do so, choose this button in the Application Status Bar, press and hold the left mouse button and then drag the drawing area. Press ESC to exit this command.

Zoom Extents

Choose one of the tools from the group to zoom the view of the drawing on the screen as per your requirement without affecting the actual size of the objects.

Orbit

This set of tools is used to rotate the view in 3D space.

ShowMotion

Choose this button to capture different views in a sequence and animate them when required.

Project Database Service

The Project Database Service (PDS) updates the scratch project database automatically. It is also used to save all non-AutoCAD Electrical attributes from block files into the project. Also, if the drawing files of a project are not found in the respective folder, an error message will be displayed in the **Project Database Service** message box.

Units

This button is used to display and control the units of drawing. It has a flyout that displays all the unit systems available for drawing.

Lock UI

 This button is used to dock/undock the toolbars, panels, and windows.

Clean Screen

 The **Clean Screen** button is located at the lower right corner of the screen. When you choose this button, all displayed toolbars, except the command window, Application Status Bar, and menu bar, disappear and the expanded view of the drawing is displayed. The expanded view of the drawing area can also be displayed by choosing **View > Clean Screen** from the **Menu Bar** or by using the CTRL+0 keys. Choose the **Clean Screen** button again to restore the previous display state.

INVOKING COMMANDS IN AutoCAD Electrical

When AutoCAD Electrical is started, you can invoke AutoCAD Electrical commands to perform any operation. For example, to draw a wire, first you need to invoke the **AEWIRE** command and then define the start point and endpoint of the wire. Similarly, if you want to trim wires, you must invoke the **AETRIM** command and then select wires for trimming. AutoCAD Electrical provides the following options to invoke a command:

Keyboard	**Ribbon**	**Application Menu**	**Toolbar**
Marking Menu	**Shortcut menu**	**Menu Bar**	**Tool Palettes**

Keyboard

You can invoke any AutoCAD Electrical command from the keyboard by entering the name of the command at the Command prompt and then pressing the ENTER key. If the **Dynamic Input** mode is on and the cursor is in the drawing area, by default the command will be entered through the **Pointer Input** box. The **Pointer Input** box is a small box displayed on the right of the cursor, as shown in Figure 1-10.

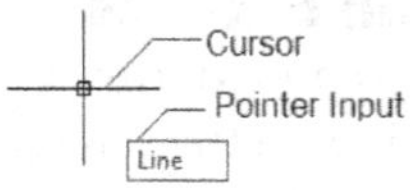

***Figure 1-10** The **Pointer Input** box displayed*

However, if the cursor is currently placed on any toolbar or menu bar, or if the **Dynamic Input** mode is turned off, the command will be entered through the Command prompt. Before you enter a command, the Command prompt is displayed as the last line in the command window area. If it is not displayed, you must cancel the existing command by pressing the ESC key. The following example shows how to invoke the **AEWIRE** command using the keyboard:

Command: **AEWIRE** 

Ribbon

Most of the commands used for creating, modifying, and annotating components are available in the **Ribbon**, as shown in Figure 1-11.

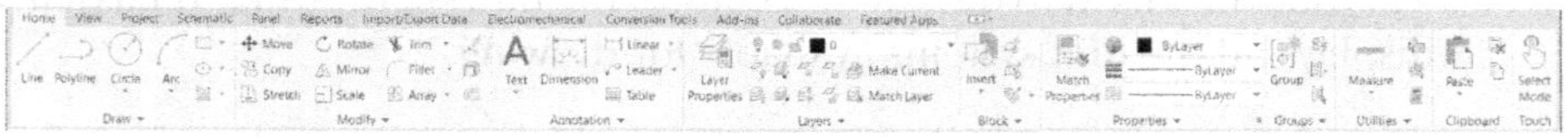

***Figure 1-11** The **Ribbon** for the **ACADE & 2D Drafting & Annotation** workspace*

When you start AutoCAD Electrical session for the first time, by default the **Ribbon** is displayed horizontally below the **Quick Access Toolbar**. The **Ribbon** consists of various tabs. These tabs have different panels, which in turn, have tools arranged in rows. Some of the panels and tools have a small black down arrow. This indicates that the corresponding panels and tools have some more buttons in the form of drop-down. Click on this down arrow to access the hidden tools. If you choose a tool from the drop-down, the corresponding command will be invoked and the tool chosen will be displayed in the panel. For example, to move components using the **Move Component** option, click on the down arrow next to the **Scoot** tool in the **Edit Components** panel of the **Schematic** tab; a drop-down will be displayed. Choose the **Move Component** tool from the drop-down and move components. Choose the down arrow to expand the panel. You will notice that a push pin is available at the left end of the panel. Click on the push pin to keep the panel in the expanded state.

You can reorder panels in a tab. To do so, press and hold the left mouse button on the panel to be moved and then drag it to the required position. To undock the **Ribbon**, right-click on the blank space in the **Ribbon** and choose the **Undock** option from the shortcut menu displayed. You can move, resize, anchor, and auto-hide the **Ribbon** using the shortcut menu that will be displayed when you right-click on the heading strip. To anchor the floating **Ribbon** vertically to the left or right of the drawing area, right-click on its heading strip; a shortcut menu will be displayed. Choose the corresponding option from this shortcut menu. The **Auto-hide** toggle button on the right of the **Express Tools** tab will hide the **Ribbon** into the heading strip and will display it only when you move the cursor over this strip.

You can customize the display of tabs and panels in the **Ribbon**. To do so, right-click on any one of the buttons in the **Ribbon**; a shortcut menu will be displayed. On moving the cursor over one of the options in the shortcut menu, a cascading menu will be displayed with a tick mark before all options indicating that the corresponding tab or panel will be displayed in the **Ribbon**. Select/clear the appropriate option to display/hide a particular tab or panel.

Application Menu

You can invoke various commands from the **Application Menu**. To do so, choose the **Application** button available at the top left corner of the AutoCAD Electrical window; the **Application Menu** will be displayed, refer to Figure 1-12.

You can search for a command using the search field on the top of the **Application Menu**. To search for a command, enter the complete or partial name of the command in the search field; a list displaying all possible commands will be displayed. If you click on a command from the list, the corresponding command will get activated.

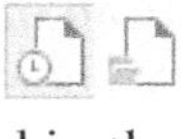

By default, the **Recent Documents** button is chosen in the **Application Menu**. As a result, the recently opened drawings will be listed. If you have opened multiple drawing files, choose the **Open Documents** button; the documents that are opened will be listed in the **Application Menu**. To set the preferences of the file, choose the **Options** button available at the bottom of the **Application Menu**. To exit AutoCAD Electrical, choose the **Exit AutoCAD Electrical** button next to the **Options** button.

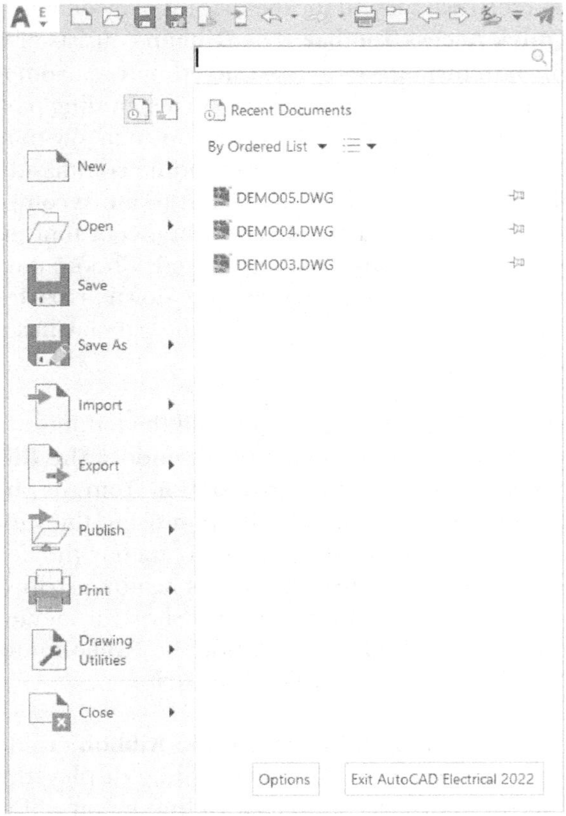

*Figure 1-12 The **Application Menu***

Menu Bar

You can also select commands from the menu bar. Menu bar is not displayed by default. To invoke the menu bar, choose the down arrow in the **Quick Access Toolbar**; a flyout will be displayed. Choose the **Show Menu Bar** option from the flyout; the menu bar will be displayed. As you move the cursor over the menu bar, different titles will be highlighted. You can choose the desired item from the menu bar by clicking on it. Once the item is chosen, the corresponding menu will be displayed directly under the title. Some of the menu items display an arrow on their right indicating that they have a cascading menu. The cascading menu provides various options to execute AutoCAD Electrical commands. You can display the cascading menu by choosing the menu item or by moving the arrow pointer to the right of that item. You can then choose any item from the cascading menu by clicking on it. For example, to insert a ladder, choose **Wires** from the menu bar and then choose the **Ladders** option; a cascading menu will be displayed, as shown in Figure 1-13. From the cascading menu, choose the **Insert Ladder** option.

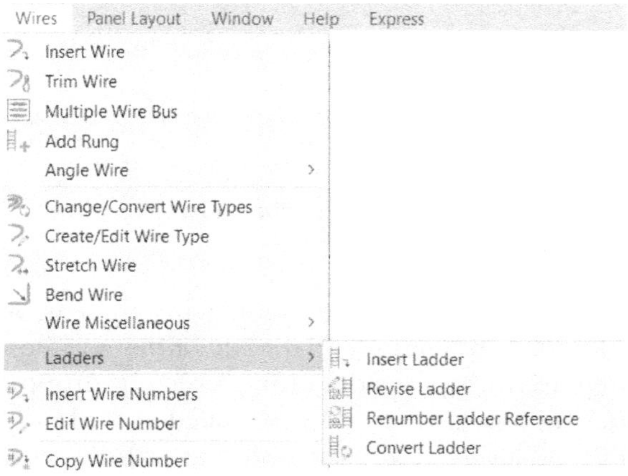

***Figure 1-13** The cascading menu*

Toolbar

Toolbars are not displayed by default. To display a toolbar, first invoke the menu bar and then choose **Tools > Toolbars > ELECTRICAL** from it; the list of toolbars will be displayed. Select the required toolbar. In a toolbar, a set of tools representing various AutoCAD Electrical commands are grouped together. When you move the cursor over a tool in a toolbar, the tool will be lifted. The tooltip (name of the tool) and a brief description related to that tool will also be displayed below the tool. Once you locate the desired tool, the command associated with it can be invoked by choosing it. For example, you can invoke the **AEWIRE** command by choosing the **Insert Wire** tool from the **ACE:Main Electrical** toolbar.

Some of the tools in a toolbar have a small triangular arrow at its lower right corner. This arrow indicates that the tool has a flyout attached to it. If you hold cursor on the triangular arrow button, a flyout containing the options for the command will be displayed, as shown in Figure 1-14. Choose the desired option from the toolbar; a command will be displayed in the command window. The **ACE:Main Electrical** and **ACE:Panel Layout** toolbars are shown in Figures 1-15 and 1-16, respectively.

***Figure 1-14** The flyout*

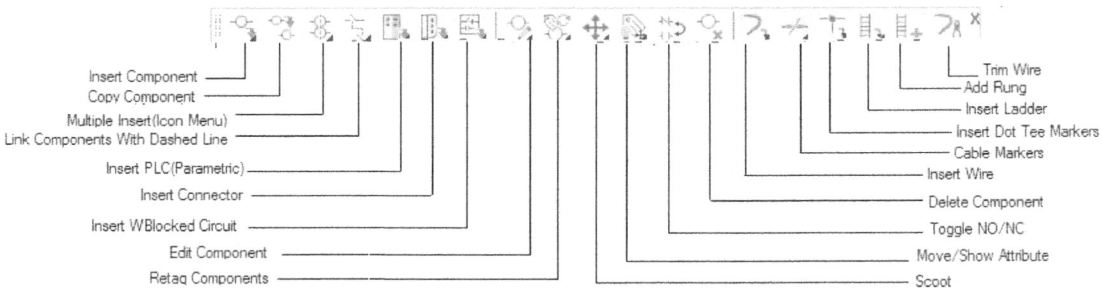

***Figure 1-15** The **ACE:Main Electrical** toolbar*

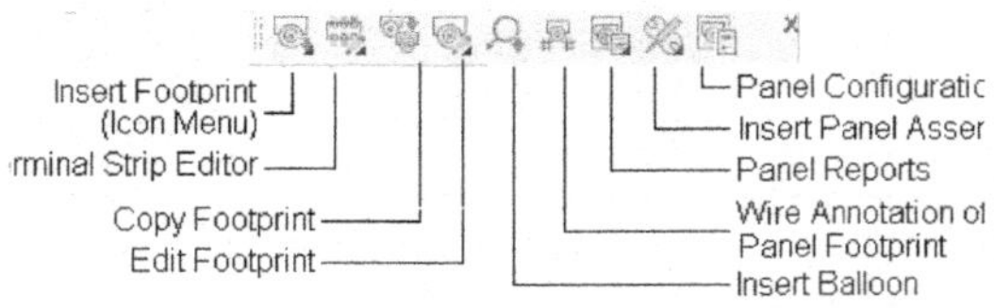

Figure 1-16 *The* ***ACE:Panel Layout*** *toolbar*

Moving and Resizing Toolbars

Toolbars can be moved anywhere on the screen by placing the cursor on the strip and then dragging it to the desired location. You must hold the cursor down while dragging. While moving toolbars, you can dock them to the top or sides of the screen by dropping them in the docking area. You can prevent docking toolbars, as and when needed, by holding the CTRL key while moving the toolbar to a desired location. You can also change the size of toolbars by placing the cursor anywhere on the border of the toolbar where it takes the shape of a double-sided arrow, as shown in Figure 1-17, and then pulling it in the required direction, as shown in Figure 1-18. You can also customize toolbars to meet your requirements.

Figure 1-17 *Resizing the* ***ACE:Main Electrical*** *toolbar*

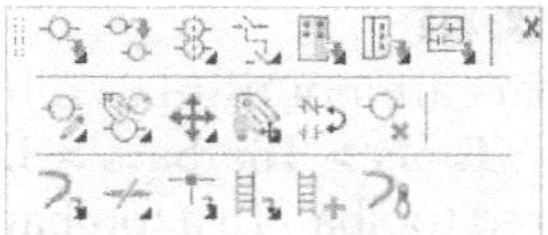

Figure 1-18 *The* ***ACE:Main Electrical*** *toolbar resized*

Marking Menu

AutoCAD Electrical provides you with marking menus which are displayed on right-clicking on an AutoCAD Electrical Object. By using the marking menus, you can easily invoke the commands, if the toolbar is not displayed in the drawing area. These marking menus are context-sensitive, which means that the commands or tools in this menu will be displayed based on the object selected. Figure 1-19 shows a marking menu that is displayed on right-clicking on a component and Figure 1-20 shows a marking menu that is displayed on right-clicking on a wire. Figure 1-21 shows a marking menu that is displayed on right-clicking on a footprint. There are two basic modes for command selection: Menu mode and Mark mode. These modes are discussed next.

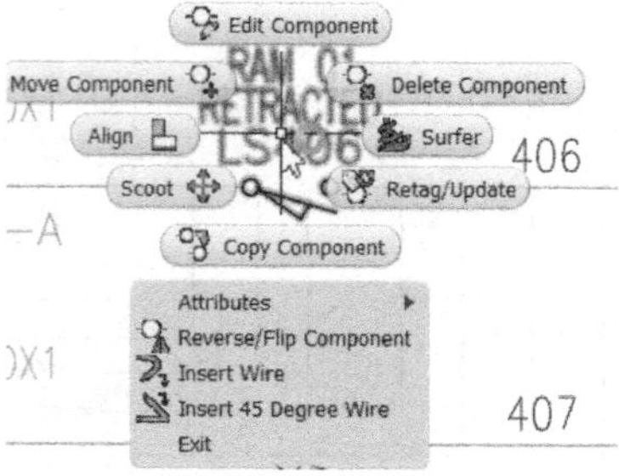

Figure 1-19 *The marking menu displayed on right-clicking on a component*

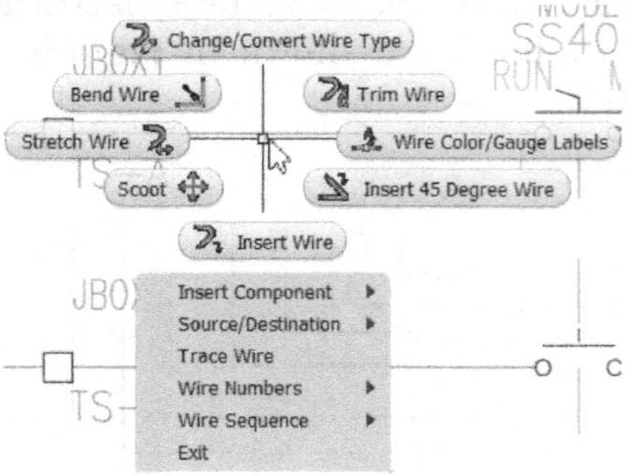

Figure 1-20 *The marking menu displayed on right-clicking on a wire*

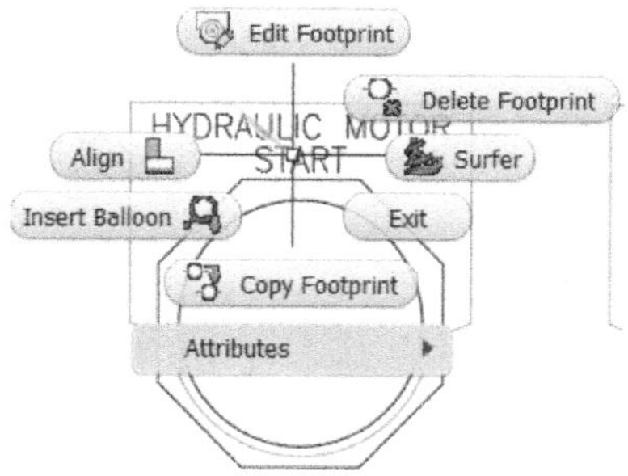

Figure 1-21 *The marking menu displayed on right-clicking on a footprint*

Menu Mode

You can invoke the Menu mode by right-clicking on an object. When you invoke this mode, a marking menu will be displayed, showing the commands related to that object. Move the cursor over the command to be executed and click on it; the selected command will be executed. To exit the marking menu, click at the center of the menu or click anywhere outside the menu. If you press the ESC key, the command in progress will be cancelled.

Mark Mode

Mark mode is similar to Menu mode. The only difference between them is that in this mode, you need to immediately move the cursor along the direction of the desired command after right-clicking on the component. On doing so, a rubber band line connected with the cursor is displayed. Release the mouse button on the command to be executed.

Shortcut Menu

If you right-click in the drawing area, the AutoCAD shortcut menu will be displayed. This shortcut menu contains the commonly used commands of Windows and an option to select the previously invoked commands again, as shown in Figure 1-22. If you right-click in the drawing area while a command is active, a shortcut menu will be displayed containing the options of that particular command. Figure 1-23 shows the shortcut menu displayed while the **AEWIRE** command is active.

When you right-click on the command window, a shortcut menu will be displayed. This shortcut menu displays the six most recently used commands and some of the window options like **Copy** and **Paste**, as shown in Figure 1-24.

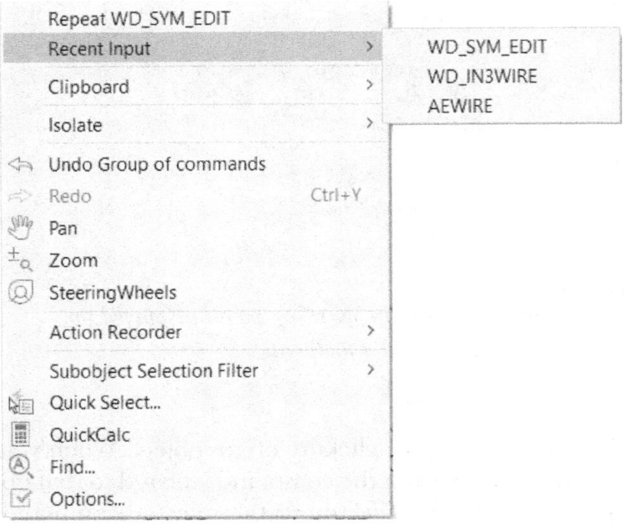

***Figure 1-22** The shortcut menu showing the recently used commands*

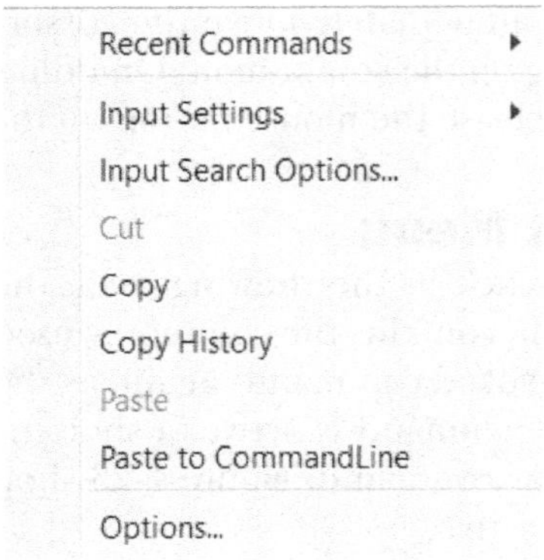

***Figure 1-23** The shortcut menu displayed on right-clicking in the drawing area with the **AEWIRE** command active*

***Figure 1-24** Command window shortcut menu*

The commands and their prompt entries are displayed in the History window (previous command lines not visible) and can be selected, copied, and pasted in the command line using the shortcut menu. As you press the up arrow key, the previously entered commands will be displayed in the command window. Once the desired command is displayed at the Command prompt, you can execute it by simply pressing the ENTER key.

When you right-click in the coordinate display area of the Status Bar, a shortcut menu will be displayed. This shortcut menu contains options to modify the display of coordinates, refer to Figure 1-25. You can also right-click on any of the toolbars to display a shortcut menu from where you can choose any toolbar to be displayed.

Tool Palettes

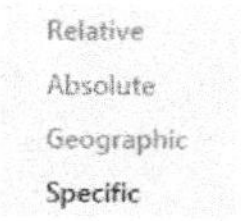

Figure 1-25 The shortcut menu displayed

The **Tool Palettes**, as shown in Figure 1-26, is an easy and convenient way of placing components in the current drawing. By default, the **Tool Palettes** is not displayed. To invoke the **Tool Palettes**, choose **Tools > Palettes > Tool Palettes** from the menu bar or choose the CTRL+3 keys to display the **Tool Palettes** as a window on the left of the drawing area. You can resize the **Tool Palettes** by using the resizing cursor that is displayed when you place the cursor on the top or bottom of the **Tool Palettes**. The **Tool Palettes** contains different commands for inserting components in Imperial and Metric units. When you move the **Tool Palettes** in the drawing area and right-click on its title bar, a shortcut menu is displayed, as shown in Figure 1-27. Using this shortcut menu, you can turn on or off the **Tool Palettes**. Also, you can move, change size, close, auto-hide, and dock the **Tool Palettes**. Also, you can create new palette, rename it, and customize palettes and commands by choosing the desired option from the shortcut menu.

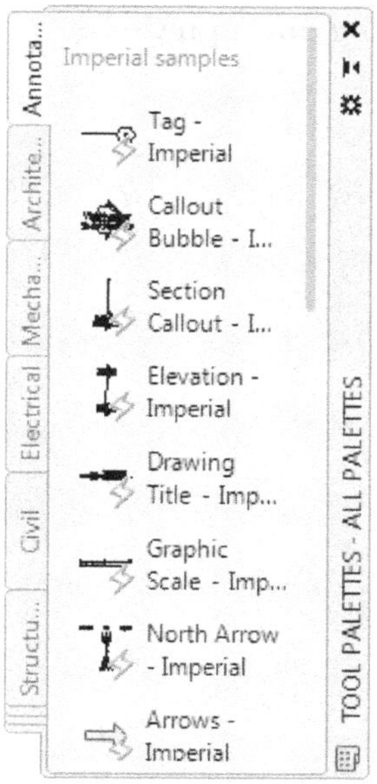

*Figure 1-26 The **Tool Palettes***

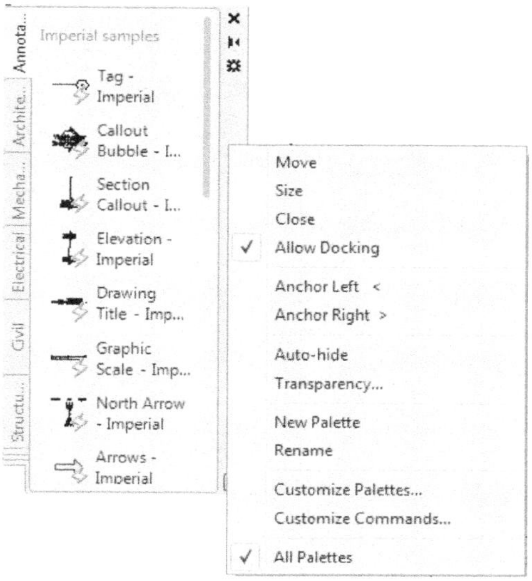

Figure 1-27** Shortcut menu displayed on right-clicking on the title bar of the **Tool Palettes

File Tabs

The **File Tabs** button is available in the **Interface** panel of the **View** tab. It is used to toggle the display of the File tab bar which displays all opened files. You can easily switch between multiple opened drawings by clicking on them.

Note

*The **View** tab is not displayed by default in the AutoCAD Electrical interface. To display it, right-click on the **Ribbon**; a shortcut menu will be displayed. Next, choose **Show Tabs > View** from the shortcut menu.*

You can also create a new drawing file by clicking on the (+) sign available at the end of the file tabs. Figure 1-28 shows the **File Tabs** button chosen in the **Ribbon** and the File tab bar displayed at the bottom of the **Ribbon**.

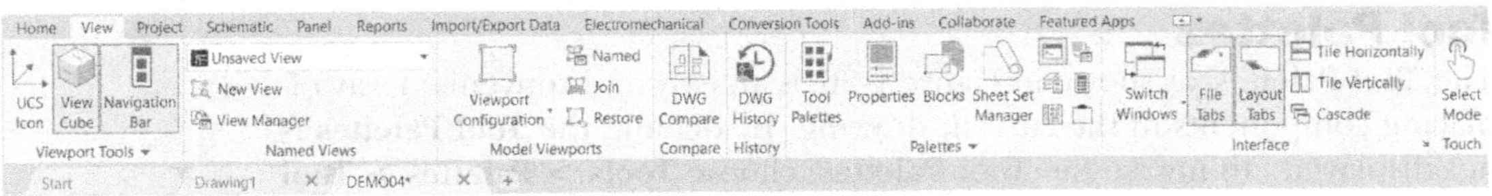

*Figure 1-28 The **File Tabs** button chosen in the **Ribbon** and File tab bar displayed at the bottom of the **Ribbon***

In the File tab bar, all the added tabs get arranged in a sequence in which the respective drawings are created. You can change the sequence of the tabs in the File tab bar by using the left mouse button. To do so, press and hold the left mouse button on any tab and drag it to the desired location. If a large number of files are opened, some of the files will not be visible in the File tab bar and therefore an overflow symbol will be displayed on the right end of the File tab bar, refer to Figure 1-29. To open a tab which is not visible in the File tab bar, click on the overflow symbol; the names of all the tabs will be displayed in a flyout, refer to Figure 1-29. Also, when you move the cursor on a tab name, previews of the Model and Layouts will be displayed, refer to Figure 1-29. You can open the desired environment by clicking on its preview.

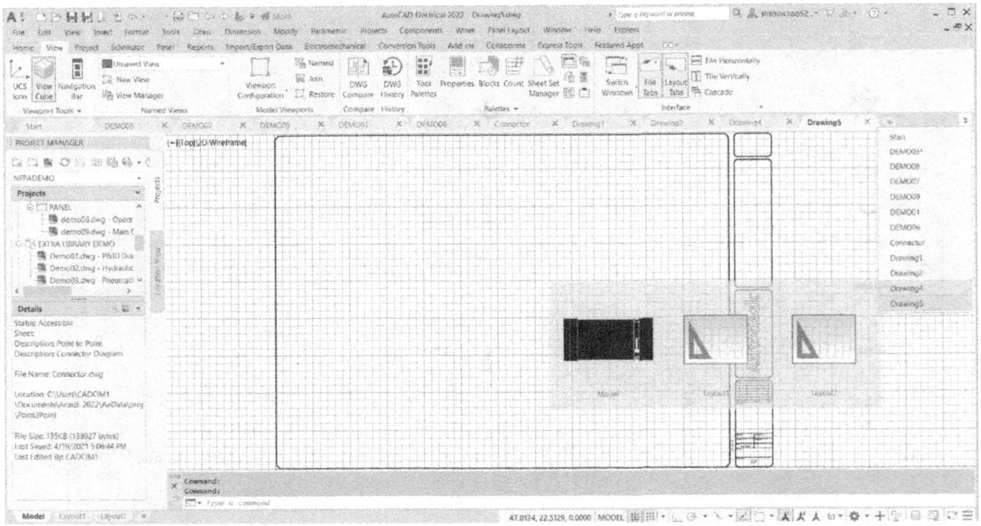

***Figure 1-29** Flyout with file tab names and preview of their respective drawings*

If you move the cursor over a file tab, the preview of the model and layout will be displayed. When you move the cursor over any preview in the file tab, the corresponding preview will be displayed in an enlarged form in the drawing area, refer to Figure 1-30.

There are two buttons available on the top of the preview window: **Plot** and **Publish**. By using **Plot**, you can plot the drawing and by using **Publish**, you can publish the drawing. When you right-click on a file tab, a shortcut menu containing various options such as **New**, **Open**, **Save**, **Save As**, **Close**, and so on will be displayed, refer to Figure 1-31. You can choose the option from the shortcut menu as per your requirement.

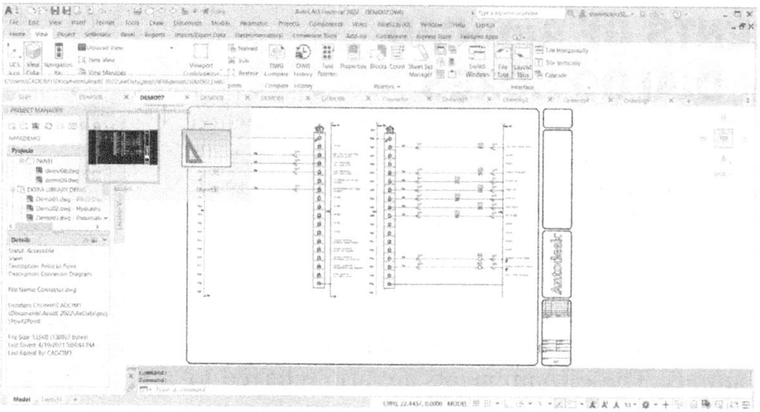

Figure 1-30 *Previews of model and layout*

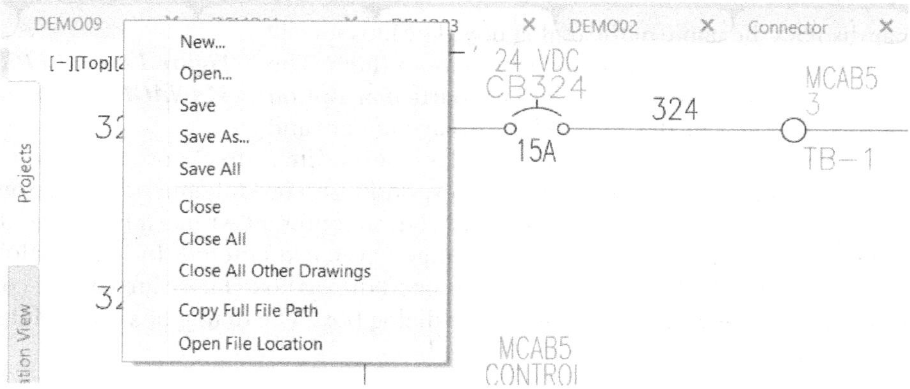

Figure 1-31 *Shortcut menu displayed on right-clicking on the File tab bar*

There are two icons displayed on the file tab: Asterisk and Lock. The Asterisk icon indicates that the file is modified but not saved. The Lock icon indicates that the file is locked and the changes cannot be saved with the original file name, although you can use the **SaveAs** tool to create another copy.

To open a drawing as a locked file, first choose the **Open** option from the shortcut menu displayed on right-clicking over the file tab; the **Select File** dialog box will be displayed. Select the desired file and then select the **Open Read-Only** option from the **Open** drop-down list. On doing so, the file will be opened as a locked file in the drawing area. You can also open the file as a locked file by using the **Open** button from the **Quick Access Bar**.

PROJECT MANAGER

The **PROJECT MANAGER** is used to create new projects, add new drawings to a project, re-order drawing files, access the existing projects, or modify the existing information in a project, refer to Figure 1-32. By default, the **PROJECT MANAGER** is opened and docked on the left of your screen. The **PROJECT MANAGER** displays a list of projects. Using the **PROJECT MANAGER**, you can open, activate, edit, and close projects. The **PROJECT MANAGER** is discussed in detail in Chapter 2.

COMPONENTS OF AutoCAD Electrical DIALOG BOXES

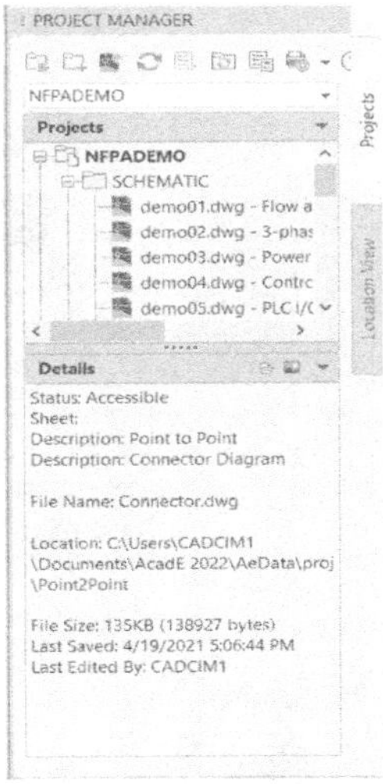

Figure 1-32 *The* ***PROJECT MANAGER***

In AutoCAD Electrical, there are certain commands, which when invoked, display a dialog box. When you choose an item with ellipses [**...**] from the menu bar, a dialog box will be displayed. For example, when you choose **Options** from the **Tools** menu, the **Options** dialog box will be displayed.

A dialog box contains a number of parts like dialog label, radio buttons, text or edit boxes, check boxes, slider bars, image boxes, and command buttons. These components are also referred to as tiles. Some of the components of a dialog box are shown in Figure 1-33.

The title bar displays the name of the dialog box. The tabs specify various sections with a group of related options under them. The check boxes are toggle buttons for making a particular option available or unavailable. The drop-down list displays an item and an arrow on the right which when selected displays a list of items to choose from. You can also select a radio button to activate the option corresponding to it. Only one radio button can be selected at a time. The text box is an area where you can enter a text like a file name. It is also called an edit box because you can make any change to the text entered. In some dialog boxes, there is the [**...**] button, which displays another related dialog box. There are certain command buttons (**OK**, **Cancel**, **Help**) at the bottom of the dialog box. The dialog box has a **Help** button for getting help on various features of the dialog box.

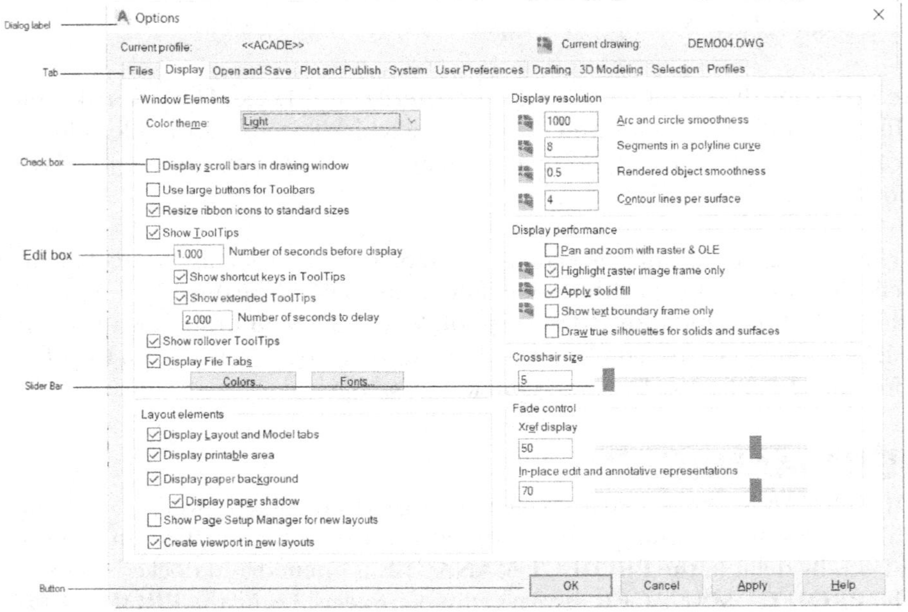

Figure 1-33 *Components of a dialog box*

SAVING THE WORK

Application Menu:	SAVE, SAVEAS
Toolbar:	Quick Access Toolbar > Save
Menu:	File > Save or Save As
Command:	QSAVE, SAVEAS, SAVE

In AutoCAD Electrical, you need to save your work before you exit from the drawing editor or turn off the system. Also, it is recommended that you save your drawings after regular time intervals, so that in the event of a power failure or an editing error, all work done by you is not lost and only the unsaved part is affected.

AutoCAD Electrical has provided the **QSAVE**, **SAVEAS**, and **SAVE** commands that allow you to save your work on the hard disk of a computer. These commands allow you to save your drawing by writing it to a permanent storage device such as a hard drive, or a diskette in any removable drive.

When you choose **Save** from the **Quick Access Toolbar** or **Application Menu**, the **QSAVE** command is invoked. If the current drawing is unnamed and you save the drawing for the first time in the present session, the **QSAVE** command will prompt you to enter the file name in the **Save Drawing As** dialog box. You can enter a name for the drawing and then choose the **Save** button. If you have saved a drawing file once and then edited it, you can use the **QSAVE** command to save it. This allows you to do a quick save.

When you invoke the **SAVEAS** command, the **Save Drawing As** dialog box will be displayed, refer to Figure 1-34. Even if the drawing has been saved with a file name, this command provides you with an option to save it with a different file name. In addition to saving the drawing, it allows you to set a new name for the drawing, which is displayed in the title bar.

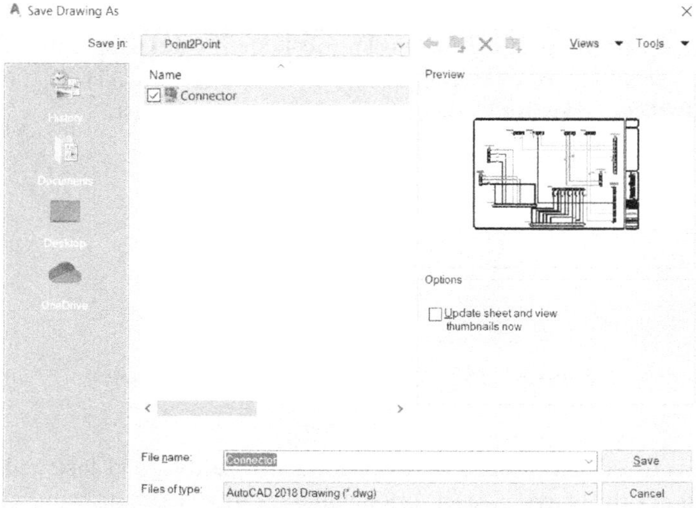

Figure 1-34 *The* ***Save Drawing As*** *dialog box*

This command is used when you want to save a previously saved drawing under a different file name. You can also use this command when you make certain changes to a template and want to save the changed template drawing without changing the original template.

The **SAVE** command is the most rarely used command and can be invoked only from the command line by entering **SAVE** at the Command prompt. This command is similar to the **SAVEAS** command and displays the **Save Drawing As** dialog box when invoked. With this command, you can save a previously saved drawing under a different file name.

Save Drawing As Dialog Box

The **Save Drawing As** dialog box displays the information related to drawing files on your system. Various options in this dialog box are described next.

Places List

A column of icons is displayed on the left in the dialog box. These icons contain shortcuts to the folders that are frequently used. You can quickly save your drawings in one of these folders. The **History** folder displays the list of the most recently saved drawings. The **FTP** folder displays the list of various FTP sites available for saving a drawing. By default, no FTP sites are shown in the dialog box. To add a FTP site to the dialog box, choose the **Tools** button at the upper-right corner of the dialog box to display a shortcut menu and select **Add/Modify FTP Locations**. The **Desktop** folder displays the list of contents on the desktop. The **Buzzsaw** icons connect you to their respective pages on a Web.

File name

To save your work, enter the name of the drawing in the **File name** edit box by typing the file name or selecting it from the drop-down list.

Files of type

The **Files of type** drop-down list, as shown in Figure 1-35, is used to specify a drawing format in which you want to save a file. For example, to save a file as an AutoCAD 2004 drawing file, select **AutoCAD 2004/LT 2004 Drawing (*.dwg)** from this drop-down list.

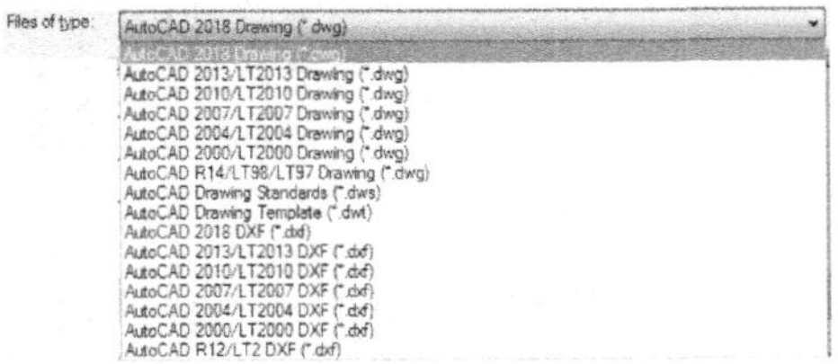

*Figure 1-35 The **Files of type** drop-down list*

Save in

The active project is listed in the **Save in** drop-down list. AutoCAD Electrical initially saves the drawing in the folder of the active project. But if you want to save the drawing in a different folder, you need to specify the path.

Views

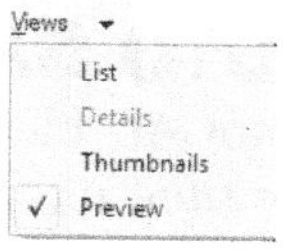

Figure 1-36 The Views flyout

The options in the **Views** flyout are used to specify how the list of files will be displayed in the **Save Drawing As** dialog box, refer to Figure 1-36. These options are discussed next.

List, Details, and Thumbnails Options

The **Files** list box displays the drawing files of a project. If you choose the **Details** option, it will display the detailed information about files (size, type, date, and time of modification) in the **Files** list box. In the detailed information, if you click on the **Name** label, the files will be listed with names in alphabetical order. If you again click on the **Name** label, the files will be listed in the reversed order. Similarly, if you click on the **Size** label, the files will be listed according to their size in ascending order. Double-clicking on the **Size** label will list the files in descending order of their size. Similarly, you can click on the **Type** label or the **Modified** label to list the files accordingly. If you choose the **List** option from the **Views** flyout, all files in the current folder will be listed in the **File** list box.

> **Tip**
> *The file name you enter to save a drawing should match its contents. This helps you to remember drawing details and makes it easier to refer to them later. Also, the file name can be 255 characters long and can contain spaces and punctuation marks.*

Create New Folder

If you choose the **Create New Folder** button, AutoCAD Electrical will create a new folder with the name **New Folder**. The new folder is displayed in the **File** list box. You can accept the name or change it as per your requirement. Alternatively, press ALT+5 to create a new folder.

Up one level

The **Up one level** button is used to display the folders that are up by one level. Alternatively, press ALT+2 to display the folder.

Tools

The **Tools** flyout has an option for adding or modifying the FTP sites, refer to Figure 1-37. These sites can then be browsed from the FTP shortcut in the **Places** list. The **Add Current Folder to Places** and **Add to Favorites** options are used to add the folder displayed in the **Save in** edit box to the **Places** list or to the **Favorites** folder.

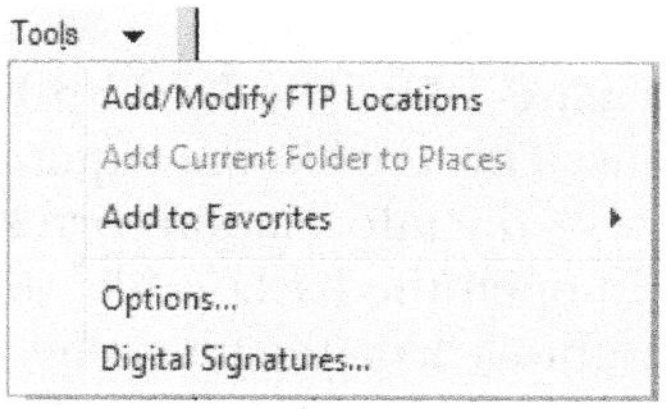

*Figure 1-37 The **Tools** flyout*

The **Options** button, when chosen, displays the **Saveas Options** dialog box where you can save the proxy images of custom objects. This dialog box has the **DWG Options** and **DXF Options** tabs. The **Display Signatures** button displays the **Security Options** dialog box that is used to configure the security options of a drawing.

AUTO SAVE

AutoCAD Electrical allows you to save your work automatically at specific intervals. To change the time intervals, you can enter the interval's duration in minutes in the **Minutes between saves** text box in the **File Safety Precautions** area of the **Options** dialog box (**Open and Save** tab). This dialog box can be invoked from the **Tools** menu. Depending on the power supply, hardware, and type of drawings, you should decide on an appropriate time and assign it to this variable. AutoCAD Electrical saves the drawing with the file extension *.sv$*. You can also change the time interval by using the **SAVETIME** system variable.

Tip

*Although the automatic save option saves your drawing after a certain time interval, you should not completely depend on it because the procedure for converting the sv$ file into a drawing file is cumbersome. Therefore, it is recommended that you save your files regularly by using the **QSAVE** or **SAVEAS** command.*

CREATING BACKUP FILES

If a drawing file already exists and you use the **SAVE** or **SAVEAS** command to update the current drawing, AutoCAD Electrical will create a backup file. AutoCAD Electrical takes the previous copy of the drawing and changes it from a file type *.dwg* to *.bak*, and the updated drawing is saved as a drawing file with the *.dwg* extension. For example, if the name of a drawing is *myproj.dwg*, AutoCAD Electrical will change the name to *myproj.bak* and save the current drawing as *myproj.dwg*.

Using the Drawing Recovery Manager to Recover Files

The automatically saved files can also be retrieved using the **Drawing Recovery Manager**. If the system crashes accidently and the automatic save operation is performed on a drawing, the **Drawing Recovery** dialog box will be displayed when AutoCAD Electrical is run the next time, as shown in Figure 1-38.

The dialog box informs you that the program unexpectedly failed and you can open the most suitable file from the backup files created by AutoCAD Electrical. To open the most suitable file among the backup files, choose the **Close** button from the **Drawing Recovery** dialog box; the **Drawing Recovery Manager** will be displayed on the left of the drawing area, as shown in Figure 1-39. The **Backup Files** rollout lists the original files, backup files, and automatically saved files. Select the required file; the preview of the file will be displayed in the **Preview** rollout. Also, the information corresponding to the selected file will be displayed in the **Details** rollout. To open the backup file, double-click on its name in the **Backup Files** rollout. Alternatively, right-click on the file name, and then choose **Open** from the shortcut menu. It is recommended that you save the backup file at the desired location before you start working on it.

CLOSING A DRAWING

You can close the current drawing file without actually quitting AutoCAD Electrical by using the **CLOSE** command. If multiple drawing files are opened, choose **Close > All Drawings** from the **Application Menu**; the drawings will be closed. If you have not saved the drawing after making the last change to it and you invoke the **CLOSE** command, AutoCAD Electrical will display a dialog box that allows you to save the drawing before closing it. This box provides you with an option to discard the current drawing or the changes made to it. It also provides you with an option to cancel the command. After closing the drawing, you are still in AutoCAD Electrical from where you can open a new or an already saved drawing file. You can also use the close button (**X**) to close the drawing.

Note
You can close a drawing even if a command is active.

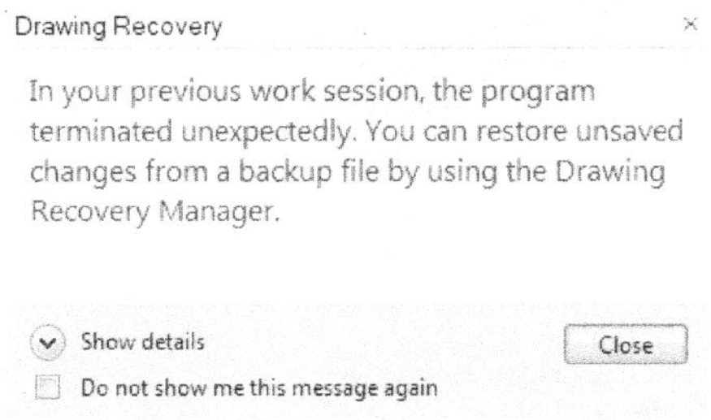

Figure 1-38 *The* ***Drawing Recovery*** *dialog box*

Figure 1-39 *The* ***Drawing Recovery Manager***

QUITTING AutoCAD Electrical

You can exit the AutoCAD Electrical program by using the **EXIT** or **QUIT** command. Even if you have an active command, you can choose **Exit AutoCAD Electrical** from the **Application Menu** to quit the AutoCAD Electrical program. In case the drawing has not been saved, the **AutoCAD** message box will be displayed. Note that if you choose **No** in this message box, all changes made in the current drawings, till they were last saved, will be lost. You can also use the close button (**X**) of the main AutoCAD Electrical window (present in the title bar) to end the AutoCAD Electrical session.

Tip

You can open the ***Drawing Recovery Manager*** *again by choosing* ***Drawing Utilities > Open the Drawing Recovery Manager*** *from the* ***Application Menu*** *or by entering DRAWINGRECOVERY at the Command prompt.*

DYNAMIC INPUT MODE

As mentioned earlier, turning the **Dynamic Input** mode on allows you to enter commands through the pointer input and dimensions using the dimensional input. When this mode is turned on, all prompts will be available at the tooltip as dynamic prompts. The settings for the **Dynamic Input** mode are made through the **Dynamic Input** tab of the **Drafting Settings** dialog box. To invoke the **Drafting Settings** dialog box, right-click on the **Dynamic Input** button in the Status Bar; a shortcut menu will be displayed. Choose the **Dynamic Input Settings** option from the shortcut menu; the **Drafting Settings** dialog box will be displayed, as shown in Figure 1-40. Alternatively, enter **DSETTINGS** at the Command prompt to display the **Drawing Settings** dialog box. The options in the **Dynamic Input** tab of the **Drafting Settings** dialog box are discussed next.

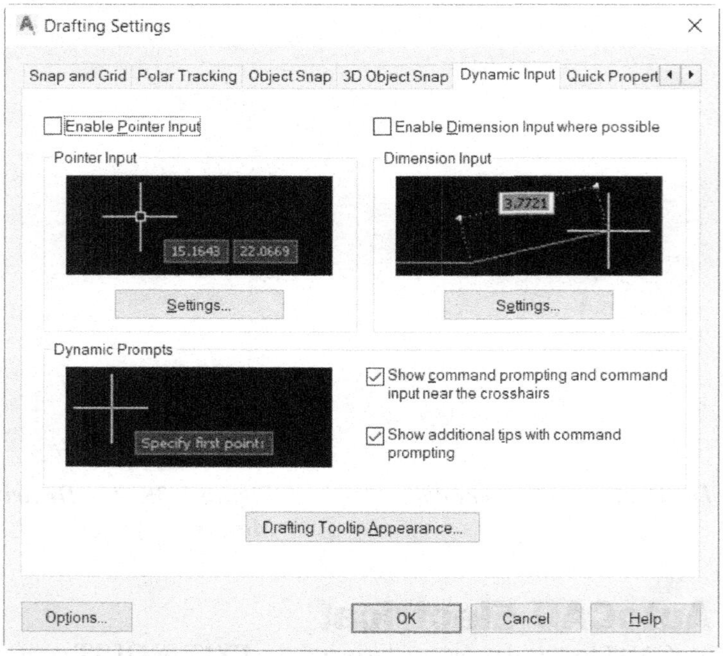

Figure 1-40 *The* ***Dynamic Input*** *tab of the* ***Drafting Settings*** *dialog box*

Enable Pointer Input

With the **Enable Pointer Input** check box selected, you can enter commands through the pointer input. Figure 1-41 shows the **AEWIRE** command entered through the pointer input. If this check box is cleared, the **Dynamic Input** will be turned off and the commands will be entered through the Command prompt in a way similar to the old releases of AutoCAD Electrical.

On choosing the **Settings** button from the **Pointer Input** area, the **Pointer Input Settings** dialog box will be displayed, as shown in Figure 1-42. The radio buttons in the **Format** area of this dialog box are used to set the default settings for specifying other points, after specifying the first point. By default, the **Polar format** and **Relative coordinates** radio buttons are selected. As a result, coordinates will be specified in the polar form, with respect to the relative coordinates system. You can select the **Cartesian format** radio button to enter coordinates in the cartesian form. Likewise, if you select the **Absolute coordinates** radio button, numerical entries will be measured with respect to the absolute coordinate system.

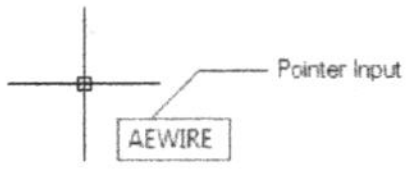

Figure 1-41 Entering a command using the pointer input

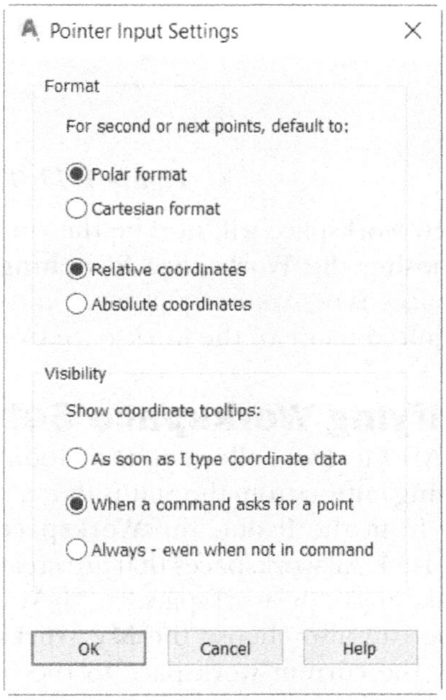

*Figure 1-42 The **Pointer Input Settings** dialog box*

The **Visibility** area in the **Pointer Input Settings** dialog box is used to set the visibility of tool tips of coordinates. By default, the **When a command ask for a point** radio button is selected. You can select the other radio buttons to modify this display.

CREATING AND MANAGING WORKSPACES

A workspace is defined as a customized arrangement of toolbars, menus, and window palettes in the AutoCAD Electrical environment. Workspaces are required when you need to customize the AutoCAD Electrical environment for a specific use, in which you need only a certain set of toolbars and menus. For such requirements, you can create your own workspaces in which only the specified toolbars, menus, and palettes will be available. By default, the **ACADE & 2D Drafting & Annotation** workspace is set as the current workspace when you start AutoCAD Electrical. You can choose any other predefined workspace from the shortcut menu that will be displayed on choosing the **Workspace Switching** button in the Status Bar or by choosing the required workspace from **Tools > Workspaces** in the menu bar. You can also choose the workspace using the **Workspaces** toolbar.

Creating a New Workspace

To create a new workspace, invoke the toolbars and window palettes that you want to display in the new workspace. Next, choose **Tools > Workspaces > Save Current As** from the menu bar; the **Save Workspace** dialog box will be displayed, as shown in Figure 1-43. Enter the name of the new workspace in the **Name** edit box and choose the **Save** button from the dialog box.

Figure 1-43 The ***Save Workspace*** *dialog box*

The new workspace will now be the current workspace in the shortcut menu that is displayed on choosing the **Workspace Switching** button on the Status Bar. Likewise, you can create workspaces as per your requirement and can switch from one workspace to the other by selecting the required name of the workspace from the drop-down list in the **Workspaces** toolbar.

Modifying Workspace Settings

AutoCAD Electrical allows you to modify workspace settings. To do so, choose the **Workspace Switching** button from the Status Bar; a flyout will be displayed. Choose the **Workspace Settings** option from the flyout; the **Workspace Settings** dialog box will be displayed, as shown in Figure 1-44. All workspaces that are created are listed in the **My Workspace** drop-down list. You can make any of the workspaces as My Workspace by selecting it in the **My Workspace** drop-down list. You can also choose the **My Workspace** tool from the **Workspace Switching** toolbar to change the current workspace to the one that was set as **My Workspace** in the **Workspace Settings** dialog box. The other options in this toolbar are discussed next.

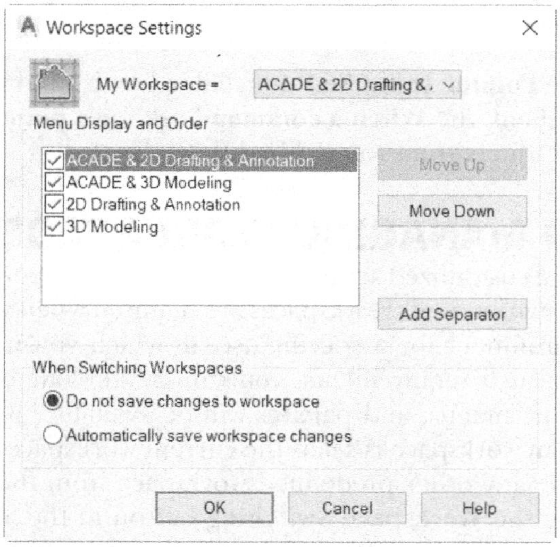

Figure 1-44 The ***Workspace Settings*** *dialog box*

Menu Display and Order Area

The options in this area are used to set the order of the display of workspaces in the drop-down list of the **Workspaces** toolbar. By default, workspaces are listed in the sequence of their creation.

To change the order, select the workspace and choose the **Move Up** or **Move Down** button. You can also add a separator between workspaces by choosing the **Add Separator** button. A separator is a line that is placed between two workspaces in the shortcut menu that is displayed on choosing the **Workspace Switching** tool in the Status Bar.

When Switching Workspaces Area

By default, the **Do not save changes to workspace** radio button is selected in this area. As a result, while switching workspaces, the changes made in the current workspace will not be saved. If you select the **Automatically save workspace changes** radio button, the changes made in the current workspace will automatically be saved when you switch to the other workspace.

WD_M BLOCK

Most of the settings of a drawing used by AutoCAD Electrical are saved in a smart block on the drawing, called WD_M.dwg. Every drawing of AutoCAD Electrical should consist of non-visible WD_M block to make the drawing compatible with AutoCAD Electrical. The drawing should consist of only one copy of WD_M block. The WD_M.dwg block is situated in the default symbol library. The WD_M.dwg consists of various attributes that define layer names, default settings, and so on. If the WD_M block is not present in the existing or new drawing and if you want to insert any electrical component or want to edit drawing properties, then the **Alert** message box will be displayed, as shown in Figure 1-45. Choose the **OK** button in this message box to insert the WD_M block at 0,0 location. By default, the **Force this drawing's configuration settings to match the project settings** check box is selected. As a result, the drawing settings will be matched to the project settings.

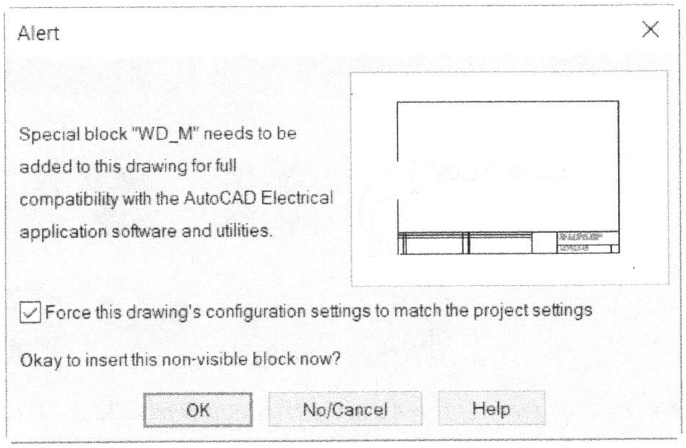

***Figure 1-45** The **Alert** message box*

Note

For inserting panel layout symbols in the drawing, you need to insert the WD_PNLM block, which will be discussed in Chapter 8.

AutoCAD Electrical HELP

You can get online help and documentation about the working of AutoCAD Electrical 2022 commands from the **Help** menu in the menu bar, refer to Figure 1-46. You can access the **Help** menu for a particular tool. To do so, place the cursor on the particular tool and then press the

F1 key; the **AutoCAD Electrical 2022 - Help** window will be displayed. This window displays the detailed description related to that particular tool. The menu bar also contains InfoCenter bar, as shown in Figure 1-47. This bar helps you search information by using certain keywords. The options in the **Help** menu and the **InfoCenter** bar are discussed next.

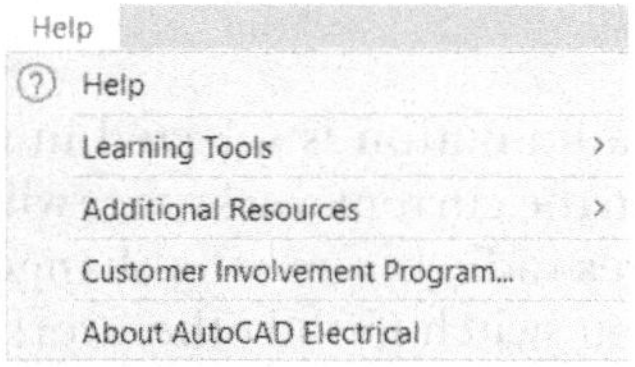

***Figure 1-46** The **Help** menu*

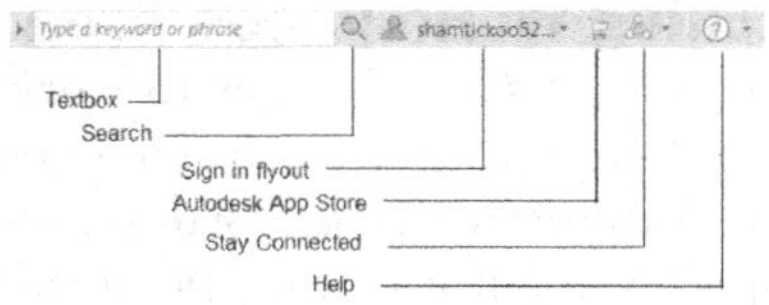

***Figure 1-47** The **InfoCenter** bar*

Help Menu

Electrical Help Topics

On choosing the **Help** option, the **AutoCAD Electrical 2022 - Help** window will be displayed. Figure 1-48 shows partial view of the **AutoCAD Electrical 2022 - Help** window.

***Figure 1-48** Partial view of the **AutoCAD Electrical 2022 - Help** window*

You can use this window to access help on different topics and commands. If you are in the middle of a command and require help regarding it, choose the **Help** button to display information about that particular command in the dialog box.

In this window, you can choose the **Help Home** button to explore new features in AutoCAD Electrical 2022. Also, you can browse videos, tutorials, and documentation for beginners and advanced users in the **Essential Skills Videos** and **Resources** sections. You can download sample files and offline help database from the **Downloads** section. You can also connect to other users of Autodesk community and discussion groups using the **Connect** section.

Click on the + sign at the left of these sections to display the list of topics under them. Next, select a topic from the list; contents of the selected topic will be displayed at the right side of the **AutoCAD Electrical 2022 - Help** window.

Search Field

When you type any word in the Search edit box and then choose the **Search** button, a list of topics related to the typed word will be displayed below the **Search** button. You can select the desired topic from the list; the information related to that topic will be displayed in the **AutoCAD Electrical 2022 - Help** window.

Learning Resources

Choose **Help > Learning Tools > Learning Resources** from the menu bar; the **AutoCAD Electrical 2022 - Help** window will be displayed. Using this window, you can access information about topics such as user interface, AutoCAD Electrical Quick Reference Guide, AutoCAD Electrical 2022 Toolset New Feature Summary, and so on. When you choose a topic, the description of the feature improvements will be displayed in a window.

Additional Resources

This utility connects you to the **Support Knowledge Base**, **Online Training Resources, Online Developer Center**, **Developer Help**, **API Help**, and **Autodesk User Group International** web pages. The **Developer Help** option provides a detailed help on customizing AutoCAD Electrical in a separate window. You can click on any link on the right of this window.

About AutoCAD Electrical

This option gives you information about the Release, Serial number, Licensed to, and also the legal description about AutoCAD Electrical.

InfoCenter Bar

InfoCenter

InfoCenter is an easy way to get the desired help documentation. Enter the keywords to be searched in the textbox and choose the **Search** button; the result will be displayed as a link in the **AutoCAD Electrical 2022 - Help** window, refer to Figure 1-49. To display the required information in this window, click on any one of the search topics from the list, the information related to the search topic will be displayed at the right of the window.

Autodesk App Store

Autodesk App Store helps you to download various applications for AutoCAD Electrical, get connected to the AutoCAD network, share information and designs, and so on. On choosing the **Autodesk App Store** button from the InfoCenter bar, the **AUTODESK APP STORE** window will be opened, refer to Figure 1-50.

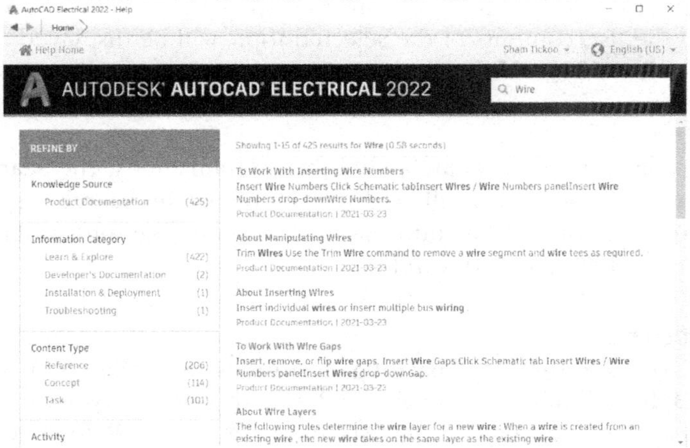

Figure 1-49 The search results displayed using ***InfoCenter***

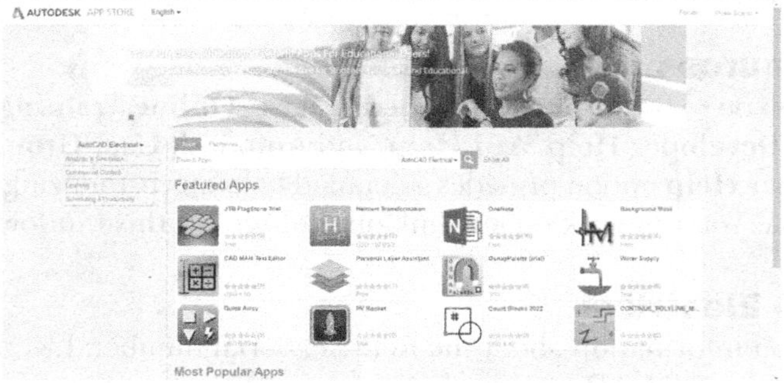

Figure 1-50 The ***AUTODESK APP STORE*** *window*

You can download various Autodesk apps from this page. Some of them are free of cost. You can also search for the apps by entering the name of the app in the **Search Apps** text box. You can download the AutoCAD apps from the **Featured Apps** panel of the **Featured Apps** tab in the **Ribbon**, refer to Figure 1-51. To download the AutoCAD apps, choose the required apps from the **Featured Apps** panel of the **Feature Apps** tab; the default browser will open with the icons of the apps to be downloaded. Now, you can download the app.

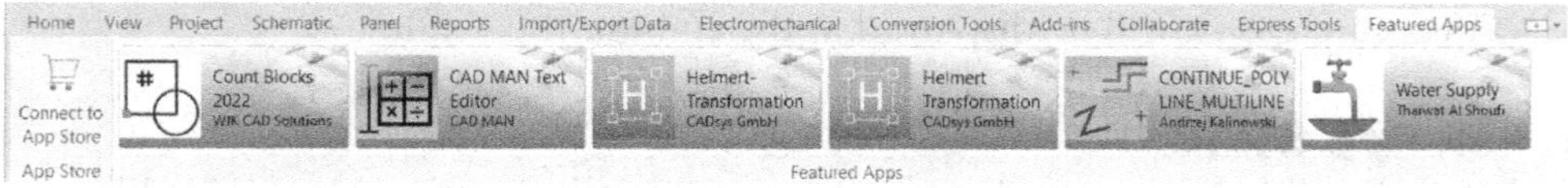

Figure 1-51 The ***Featured Apps*** *panel of the* ***Feature Apps*** *tab in the* ***Ribbon***

You can also download apps other than the apps available in the **Featured Apps** panel. To do so, choose the **Connect to App Store** button from the **App Store** panel of the **Featured Apps** tab; the **AUTODESK APP STORE** window will be displayed. Next, choose **Show All** from the **AUTODESK APP STORE** window. Now, you can download the required apps from the window. You can also search apps by entering the name of the required app in the **Search Apps** text box. Some of the apps are paid and some of them are free to download.

The downloaded and installed apps will be available in the **Add-Ins** tab. You can choose the required app to work with and can manage it by choosing the **Exchange App Manager** button available in the **App Manager** panel of the **Add-Ins** tab. To manage the app, choose the **App Manager** button from the **App Manager** panel in the **Add-ins** tab; the **Autodesk App Manager** dialog box with all the installed apps will be displayed. To manage a particular app, right-click on it; a shortcut menu will be displayed, as shown in Figure 1-52.

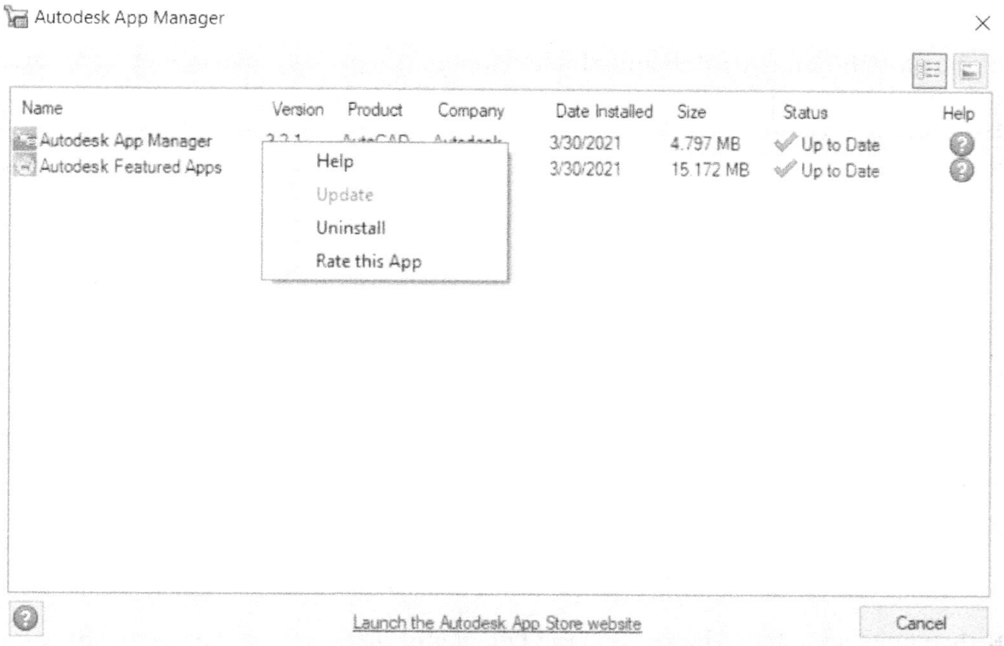

Figure 1-52 *The **Autodesk App Manager** dialog box*

Now, you can choose the required option to manage the app.

Stay Connected

When you choose this button, a flyout is displayed, as shown in Figure 1-53. The options in this flyout provide you quick access to the subscription center and social media.

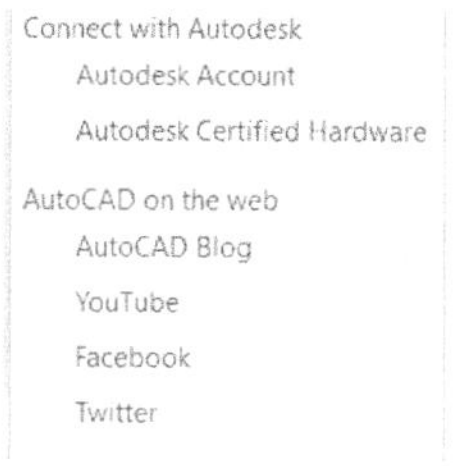

Figure 1-53 *The flyout displayed*

SAVE TO WEB & MOBILE

Application Menu: Save As > Drawing to AutoCAD Web & Mobile
Quick Access Toolbar: Save to Web & Mobile **Command:** SAVETOWEBMOBILE

Using the **Save to Web & Mobile** tool, you can save copy of your drawings in your Autodesk web & mobile account from any remote location in the world using any device such as desktop or mobile having internet access. You can access this tool from the **Quick Access Toolbar** or **Application Menu**. When you choose this tool for the first time, you will be prompted to install the Save to AutoCAD Web and Mobile Plug-in. Choose the **Save to Web & Mobile** tool from the **Quick Access Toolbar**; the **Save in AutoCAD Web & Mobile** dialog box will be displayed, refer to Figure 1-54.

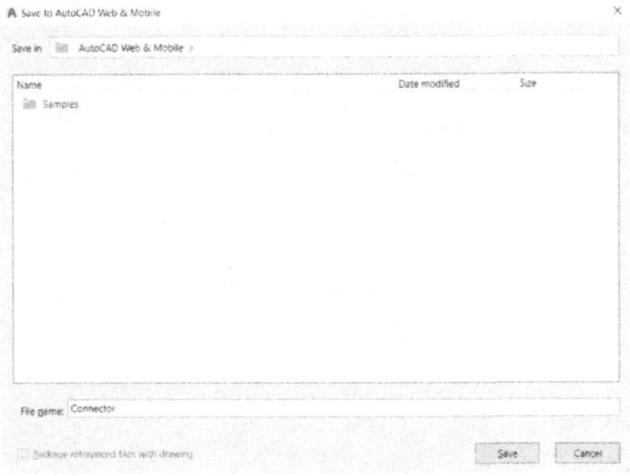

***Figure 1-54** The **Save in AutoCAD Web & Mobile** dialog box*

Enter the name of the file to be saved in the **File name** edit box. Next, choose the **Save** button.

Note

*1. The name of the saved files appear in the **Name** column of the list box available in the **Save in AutoCAD Web & Mobile Cloud Files** dialog box when you invoke this dialog box again.*

2. You need to first sign in to your Autodesk account to use this tool. If you are not signed in and choose this tool, the Autodesk sign-in window will appear and will prompt you to sign in first.

*3. The drawing files saved in your web and mobile account are saved to the cloud and utilise the cloud space. You can save them to your device by using the **Save As** tool.*

4. You can also share the saved files with any of the co-workers or clients across the world. They can review or edit the drawing files depending upon the permissions you grant them.

After saving the file to your web and mobile account, you can access it from anywhere across the world using any device (mobile, tablet, etc) having Wi-fi or internet connection.

Note

*You can access the saved web and mobile files using the **Open from Web & Mobile** tool available in the **Quick Access Toolbar**.*

ADDITIONAL HELP RESOURCES

1. You can get help for a command by pressing the F1 key. On doing so, the **AutoCAD Electrical 2022 - Help** window containing information about the command will be displayed. You can exit the dialog box and continue working with the command.

2. You can get help about a dialog box by choosing the **Help** button in that dialog box.

3. Autodesk has provided several resources that can be used to seek assistance for your AutoCAD Electrical questions. The following is a list of some of the resources:

   a. Autodesk website: *https://www.autodesk.com*
   b. AutoCAD Electrical Technical Assistance website: *https://knowledge.autodesk.com*
   c. AutoCAD Electrical Discussion Groups website

4. CADCIM Technologies provides technical support for the study material discussed in our textbooks. Send us an email at *techsupport@cadcim.com* to seek technical assistance.

Note
For the printing purpose, this textbook will follow the white background.

Self-Evaluation Test

Answer the following questions and then compare them to those given at the end of this chapter:

1. Which of the following combinations of keys should be pressed to turn the display of the **Tool Palettes** window on or off?

   (a) CTRL+3 (b) CTRL+0
   (c) CTRL+5 (d) CTRL+2

2. If WD_M block is not present in a drawing and you insert a component in it, then the __________ message box will be displayed.

3. The __________ option in the **Workspace Switching** flyout is used to save the current workspace settings as a new workspace.

4. You can use the __________ command to close the current drawing file without actually quitting AutoCAD Electrical.

5. You can retrieve the automatically saved files by using the __________.

6. The __________ button in the Application Status Bar is used to display the expanded view of the drawing.

7. You can press the F3 key to display the **AutoCAD Electrical** text window that displays previously used commands and prompts. (T/F)

8. You cannot create a new drawing using the **PROJECT MANAGER**. (T/F)

9. AutoCAD Electrical marking menu will be displayed when you right-click on the AutoCAD Electrical object such as component, wire, and so on. (T/F)

10. You can press the F1 key to display the **AutoCAD Electrical 2022 - Help** window. (T/F)

Review Questions

Answer the following questions:

1. Which of the following combinations of keys needs to be pressed to toggle display of all toolbars displayed on the screen?

   (a) CTRL+3 (b) CTRL+0
   (c) CTRL+5 (d) CTRL+2

2. Which of the following commands is used to exit the AutoCAD Electrical program?

   (a) **QUIT** (b) **END**
   (c) **CLOSE** (d) **EXIT**

3. Which of the following commands is invoked when you choose **Save** from the **File** menu or choose the **Save** button in the **Quick Access Toolbar**?

   (a) **SAVE** (b) **LSAVE**
   (c) **QSAVE** (d) **SAVEAS**

4. The __________ button is used to add or remove buttons in the Application Status Bar.

5. The __________ button is used to toggle the display of the File tab bar which displays all opened files.

6. The __________ contains different commands for inserting components in Imperial and Metric units.

7. The shortcut menu invoked by right-clicking in the command window displays the most recently used commands and some of the window options such as **Copy**, **Paste**, and so on. (T/F)

8. The F12 function key is used to toggle the **Dynamic Input** mode. (T/F)

Answers to Self-Evaluation Test

1. CTRL+3, **2. Alert**, **3. Save Current As**, **4. CLOSE**, **5. Drawing Recovery Manager**, **6. Clean Screen**, **7.** F, **8.** F, **9.** T, **10.** T

Chapter 2

Working with Projects and Drawings

Learning Objectives

After completing this chapter, you will be able to:

- *Create new projects and drawings*
- *Open existing projects*
- *Add existing and new drawings to the current project*
- *Copy existing projects*
- *Understand the working of the PROJECT MANAGER*

INTRODUCTION

AutoCAD Electrical is a project-based software in which wiring diagrams related to each other are grouped under a project. A project is a set of electrical wiring diagrams that form a project file *<project_name>.wdp*. Each project is defined by an ASCII text file with *.wdp* extension. These project files contain a list of project information such as project settings, project or drawing properties, names and descriptions of drawing files, symbol library paths, and so on. You can have an unlimited number of projects. However, only one project can be active at a time. The list of these projects is displayed in the **PROJECT MANAGER**.

PROJECT MANAGER

Command: AEPROJECT

The **PROJECT MANAGER** is used to create new projects, open existing projects, add new drawings to a project, re-order drawing files, access existing projects, and modify existing information in a project. By default, the **PROJECT MANAGER** is displayed and docked on the left of the screen, refer to Figure 2-1. If the **PROJECT MANAGER** is not displayed by default, choose the **Manager** tool from the **Project Tools** panel of the **Project** tab; the **PROJECT MANAGER** will be displayed.

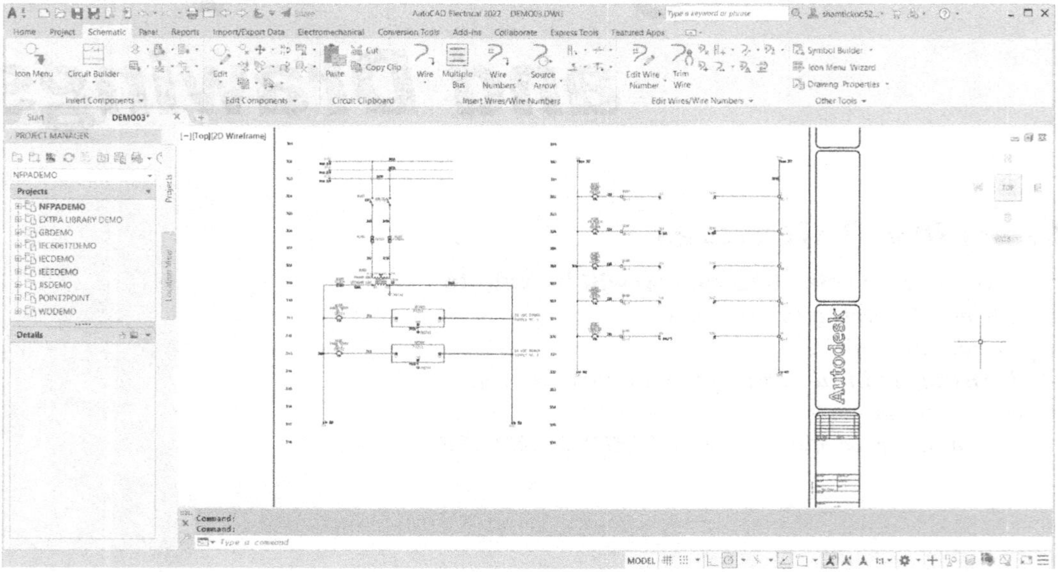

Figure 2-1** AutoCAD Electrical screen with the **PROJECT MANAGER

Alternatively, choose the **Project Manager** tool from the **ACE:Main Electrical 2** toolbar to display the **PROJECT MANAGER**. The **PROJECT MANAGER** is similar to other AutoCAD Electrical tool palettes. You can dock the **PROJECT MANAGER** at a specific location on the screen. Also, if you do not want to use the project tools, you can hide the **PROJECT MANAGER**.

The **PROJECT MANAGER** is divided into two tabs: **Projects** and **Location View**. These tabs are discussed in brief next.

Projects TAB

In the **PROJECT MANAGER**, the **Projects** tab is chosen by default, refer to Figure 2-1. When you double-click on the title bar of the **PROJECT MANAGER**, it gets undocked and is displayed separately on the screen, refer to Figure 2-2. When you right-click on the title bar of the undocked **PROJECT MANAGER**, a shortcut menu is displayed, as shown in Figure 2-3. You can change the appearance, location, and display settings of the **PROJECT MANAGER** by choosing the respective options from the shortcut menu.

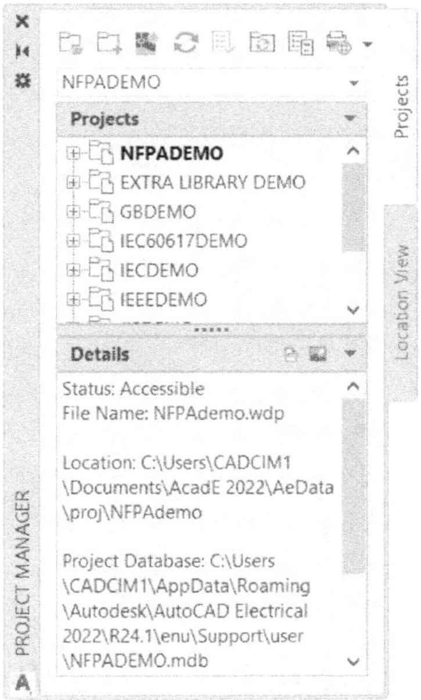

*Figure 2-2 The undocked **PROJECT MANAGER***

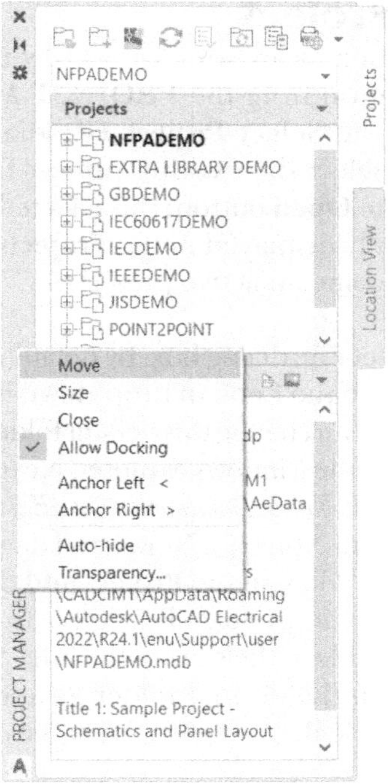

*Figure 2-3 The shortcut menu displayed by right-clicking on the title bar of the **PROJECT MANAGER***

When you double-click on a project name in the **Projects** rollout, a list of drawings associated with the project will be displayed. Also, the details of the selected project will be displayed in the **Details** rollout of the **PROJECT MANAGER**, as shown in Figure 2-4. The name of the active project will appear in bold text in the **Projects** rollout of the **PROJECT MANAGER**.

In AutoCAD Electrical 2022, you will find five new projects in the Project Manager which have been added based on various drafting standards. Names of these projects are **IECDEMO, JISDEMO, GBDEMO**, **IEC60617DEMO**, and **IEEEDEMO**. These projects are based on respective symbol libraries. For example, in the **IECDEMO** project, the drawings are based on IEC symbol library whereas in the **IEEEDEMO** project, the drawings are based on IEEE symbol library. Detailed information about various drafting standards is provided later in this chapter.

When you right-click on a drawing name, a shortcut menu will be displayed. You can use the options in this shortcut menu to open, close, copy, remove, replace, rename, or access other editing options to modify the drawing file. You can open a drawing file by double-clicking on it in the **PROJECT MANAGER** and the corresponding drawing file name will appear in bold text in the **Projects** rollout of the **PROJECT MANAGER**.

Note

You cannot create two projects with the same name in the ***PROJECT MANAGER****. Moreover by using the* ***PROJECT MANAGER****, you can switch to different projects and change their settings.*

Opening a Project

You can open an existing project by using the **Open Project** button from the **PROJECT MANAGER**. On doing so, the **Select Project File** dialog box will be displayed. Next, select the existing project from this dialog box and choose the **Open** button; the selected project's name will be automatically displayed in the **Projects** rollout in bold text and it will become an active project.

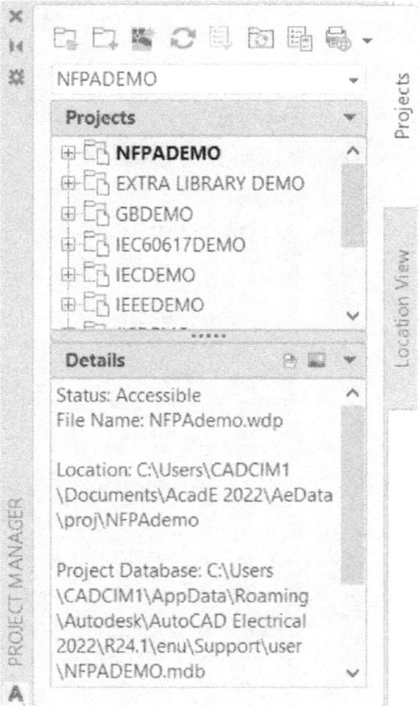

Figure 2-4 The ***PROJECT MANAGER*** *displaying the* ***Details*** *rollout*

In the **Select Project File** dialog box, by default, the **NfpaDemo** folder is displayed in the **Look in** drop-down list. In AutoCAD Electrical 2022, you can change this default folder. To change the default folder, open the windows explorer. Next, navigate to the location c:*\Users\User Name\Documents\AcadE 2022\AeData* and open the *wd.env* file in Notepad. Now, scroll down in the file and go to the highlighted line in the **Project and drawing-related paths and settings** section, refer to Figure 2-5. In this line, remove the '*' symbol available at the start of the line. Also, replace the 'x:/some path/' with the desired folder location; the selected text is replaced by the desired path. Next, save the *wd.env* file.

Now, choose the **Open Project** button from the **PROJECT MANAGER**; the modified **Select Project File** dialog box will be displayed. You will notice that the name of the folder is changed.

Note

If you put any extra space at the start, in between, or at the end of the highlighted line, the desired path will not change.

Creating a New Project

Command: ACENEWPROJECT

You can create a new project by choosing the **New Project** button from the **PROJECT MANAGER**. On doing so, the **Create New Project** dialog box will be displayed. Alternatively, use the **ACENEWPROJECT** command to create a new project.

Working with Drawings

You can add any number of drawings to your project at any time. A project file can have drawings located in different directories. But, it is recommended that you save your drawings and project file (*.wdp*) in the same folder. When you create a new drawing, an invisible smart block, WD_M, will automatically be added to your drawing at location 0,0. The WD_M block defines drawing settings. These settings may be different from the project settings. Thus, you can have different settings for different drawings in a single project. Also, each AutoCAD Electrical drawing should contain only one copy of invisible WD_M block. If multiple WD_M blocks are present in the drawing, the settings may not be stored and read consistently. Also, note that the drawing to be created will automatically get added to the active project.

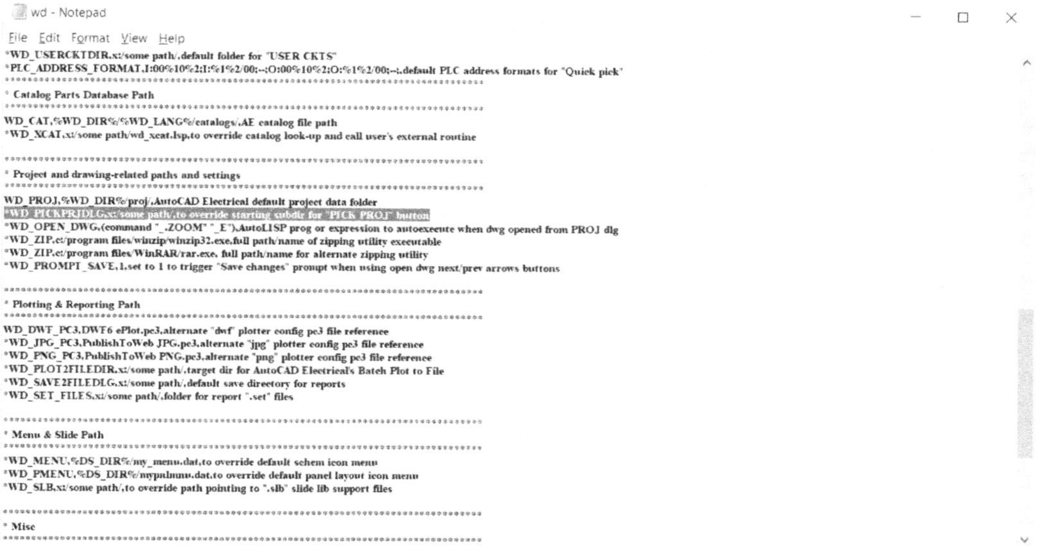

Figure 2-5 *The line highlighted in the wd.env file*

Creating a New Drawing

Command: ACENEWDRAWING

The **New Drawing** button is used to create a new drawing file. To do so, choose the **New Drawing** button from the **PROJECT MANAGER**; the **Create New Drawing** dialog box will be displayed.

Alternatively, invoke this dialog box by using the **ACENEWDRAWING** command or by right-clicking on the **Projects** rollout of the **PROJECT MANAGER** and then choosing the **New Drawing** option from the shortcut menu displayed.

Note

You can have the same drawing in multiple projects.

Adding Existing Drawings to the Current Project

You can also add the existing drawings to the current project. To do so, right-click on the project name in the Projects rollout of the **PROJECT MANAGER**; a shortcut menu will be displayed, as shown in Figure 2-6. Choose the **Add Drawings** option from the shortcut menu; the **Select Files to Add** dialog box will be displayed. In this dialog box, select the desired project folder from the **Look in** drop-down list and then select drawings to be added to the project. Next, choose the **Add** button; the **Apply Project Defaults to Drawing Settings** message box will be displayed.

If you choose the **Yes** button, the default values of the project will be added to the WD_M block definition of the newly added drawing. But if you choose the **No** button, the new drawing will retain its existing settings. Also, the selected drawings will be added to your project and will appear at the end of the Project Drawing list.

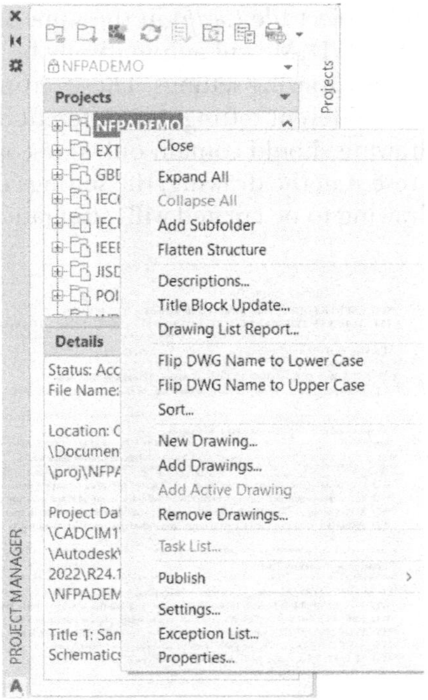

*Figure 2-6 The shortcut menu displayed by right-clicking on the active project name in the **PROJECT MANAGER***

> **Note**
>
> *You can select multiple drawing files at a time from the **Select Files to Add** dialog box by pressing the SHIFT or CTRL key.*

You can group drawings, remove drawings, assign description to drawings, preview drawings, configure the drawing list display, and so on by using the **PROJECT MANAGER**. Also, you can access the existing project and modify its related information using the **PROJECT MANAGER**.

Configuring the Drawing List Display

The **Drawing List Display Configuration** button is used to configure the information related to various drawings in the **Projects** rollout of the **PROJECT MANAGER**. This button is used to display the required information such as drawing number, description, and so on. By default, only the name of the drawing is displayed in the **Projects** rollout of the **PROJECT MANAGER**. Choose the **Drawing List Display Configuration** button from the **PROJECT MANAGER**; the **Drawing List Display Configuration** dialog box will be displayed, as shown in Figure 2-7.

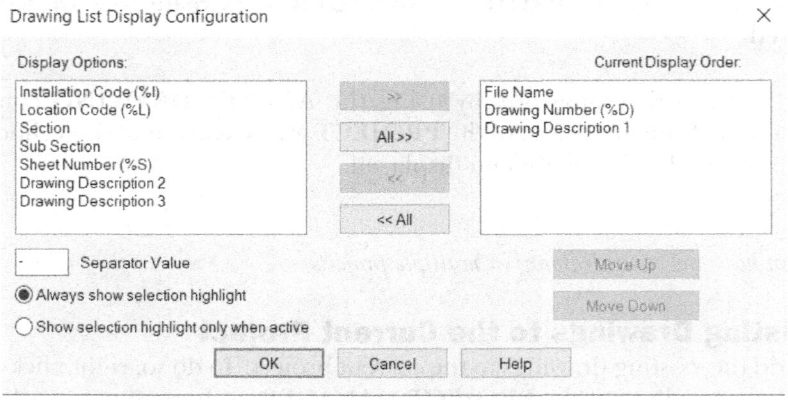

*Figure 2-7 The **Drawing List Display Configuration** dialog box*

Copying a Project

Command: AECOPYPROJECT

The **Copy** tool is used to copy the entire project as well as copy the drawings present within that project. To do so, choose the **Copy** tool from the **Project Tools** panel of the **Project** tab; the **Copy Project: Step 1 - Select Existing Project to Copy** wizard will be displayed.

Note

Before copying drawings to a new project, you need to close all drawings of that project.

Next, enter the name and path of the existing project in the **Enter existing project name** edit box. Alternatively, choose the **Browse** button from this dialog box to select the existing project; the **Select existing project to copy** dialog box will be displayed. Next, double-click on the project folder that you want to copy. Select the project's *.wdp* file and choose the **Open** button; the name and path of the existing project will be displayed in the **Enter existing project name** edit box. You can also choose the **Copy Active Project** button available in this dialog box to copy the currently active project.

Choose the **OK** button; the **Copy Project: Step 2 - Select path and name for new project** wizard will be displayed. Enter a name for the new project in the **File name** edit box and select the path for the new project from the **Save in** drop-down list. By default, the *.wdp* extension will be displayed in the **Save as type** edit box. Next, choose the **Save** button; the **Select Drawings to Process** dialog box will be displayed.

Select the drawing files that you want to copy in the project from the top list of this dialog box. Next, choose the **Process** button; the selected drawings will be transferred from the top list of the **Select Drawings to Process** dialog box to the bottom list. The other options in the **Select Drawings to Process** dialog box have already been discussed. Choose the **OK** button from the **Select Drawings to Process** dialog box; the **Copy Project: Step 4 -- Enter Base Path for Project Drawings** wizard will be displayed.

In this dialog box, enter the directory path where the new project will be saved. The directory path will be created, if it does not exist. In the **Select Project-Related Files to be Copied** area, you can specify the project related files that need to be copied to the project that you want to create. Note that the project related files to be copied will be activated in the **Select Project-Related Files to be Copied** area. To copy these files, select the check boxes on the left of the respective files and then choose the **OK** button; the **Copy Project: Step 5 -- Adjust new drawing file names** wizard will be displayed.

Choose the **Edit** button in this wizard to edit the file name and path of the selected drawing, if needed. Similarly, choose the **Find/Replace** button to find or replace a drawing file name and its path. Next, choose the **OK** button; the name of the new project will be displayed at the top in the **Projects** rollout in bold text and will become the active project.

Deleting a Project

Command: AEDELETEPROJECT

The **Delete** tool or the **AEDELETEPROJECT** command is used to delete an existing project and its drawings permanently. To do so, choose the **Delete** tool from the **Project Tools** panel of the **Project** tab; the **Select Existing Project to Delete** dialog box will be displayed.

Enter the name of the project file to be deleted in the **File name** edit box. Alternatively, select a project name from the list displayed in the **Select Existing Project to Delete** dialog box. Next, double-click on the name of the project folder, if it exists. Select the project definition file(*.wdp*) and choose the **Open** button; the **Project File Delete Utility** dialog box will be displayed. In this dialog box, the **Delete ".wdp" project list file** check box is used to permanently delete the selected project file with *.wdp* extension. The **Delete project's AutoCAD drawing files** check box is used to delete only the drawing files of a Project. Note that this check box will be activated only if the related project consists of drawing files.

The **List** button will be activated only if you select the **Delete project's AutoCAD drawing files** check box. Choose the **List** button; the **Select Drawings to Process** dialog box will be displayed. Select the drawings to be deleted from the drawing file list of the project. Next, choose the **OK** button from the **Select Drawings to Process** dialog box; the **Project File Delete Utility** dialog box will be displayed again. Choose the **Delete Files** button from the **Project File Delete Utility** dialog box; the selected files will be deleted permanently, and you cannot retrieve them.

Location View TAB

The options in the **Location View** tab of the **PROJECT MANAGER** are used to display and filter the components based on the installation and location codes, display details and connections of the components, and so on. Figure 2-8 shows the **PROJECT MANAGER** with the **Location View** tab chosen.

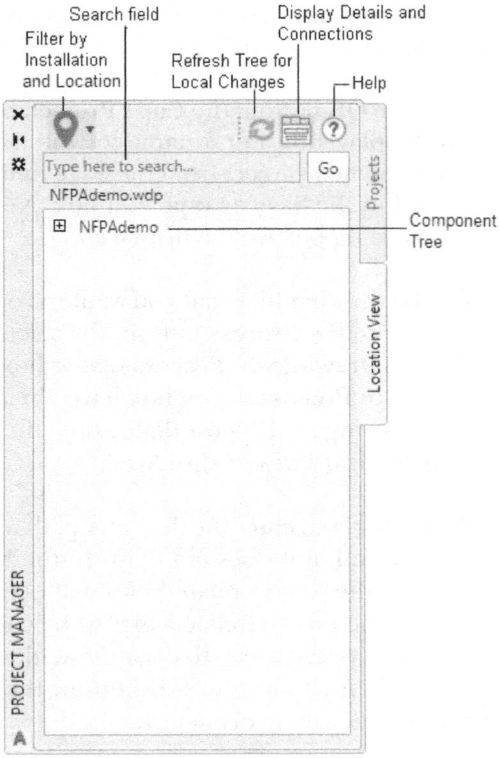

***Figure 2-8** The **PROJECT MANAGER** with the **Location View** tab chosen*

TUTORIALS

Tutorial 1

In this tutorial, you will create a new project and add project description to it, refer to Figure 2-9. You will also create a new drawing in the project. **(Expected time: 10 min)**

The following steps are required to complete this tutorial:

a. Create a new project.
b. Open the CADCIM project.
c. Create a new drawing.

Creating a New Project

1. Start AutoCAD Electrical 2022 by double-clicking on the shortcut icon of AutoCAD Electrical 2022 on the desktop of your computer.

2. In the AutoCAD Electrical 2022 interface, click on the **Start Drawing** window. The **PROJECT MANAGER** is displayed by default on the left of the screen. If it is not displayed, choose the **Manager** tool from the **Project Tools** panel of the **Project** tab. Alternatively, choose the **Project Manager** tool from the **ACE:Main Electrical 2** toolbar to display the **PROJECT MANAGER**.

3. In the **PROJECT MANAGER**, make sure the **Projects** tab is chosen. Next, choose the **New Project** button; the **Create New Project** dialog box is displayed, as shown in Figure 2-10.

4. Enter **CADCIM** in the **Name** edit box.

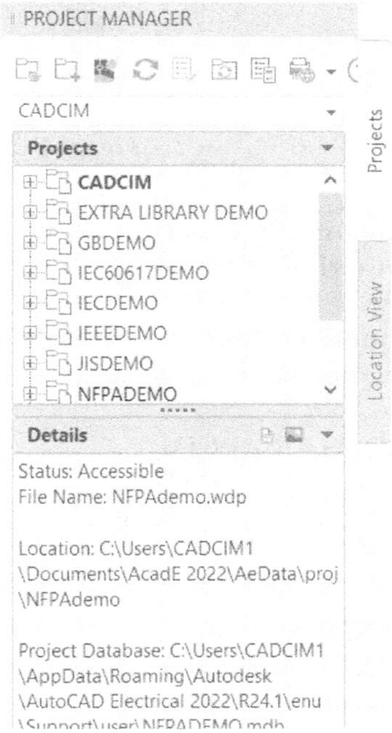

***Figure 2-9** The **PROJECT MANAGER** displaying the **CADCIM** project and its description*

5. Choose the **Browse** button; the **Browse For Folder** dialog box is displayed. By default, the **Proj** folder is selected in this dialog box. Choose the **OK** button; the location of the project is automatically displayed in the **Location** edit box.

6. Select the **Create Folder with Project Name** check box, if not selected.

7. By default, the name and path of the existing project is displayed in the **Copy Settings from Project File** edit box as *c:\Users\User Name\Documents\AcadE 2022\AeData\proj\NFPAdemo\NFPAdemo.wdp*. If not displayed, choose the **Browse** button; the **Select Project File** dialog box is displayed. Select the *NFPADemo* folder and then select the **NFPAdemo.wdp** project file.

Next, choose the **Open** button; the path and name of the existing project is displayed in the **Copy Settings from Project File** edit box.

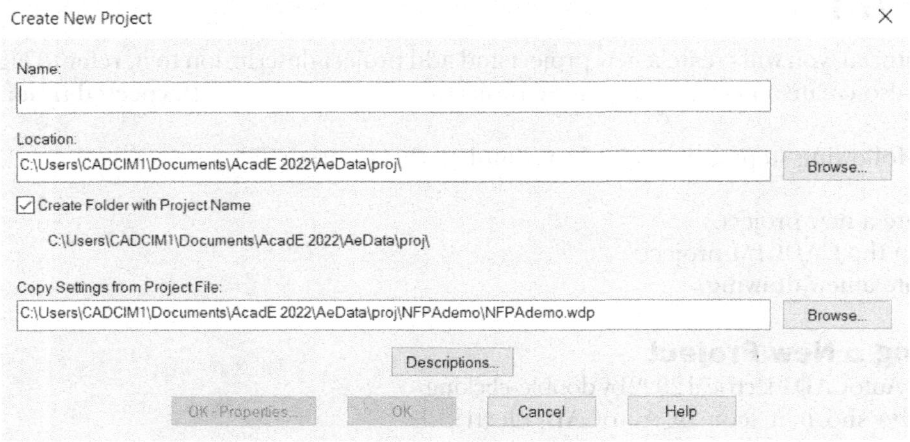

***Figure 2-10** The **Create New Project** dialog box*

8. Choose the **Descriptions** button from the **Create New Project** dialog box; the **Project Description** dialog box is displayed.

9. Enter the following information in the **Project Description** dialog box, as shown in Figure 2-11.

   LINE1= CADCIM
   LINE2= AutoCAD Electrical
   LINE3= Sample Project
   LINE4= 1
   LINE5= 07/4/2021
   LINE6= Sham
   LINE7= John
   LINE8= Crystal
   LINE9= 1.00

   Next, select the **in reports** check box corresponding to each of these lines to include this description in the reports, refer to Figure 2-11.

Note

*The information that you enter in the description lines of the **Project Description** dialog box will be displayed at a particular location in the title block of the drawing files of the **CADCIM** project after updating the title block, as shown in Figure 2-12. Updation of title block is explained in detail in Chapter 12.*

10. Choose the **OK** button from the **Project Description** dialog box to save the changes made in this dialog box.

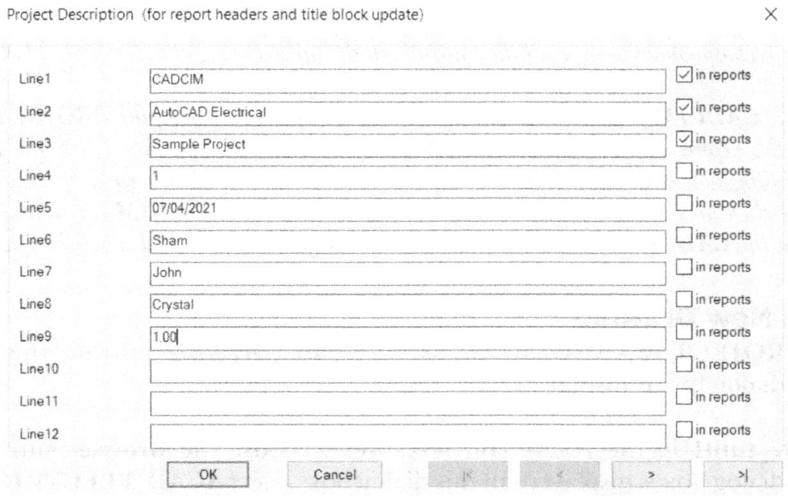

***Figure 2-11** The **Project Description** dialog box*

11. Choose the **OK** button from the **Create New Project** dialog box. You will notice that the **CADCIM** project appears in bold text at the top of the project list displayed in the **PROJECT MANAGER**, refer to Figure 2-13.

12. Next, select the **CADCIM** project from the **PROJECT MANAGER**.

13. Choose the **Details** button from the **Details/Preview** rollout. You will notice that the project description/information entered in the first nine lines of the **Project Description** dialog box is displayed in the **Details/Preview** area, as shown in Figure 2-13.

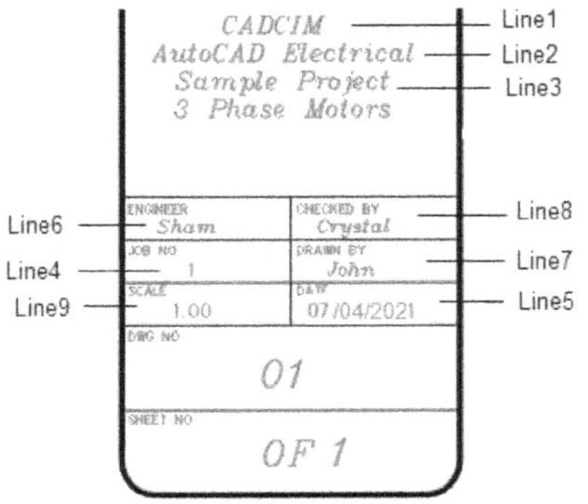

***Figure 2-12** The title block showing the description entered in the **Project Description** dialog box*

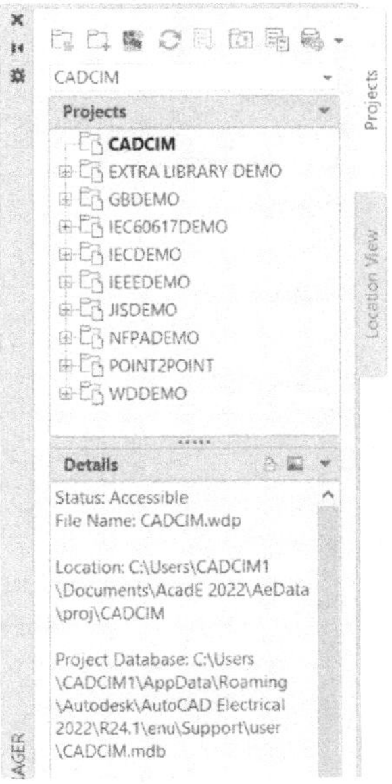

***Figure 2-13** The **PROJECT MANAGER** displaying the **CADCIM** project and its description*

Note

*1. It is recommended to save all the tutorials in the forthcoming chapters in the **CADCIM** project.*

*2. If the **CADCIM** project is not displayed in the **Projects** rollout of the **PROJECT MANAGER**, select the **Open Project** option from the Project selection drop-down list; the **Select Project File** dialog box is displayed. Select the **Proj** folder from the **Look in** drop-down list and then double-click on the **CADCIM** folder name. Next, select the **CADCIM** file and choose the **Open** button; the **CADCIM** project is displayed in the **Projects** rollout and it becomes the active project.*

Creating a New Drawing

1. In the **PROJECT MANAGER**, choose the **New Drawing** button; the **Create New Drawing** dialog box is displayed.

2. Enter **C02_tut01** in the **Name** edit box. Next, choose the **Browse** button; the **Select template** dialog box is displayed. In this dialog box, select **ACAD_ELECTRICAL.dwt** from the list displayed and then choose the **Open** button; the name and path of the template is displayed in the **Template** edit box, refer to Figure 2-14.

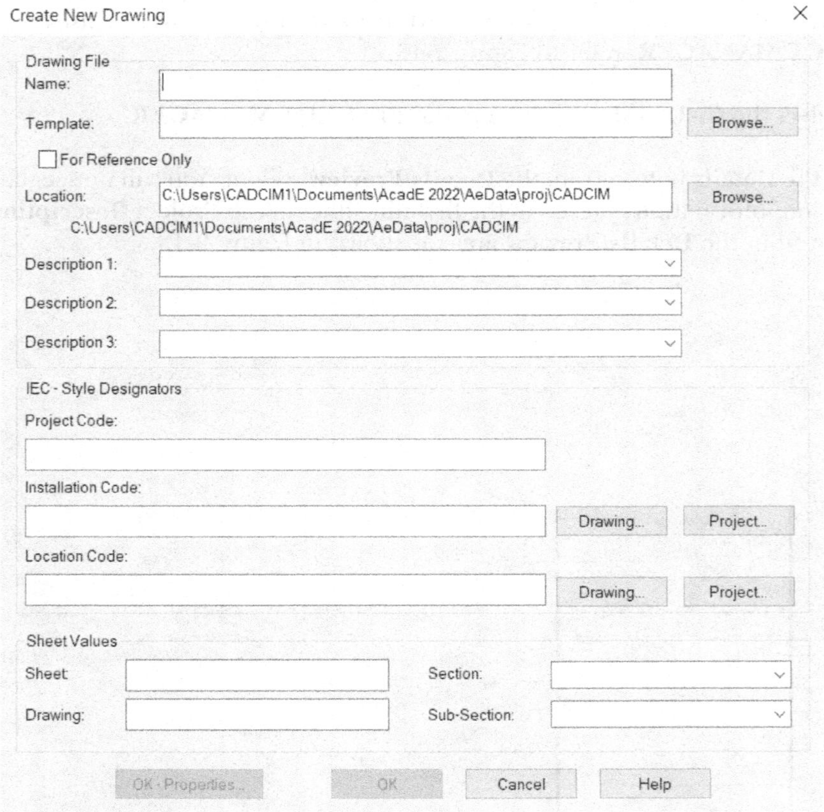

***Figure 2-14** The **Create New Drawing** dialog box*

3. Clear the **For Reference Only** check box, if selected. Next, choose the **Browse** button available on the right of the **Location** edit box; the **Browse For Folder** dialog box is displayed. By default, the **CADCIM** project is selected. Choose the **OK** button; the location of the drawing is automatically displayed in the **Location** edit box.

4. Enter **3 Phase Motors** in the **Description 1** edit box.

5. Enter **Motor Control Circuit** in the **Description 2** edit box and enter **01** in the **Drawing** edit box of the **Sheet Values** area.

6. Choose the **OK** button from the **Create New Drawing** dialog box; you will notice that the drawing that you created gets added to the active project. To view the drawing that you created, double-click on **CADCIM** in the **PROJECT MANAGER**; the drawing list is displayed and the drawing *C02_tut01.dwg* appears in bold text in the **PROJECT MANAGER**. In this case, the active project is **CADCIM**; and therefore, the drawing gets added to this project. Click on the drawing, the details and the preview of the drawing is displayed in the **PROJECT MANAGER**, as shown in Figures 2-15 and 2-16.

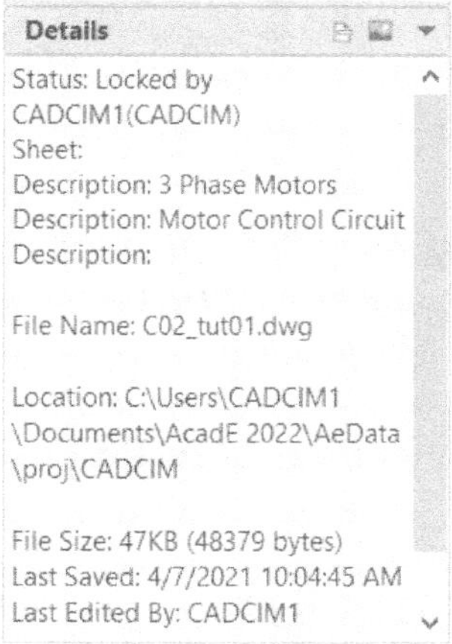

Figure 2-15 *The **Details** rollout displaying the description of C02_tut01.dwg*

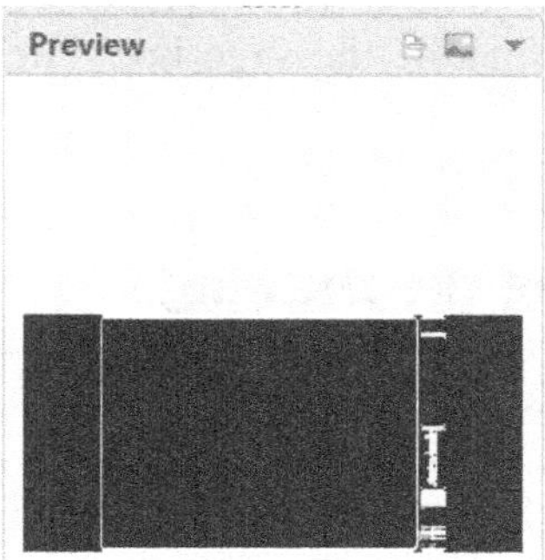

Figure 2-16 *The **Preview** area displaying the preview of C02_tut01.dwg*

Tutorial 2

In this tutorial, you will add and remove drawings from the **CADCIM** project and replace the drawings of the **POINT2POINT** project with the **CADCIM** project that you created in Tutorial 1 of this chapter. **(Expected time: 15 min)**

The following steps are required to complete this tutorial:

a. Add drawings from the existing project.
b. Remove the drawing.
c. Replace the drawing.

Adding Drawings from the Existing Project

1. Right-click on the **CADCIM** project; a shortcut menu is displayed. Choose the **Add Drawings** option from the shortcut menu; the **Select Files to Add** dialog box is displayed.

2. Select the **Proj** folder from the **Look in** drop-down list; a list of all the projects saved in this folder is displayed. Next, double-click on the **NfpaDemo** folder from the **Select Files to Add** dialog box; the drawings present in this folder are displayed.

3. Press SHIFT/CTRL and select the *DEMO01, DEMO02, DEMO03, DEMO04,* and *DEMO05* drawings. Next, choose the **Add** button; the **Apply Project Defaults to Drawing Settings** message box is displayed.
4. Choose the **Yes** button from this message box; the selected drawings are added to the **CADCIM** project, as shown in Figure 2-17.

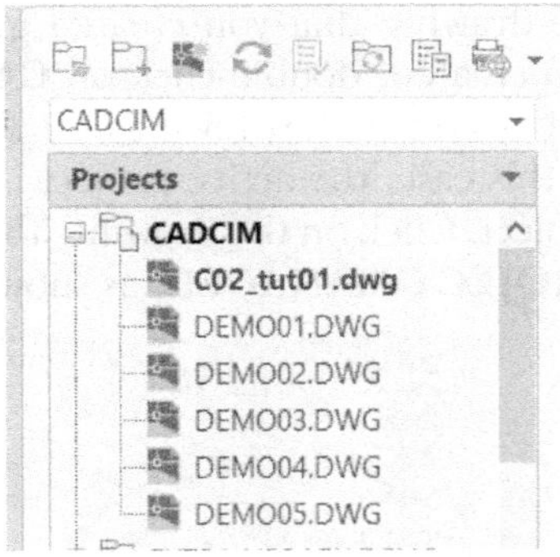

***Figure 2-17** The **PROJECT MANAGER** displaying the **CADCIM** project with the drawings added to it*

Removing the Drawing

1. Right-click on the *DEMO01.dwg* drawing and then choose the **Remove** option from the shortcut menu displayed; the **PROJECT MANAGER - Remove Files** message box is displayed. Choose the **Yes** button in this message box; the *DEMO01.dwg* drawing is removed from the drawing list, see Figure 2-18.

Replacing the Drawing

In this section, you will replace the *DEMO02.dwg* drawing of the **CADCIM** project with the *Connector.dwg* of the **POINT2POINT** project.

1. Right-click on the *DEMO02.dwg* drawing; a shortcut menu is displayed. Choose the **Replace** option from the shortcut menu; the **Select Replacement Drawing** dialog box is displayed.

2. Select the **Proj** folder from the **Look in** drop-down list. Next, double-click on the **Point2Point** folder; the *Connector.dwg* drawing is displayed in the **Select Replacement Drawing** dialog box.

3. Select the *Connector.dwg* drawing from this dialog box and choose the **Select** button; the **Apply Project Defaults to Drawing Settings** message box is displayed. Next, choose the **Yes** button; the *DEMO02.dwg* is replaced with the *Connector.dwg* drawing in the **CADCIM** project, as shown in Figure 2-19.

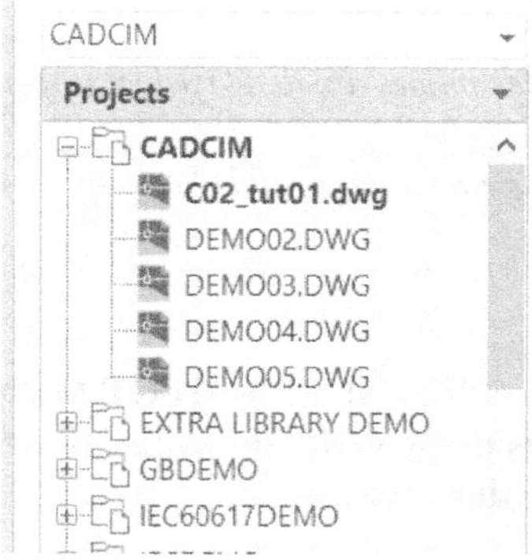

***Figure 2-18** The **PROJECT MANAGER** displaying the **CADCIM** project after removing the DEMO01.dwg drawing*

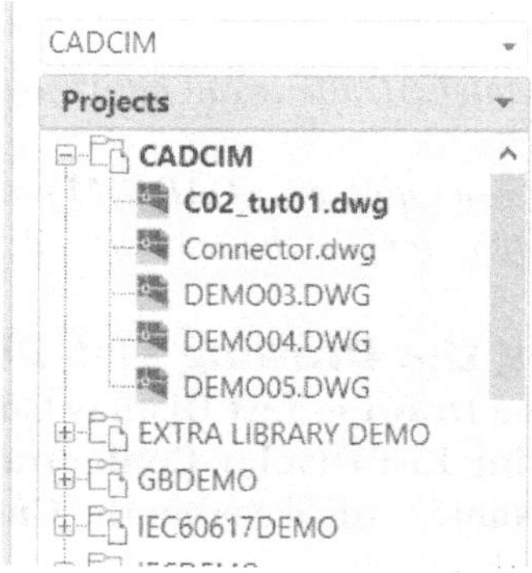

***Figure 2-19** The **PROJECT MANAGER** displaying the **CADCIM** project with the DEMO02.dwg drawing replaced by the Connector.dwg drawing*

Tutorial 3

In this tutorial, you will create a subfolder within a **CADCIM** project and configure the drawing list display for the **CADCIM** project. **(Expected time: 15 min)**

The following steps are required to complete this tutorial:

a. Create a subfolder.
b. Configure the drawing list display.

Creating a Subfolder

In this section, you will create a subfolder in the **CADCIM** project that is created in Tutorial 1.

1. Make sure that the **CADCIM** project is activated in the **PROJECT MANAGER**. Next, right-click on it; a shortcut menu is displayed.

2. Choose the **Add Subfolder** option from the shortcut menu; a subfolder with the name **NEW FOLDER** is created within the **CADCIM** project. Rename it as *TUTORIALS*.

3. Drag and drop the *C02_tut01.dwg* drawing created in Tutorial 1 on the *TUTORIALS* subfolder; the *C02_tut01.dwg* drawing is moved to the *TUTORIALS* subfolder, refer to Figure 2-20.

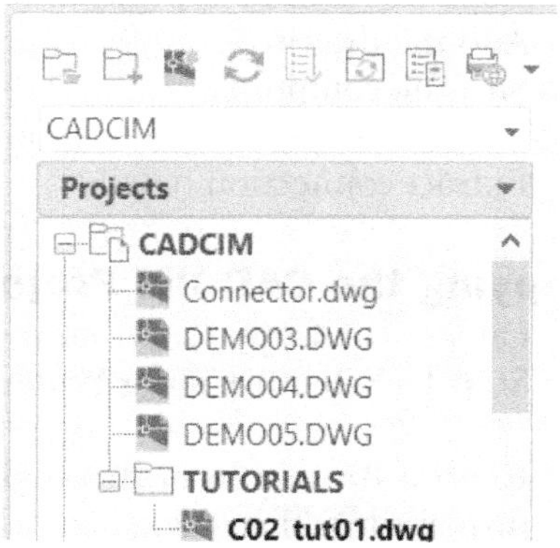

***Figure 2-20** The C02_tut01.dwg moved to the TUTORIALS subfolder*

Note

1. In AutoCAD Electrical, all projects and their subfolder names are displayed in uppercase.

*2. You can collapse the **CADCIM** project by right-clicking on its name and choosing the **Collapse All** option from the shortcut menu displayed.*

Configuring the Drawing List Display

1. Choose the **Drawing List Display Configuration** button from the **PROJECT MANAGER**; the **Drawing List Display Configuration** dialog box is displayed. You will notice that only the **File Name** is displayed in the **Current Display Order** area.

2. Select **Drawing Number(%D)** from the **Display Options** area and then choose the **>>** button; **Drawing Number(%D)** is moved to the **Current Display Order** area. Similarly, select **Drawing Description 1** from the **Display Options** area and choose the **>>** button; **Drawing Description 1** is moved to the **Current Display Order** area. Next, choose the **OK** button to close the **Drawing List Display Configuration** dialog box.

   You will notice that drawing numbers are added to the first four drawings in the **CADCIM** project and description is added to the *C02_tut01.dwg*.

Tutorial 4

In this tutorial, you will copy an existing project, name it **Copy_CADCIM**, and add some drawings into it. You will also surf a component and extract component information and connection details of the **CADCIM** project to an external file using the **Location View** tab in the **PROJECT MANAGER**. **(Expected time: 15 min)**

The following steps are required to complete this tutorial:

a. Copy the **CADCIM** project.
b. Add the drawings.
c. Surf the component.
d. Extract component information.
e. Extract connection details.

Copying the CADCIM Project

1. Choose the **Copy** tool from the **Project Tools** panel of the **Project** tab; the **Copy Project: Step 1 - Select Existing Project to Copy** dialog box is displayed, as shown in Figure 2-21.

2. Choose **Browse** from this dialog box; the **Select existing project to copy** dialog box is displayed. In this dialog box, select the *c:\Users\User Name\Documents\AcadE 2022\AeData\Proj* folder and then double-click on the **CADCIM** folder and then select the **CADCIM.wdp** file displayed, refer to Figure 2-22. Next, choose the **Open** button; path of the *CADCIM.wdp* file is displayed in the edit box of the **Copy Project: Step 1 - Select Existing Project to Copy** dialog box.

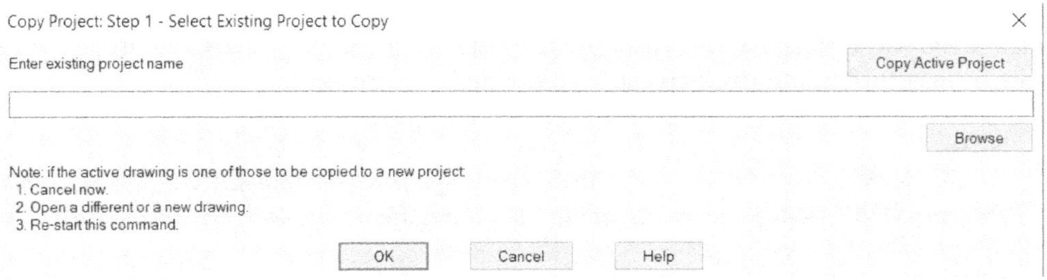

***Figure 2-21** The **Copy Project Step 1 - Select Existing Project to Copy** dialog box*

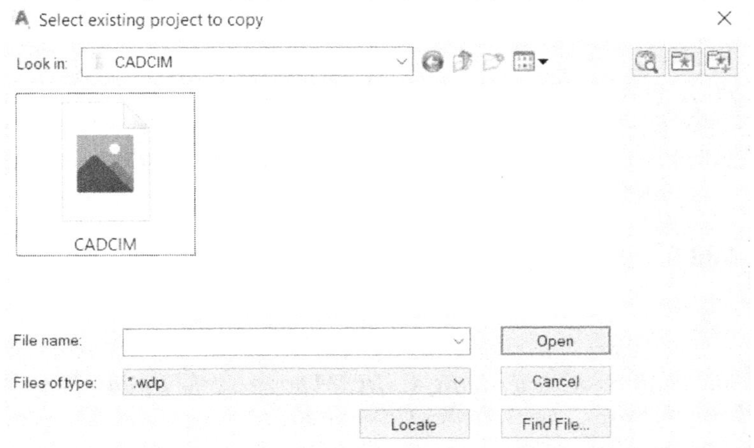

***Figure 2-22** The **Select existing project to copy** dialog box*

Note

*If the **CADCIM** project is activated, then you can choose the **Copy Active Project** button instead of following step 2.*

3. Now, choose **OK** from this dialog box; the **Copy Project: Step 2 -- Select path and name for new project** dialog box is displayed. Choose the **Up One Level** button from this dialog box and then choose the **Create New Folder** button; the **New Folder** is added to the **Proj** folder. Next, rename the new folder as *Copy_CADCIM*.

4. Double-click on the **Copy_CADCIM** folder in the **Copy Project: Step 2 -- Select path and name for new project** dialog box and then enter **Copy_CADCIM** in the **File name** text box.

5. Now, choose the **Save** button; the **Select Drawings to Process** dialog box is displayed. In this dialog box, choose all the drawings except the *c02_tut01* file and choose the **Process** button. Next, choose the **OK** button; the **Copy Project: Step 4 -- Enter Base Path for Project Drawings** dialog box is displayed.

6. In this dialog box, the edit box located at the top displays default path for the project drawings. Choose **OK** from this dialog box; the **Copy Project: Step 5 -- Adjust new drawing file names** dialog box is displayed.

7. Choose **OK** from this dialog box; the **Copy_CADCIM** project is added to the **Projects** rollout of the **PROJECT MANAGER**. Also, the drawings in the **Copy_CADCIM** project get saved in its project folder and this project becomes an active project.

Note
To change the name of drawings in the copied project, select drawings one by one in the ***Copy Project: Step 5 -- Adjust new drawing file names*** *dialog box and then choose* ***Edit****.*

Adding the Drawings

1. Right-click on the **Copy_CADCIM** project in the **PROJECT MANAGER**; a shortcut menu is displayed. Choose the **Add Drawings** option from the shortcut menu; the **Select Files to Add** dialog box is displayed.

2. In this dialog box, choose the **Up one level** button; the **Proj** folder is displayed in the **Look in** drop-down list. Next, double-click on the **NfpaDemo** folder; the drawings present in this folder are displayed. Press SHIFT/CTRL and select the *DEMO07, DEMO08,* and *DEMO09* drawings.

3. Choose the **Add** button; the **Apply Project Defaults to Drawing Settings** message box is displayed.

Note
While adding the drawings to the ***Copy_CADCIM*** *project, if the* ***Update Terminal Associations*** *message box is displayed, choose the* ***Yes*** *button in it.*

4. Choose the **Yes** button from this message box; the selected drawings are added to the **Copy_CADCIM** project. If the drawings are not displayed in the **Copy_CADCIM** project, choose **Refresh** from the **PROJECT MANAGER**.

Surfing the Component

1. Activate the **CADCIM** project. Choose the **Location View** tab from the **PROJECT MANAGER**. Next, choose the **CADCIM** node from the **PROJECT MANAGER**. Next, choose the **Display Details and Connections** button; the **PROJECT MANAGER** is expanded with the **Details** tab activated, as shown in Figure 2-23.

   You will notice that the component information is displayed in the **Details** pane in the grid.

2. Scroll down in the **PROJECT MANAGER** and select **LS406** from the **Tag** column. Next, right-click on it; a shortcut menu is displayed, as shown in Figure 2-24.

3. Choose **Surf** from the shortcut menu; the **Surf** dialog box is displayed, as shown in Figure 2-25. Choose **Go To** from the **Surf** dialog box; the *demo004.dwg* file is opened with **LS406** zoomed in.

Note
The options in the ***Surf*** *dialog box are discussed in detail in Chapter 6.*

*Figure 2-23 Expanded **PROJECT MANAGER***

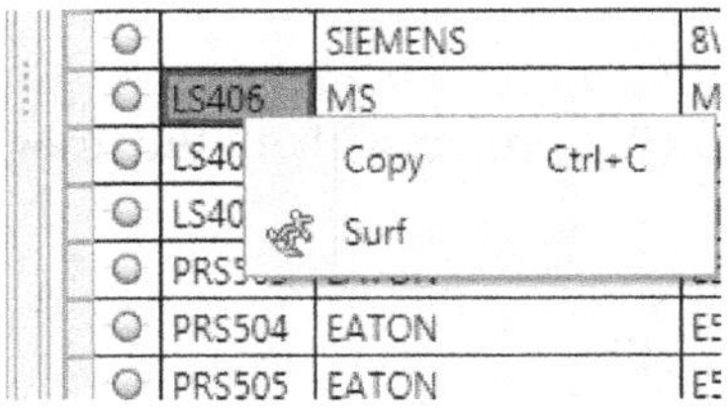

Figure 2-24 Shortcut menu displayed

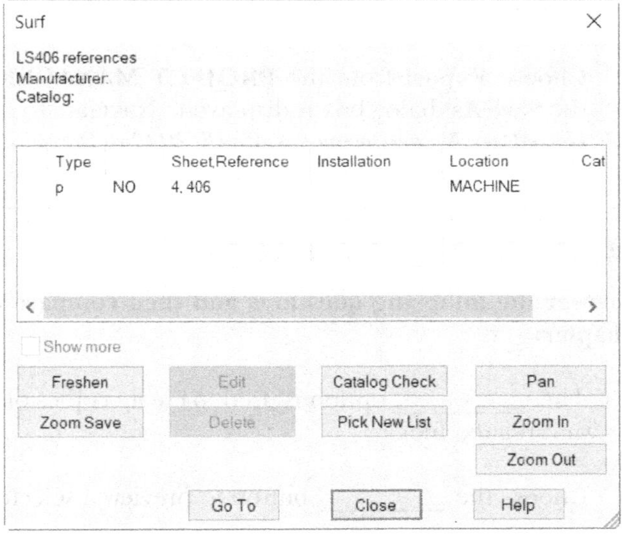

*Figure 2-25 The **Surf** dialog box*

Extracting the Component Information

1. Right-click on the **Tag** column in the grid; a shortcut menu is displayed. Choose **More** from this shortcut menu; the **Columns to display** dialog box is displayed.

2. Scroll down in the **Columns to display** dialog box and select the **Category**, **Rating1**, **Location**, and **Terminal Number** check boxes, as shown in Figure 2-26. Next, choose **OK**; the **Category**, **Rating1**, **Location**, and **Terminal Number** columns are added to the grid in the **Details** tab.

3. Choose **Export** from the **PROJECT MANAGER**; the **Save As** dialog box is displayed. Browse to *c:\Users\User Name\Documents\AcadE 2022\AeData\proj\CADCIM* and choose **Save**; the *ComponentDetails.csv* file is saved at the specified location.

Extracting the Connection Details

1. Choose the **Connections** tab in the **PROJECT MANAGER**; the **PROJECT MANAGER** is modified with connection details of the **CADCIM** project displayed in the grid.

2. Right-click on the **Wire Number** column in the grid; a shortcut menu is displayed. Choose **More** from this shortcut menu; the **Columns to display** dialog box is displayed.

3. Scroll down in the **Columns to display** dialog box and select the **Wire Type**, **Wire Color**, and **Wire Gauge** check boxes. Next, choose **OK**; the **Wire Type**, **Wire Color**, and **Wire Gauge** columns are added to the grid in the **Connections** tab.

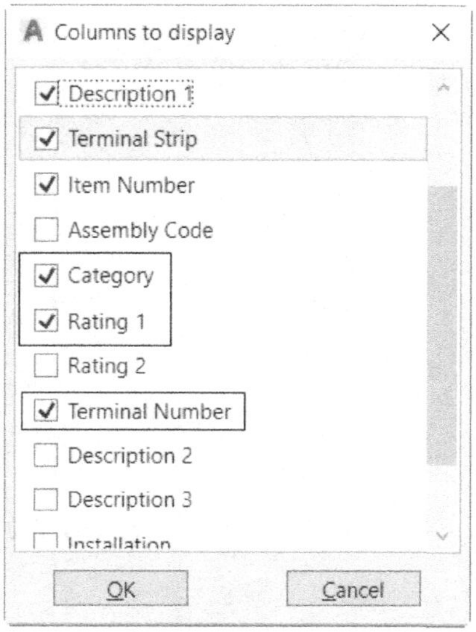

***Figure 2-26** The **Columns to display** dialog box*

4. Choose **Export** from the **PROJECT MANAGER**; the **Save As** dialog box is displayed. Navigate to *c:\Users\User Name\Documents\AcadE 2022\AeData\proj\CADCIM* and choose **Save**; the *ConnectionDetails.csv* file is saved at the specified location.

Self-Evaluation Test

Answer the following questions and then compare them to those given at the end of this chapter:

1. The __________ option is used to create copies of an existing project with all the drawings and related files.

2. Choose the __________ button to preview a selected drawing.

3. A project file is an ASCII text file. (T/F)

4. The **PROJECT MANAGER** is used to manage the project files only. (T/F)

Review Questions

Answer the following questions:

1. Which of the following buttons is used to change the appearance of the drawing list displayed in the **PROJECT MANAGER**?

   (a) **Plot Project** (b) **Project Task List**
   (c) **Move Up** (d) **Drawing List Display Configuration**

2. Which of the following options is used to add existing drawing(s) to a project?

(a) **Descriptions** (b) **Add Active Drawing**
(c) **Add Drawings** (d) **Title Block Update**

3. The **Remove Drawings** option is used both for removing and deleting a drawing file. (T/F)

4. You can close only the non-active projects. (T/F)

5. Projects can be closed by using the Project selection drop-down list. (T/F)

EXERCISES

Exercise 1

Create a new project with the project name **NEW_PROJECT** and add appropriate description to it. **(Expected time: 10 min)**

Note

It is recommended to save all exercises of the forthcoming chapters in the ***NEW_PROJECT*** *project.*

Exercise 2

Create a new drawing with the name *C02_exer02.dwg* in the project created in Exercise 1 and add appropriate description to it. **(Expected time: 10 min)**

Exercise 3

Using the **Copy** button, create a new project with the name **COPY_PROJECT** containing the drawings and settings of any existing project. **(Expected time: 10 min)**

Answers to Self-Evaluation Test

1. Copy Project, **2. Preview**, **3.** T, **4.** F

This page is intentionally left blank

Chapter 3

Working with Wires

Learning Objectives

After completing this chapter, you will be able to:

- *Insert different types of wires*
- *Modify inserted wires*
- *Create and manage wire types and their properties*
- *Insert, erase, hide, fix, and reposition wire numbers*
- *Insert special wire numbering*
- *Insert in-line wire markers*
- *Check line entities on non-wire layers*

INTRODUCTION

In this chapter, you will learn in detail about the types of wire and wire layers. Also, you will learn how to insert wires in a drawing, modify wires using commands such as trim, stretch, create different wire types in drawings, insert in-line wire markers in wires, insert wire numbers in wires and reposition them, and troubleshoot wires. Later in this chapter, you will learn how to check, repair, trace wire gaps and pointers, and manipulate wire gaps. You will also learn about the source and destination signal arrows in this chapter.

WIRES

A wire is a stretched out strand of drawn metal and is usually cylindrical in shape. It is used to carry electricity and telecommunication signals. The standard size of a wire is determined by wire gauge.

Wires are AutoCAD lines when they are placed on an AutoCAD Electrical defined wire layer. By default, the wire layer, WIRES is present in a drawing. In AutoCAD Electrical, you can create as many wire layers as you want. You can also assign wire numbers to the wires that will be included in various wire connection reports. Two wire segments or a wire segment and a component are said to be connected if they fall within a trap distance of any part of the other wire segment or component. If more than one wire is connected together, it forms a wire network. In the next section, you will learn about the types of wires and the insertion of wires in drawings.

Inserting Wires into a Drawing

In this section, you will learn about different types of wires and will also learn to insert them in a drawing. You can insert wire segments on a wire layer horizontally, vertically, or angled at 22.5, 45, or 67.5 degrees.

Inserting Single Wire

Command: AEWIRE

The **Wire** tool is used to insert wires. To insert a wire, choose the **Wire** tool from the **Insert Wires/Wire Numbers** panel of the **Schematic** tab; you will be prompted to specify the start point of the wire. You can either select a point using the pointing device or enter its coordinates at the Command prompt. After the start point of the wire is selected, AutoCAD Electrical will prompt you to enter the endpoint of the wire. Specify the endpoint where you want to terminate the wire; a wire will be drawn between the two specified points. You can specify the endpoint horizontally or vertically by entering H or V at the Command prompt. At this point, you may continue to select points or exit the **Wire** tool by pressing ESC or pressing the ENTER key twice. The command sequence that is displayed when you invoke the **Wire** tool is given next.

Choose the **Wire** tool
Current wiretype: "WIRES"
Specify wire start or [wireType/X=show connections]: *Specify the wire start point or enter an option (the options are discussed next).*
Specify wire end or [V=start Vertical H=start Horizontal TAB: Collision off Continue]: Enter.
Specify wire start or [Scoot/wireType/X=show connections]: *Press ESC to exit the command.*

Scoot/wireType/X=show connections

This Command prompt is displayed when you press ENTER after specifying the endpoint of the wire. Enter **S** at the Command prompt; the cursor will change into a selection box on the screen and you will be prompted to select the component, wire, or wire number. Select the wire, wire number, or component from your drawing to move them as per your requirement. The **Scoot** tool is used to move components to the wire it is currently located on. This tool is discussed in detail in the later chapters.

Inserting Wires at Angles

You can also insert wires at certain angles. To do so, click on the down arrow on the **Wire** tool from the **Insert Wires/Wire Numbers** panel of the **Schematic** tab; the **Wire** drop-down will be displayed, as shown in Figure 3-1.

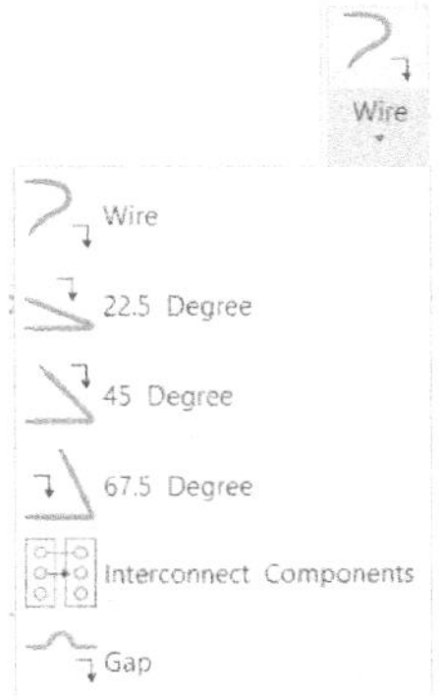

*Figure 3-1 The **Wire** drop-down for inserting different types of angled wires*

Choose the **22.5 Degree** tool from the **Wire** drop-down in the **Insert Wires/Wire Numbers** panel of the **Schematic** tab; you will be prompted to select a component or the branch for the 22.5 degree wire. You can specify a point by using the pointing device or entering its coordinates at the Command prompt. After the start point of the wire is selected, you will be prompted to specify the endpoint of the wire. Specify the endpoint where you want to terminate the wire; a wire will be drawn between the two specified points. At this point, you may continue selecting points or exit the **22.5 Degree** tool by pressing ESC. Similarly, choose the **45 Degree** and the **67.5 Degree** tools to insert wires at 45 degree and 67.5 degree, respectively.

Inserting Multiple Bus Wiring

Command: AEMULTIBUS

The **Multiple Bus** tool is used to draw multiple wire bus. This tool is very useful when you want to create 3-phase circuits, point-to-point wiring diagrams, and so on. To draw multiple wire bus, choose the **Multiple Bus** tool from the **Insert Wires/Wire Numbers** panel of the **Schematic** tab; the **Multiple Wire Bus** dialog box will be displayed, as shown in Figure 3-2.

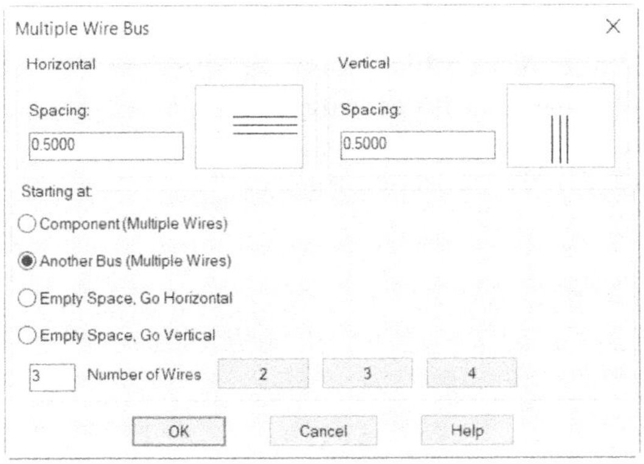

*Figure 3-2 The **Multiple Wire Bus** dialog box*

Specify the horizontal spacing between two adjacent wires of a bus in the **Spacing** edit box of the **Horizontal** area. By default, 0.5000 is displayed in this edit box. Specify the vertical spacing between two adjacent wires of a bus in the **Spacing** edit box of the **Vertical** area. By default, 0.5000 is displayed in this edit box.

The **Starting at** area is used to specify the start point of the wire. The **Component (Multiple Wires)** radio button is used to start bus wires at wire connection points of the component. Note that you can select the connection points of a component using the window selection method. After selecting this radio button, the **Number of Wires** edit box will not be available. The **Another Bus (Multiple Wires)** radio button is used to start branching off the bus from the existing bus or set of wires. The **Empty Space, Go Horizontal** radio button is used to start a horizontal bus in the empty space. Select the **Empty Space, Go Vertical** radio button to start the vertical bus from the point you specify in the empty space.

The **Number of Wires** edit box is used to specify the number of wires that you want to insert into the drawing.

Trimming a Wire

Command: AETRIM

The **Trim Wire** tool is used to erase the specified portion of a wire segment and wire tees. You can use this tool to trim multiple wires by selecting a single wire and drawing a fence through wires. To do so, choose the **Trim Wire** tool from the **Edit Wires/Wire Numbers** panel of the **Schematic** tab; the selection box will appear on the screen and you will be prompted to select the wire to trim. Select the wire segment to remove it from the drawing. Next, press ENTER or ESC to terminate the tool. Alternatively, click on the blank space in the drawing area to exit the tool.

Stretching Wires

Command: AESTRETCTWIRE

The **Stretch Wire** tool is used to stretch a wire to connect it to another wire or to a component. To do so, choose the **Stretch Wire** tool from the **Modify Wire** drop-down in the **Edit Wires/Wire Numbers** panel of the **Schematic** tab, as shown in Figure 3-3; the cursor will change into a selection box on the screen and you will be prompted to select the end of wire. Select the end of the wire that you want to connect to another wire or to a component. Next, press ESC or ENTER to exit the tool.

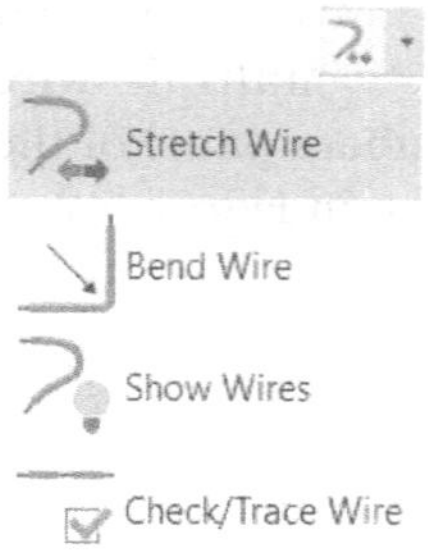

***Figure 3-3** The **Modify Wire** drop-down*

Note

*You can use the **Stretch Wire** tool only if a drawing consists of a component or a wire up to which the selected wire will be stretched.*

Creating Wire Types

Command: AEWIRETYPE

The **Create/Edit Wire Type** tool is used to create new wire types, edit existing wire types, configure wire types, and manage wire types. To create a wire type, choose the **Create/Edit Wire Type** tool from the **Modify Wire Type** drop-down in the **Edit Wires/Wire Numbers** panel of the **Schematic** tab, as shown in Figure 3-4; the **Create/Edit Wire Type** dialog box will be displayed, as shown in Figure 3-5.

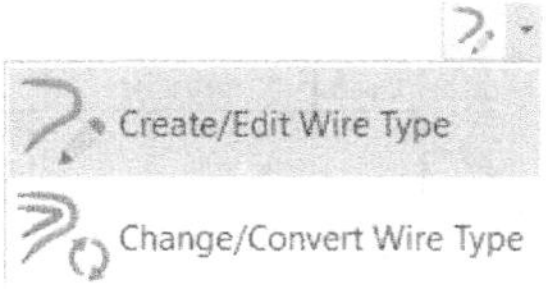

*Figure 3-4 The **Modify Wire Type** drop-down*

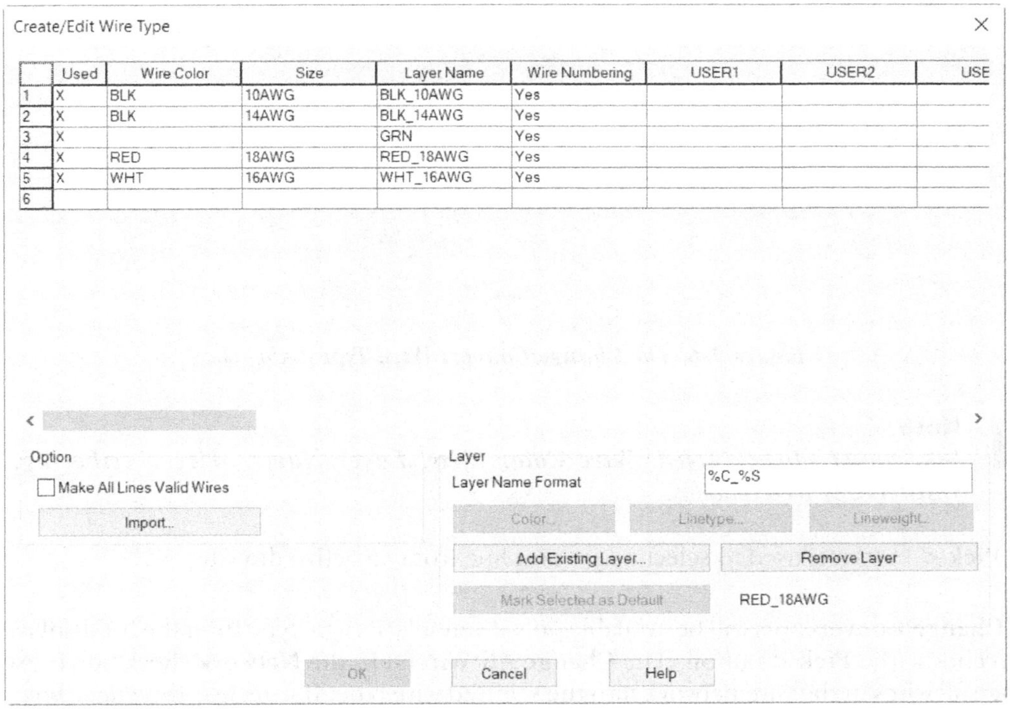

*Figure 3-5 The **Create/Edit Wire Type** dialog box*

This dialog box displays the list of wire types used in the active drawing and the list of layer names. It also displays the properties of wires such as color, size, and user-defined properties of wire in different columns of the grid area. This dialog box can also be used to specify whether the wire numbers should be assigned to the wires of a selected wire layer or not.

Changing and Converting Wire Types

Command: AECONVERTWIRETYPE

The **Change/Convert Wire Type** tool is used to change the wire type of existing wires. This tool is also used to convert lines into wires. To change the wire type of existing wires, choose the **Change/Convert Wire Type** tool from the **Modify Wire Type** drop-down in the **Edit Wires/Wire Numbers** panel of the **Schematic** tab; the **Change/Convert Wire Type** dialog box will be displayed, as shown in Figure 3-6. This dialog box displays all valid wire type layers present in a drawing. Most of the options in this dialog box are same as those in the **Create/Edit Wire Type** dialog box.

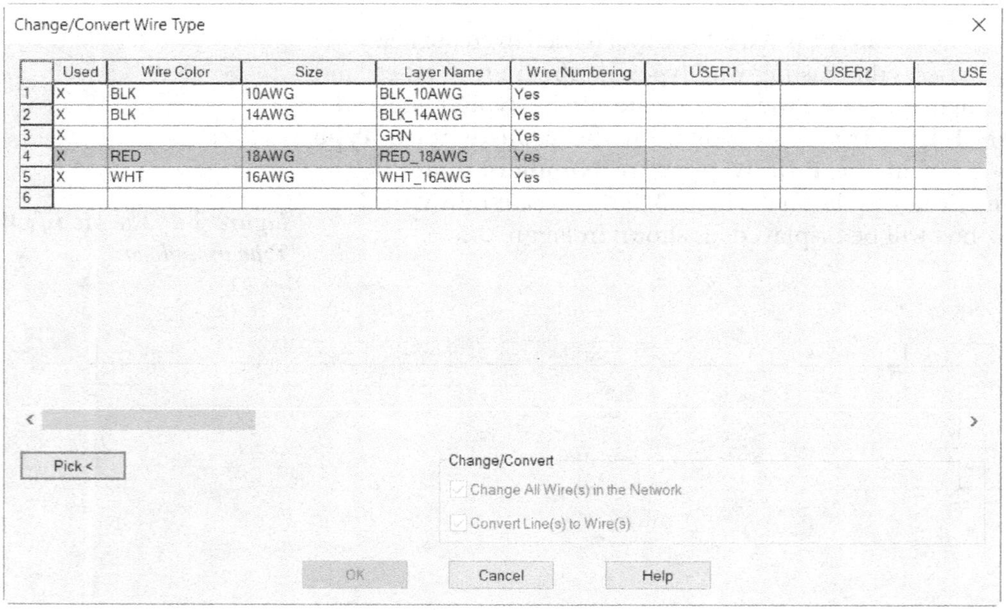

***Figure 3-6** The **Change/Convert Wire Type** dialog box*

Note
*You cannot edit the **Used, Wire Color, Size, Layer Name, Wire Numbering,** and **USER1-USER20** columns.*

The **Pick <** button is used to select a wire or a line from an active drawing.

The **Change/Convert** area will be available only if you select a wire type from the list displayed or if you choose the **Pick <** button. The **Change All Wire(s) in the Network** check box is used to change all wires in the wire network into the selected wire type. If you clear this check box, only the single wire that you select will be changed to the selected wire type. The **Convert Line(s) to Wire(s)** check box is used to convert the selected line(s) to wire(s).

Select a wire type from the list displayed in the **Change/Convert Wire Type** dialog box. Next, choose the **OK** button; the cursor will change to the selection box and you will be prompted to select objects. Select wires and lines (if any) and press ENTER; the selected wires and lines (if found) will automatically be changed or converted to the wire type that you have selected from the **Change/Convert Wire Type** dialog box. Note that the row of the default wire type is highlighted in gray and the row of the selected wire type is highlighted in blue.

Setting Wire Types

You can change the wire type while inserting a wire into a drawing. To do so, choose the **Wire** tool from the **Insert Wires/Wire Numbers** panel of the **Schematic** tab; you will be prompted to specify the wire start. Enter **T** at the Command prompt and press ENTER; the **Set Wire Type** dialog box will be displayed, as shown in Figure 3-7.

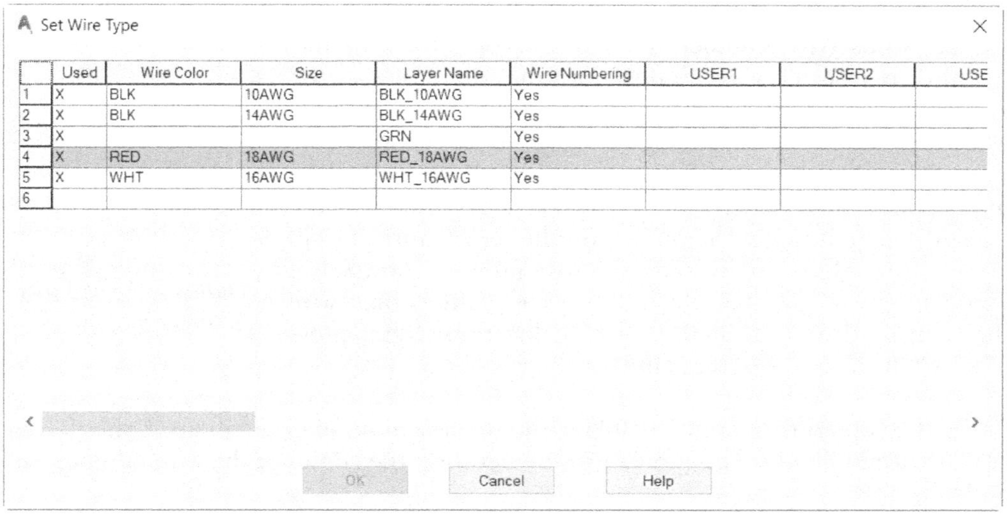

Figure 3-7 The ***Set Wire Type*** *dialog box*

The **Used** column displays the layers or the wire types that are already used in a drawing. If the **Used** column is blank, it indicates that the drawing consists of the layer name but currently no wires are found on that layer. However, if 'X' is displayed in the **Used** column, it indicates that the particular wire type is currently used in the drawing.

The **USER1 -USER20** columns are user-defined columns. Note that the row of the default wire type will be highlighted in gray color. Select the wire type; the **OK** button will be activated. Choose the **OK** button; the selected wire type will become the current wire type and the wires that you insert will be created with the current wire type unless and until the wire type has been changed again.

Inserting Wire Numbers

Command: AEWIRENO

The **Wire Numbers** tool is used to assign wire numbers to wires in a drawing. To assign wire numbers to wires, choose the **Wire Numbers** tool from the **Insert Wire numbers** drop-down in the **Insert Wires/Wire Numbers** panel of the **Schematic** tab, as shown in Figure 3-8; the **Wire Tagging** dialog box will be displayed. Using this dialog box, you can assign wire numbers to individual wires, wires in a single drawing, or wires of a whole project.

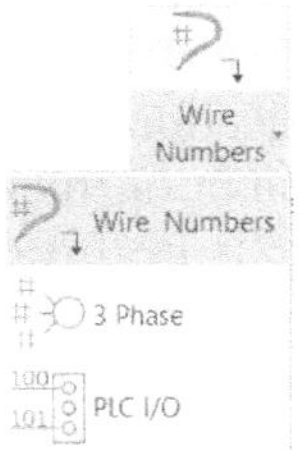

Figure 3-8 *The* ***Insert Wire numbers*** *drop-down*

Copying Wire Numbers

Command: AECOPYWIRENO

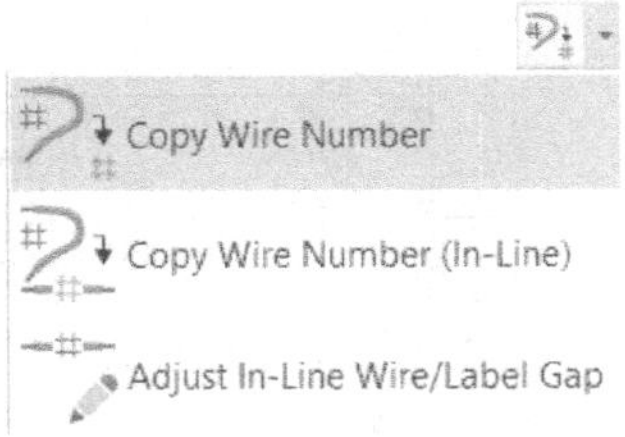

***Figure 3-9** The **Copy Wire Number** drop-down*

The **Copy Wire Number** tool is used to copy and insert the already inserted wire number into the wire. You can use this tool to place an extra copy of the wire number anywhere in a wire network. To do so, choose the **Copy Wire Number** tool from the **Copy Wire Number** drop-down in the **Edit Wires/Wire Numbers** panel of the **Schematic** tab, as shown in Figure 3-9; the cursor will change into a selection box on the screen and you will be prompted to select wire for the extra wire number copy. Select the wire whose wire number you want to copy; the copied wire number will be displayed at the selected location. Next, press ENTER or ESC to exit the command.

Note that the copied wire numbers follow the main wire number attribute of the network. If you update or modify the wire number in the network, the copies of the wire number will also get updated.

Deleting Wire Numbers

Command: AEERASEWIRENUM

The **Delete Wire Numbers** tool is used to delete a wire number. Using this tool, you can delete the main wire number of the wire network as well as you can delete the copies of the wire number. To delete the wire number, choose the **Delete Wire Numbers** tool from the **Edit Wires/Wire Numbers** panel of the **Schematic** tab; the cursor will change into a selection box and you will be prompted to select objects. Select the wire numbers to be deleted and press ENTER; the selected wire numbers will be deleted automatically.

Editing Wire Numbers

Command: AEEDITWIRENO

The **Edit Wire Number** tool is used to edit an existing wire number and assign a new wire number to the wires that do not have wire numbers. To edit an existing wire number, choose the **Edit Wire Number** tool from the **Edit Wire Number** drop-down in the **Edit Wires/Wire Numbers** panel of the **Schematic** tab, as shown in Figure 3-10; the cursor will change into a selection box and you will be prompted to select the wire or the component. Select the wire number or the wire; the **Edit Wire Number/Attributes** dialog box will be displayed. Using the options in this dialog box, you can modify, fix, and unfix the wire number and add/edit attributes in it.

Fixing Wire Numbers

Command: AEFIXWIRENO

The **Fix** tool is used to lock wire numbers in a drawing to their current values. To do so, choose the **Fix** tool from the **Edit wire Number** drop-down in the **Edit Wires/Wire Numbers** panel of the **Schematic** tab; the cursor will change into a selection box and you will be prompted to select objects. Select the wire numbers that you want to lock and press ENTER; the wire numbers will be fixed and moved to the WIREFIXED layer automatically. Note that if you update wire numbers, the fixed wire numbers will not be updated.

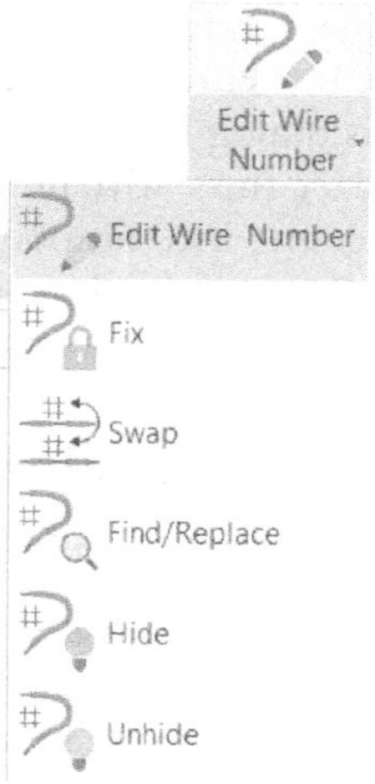

***Figure 3-10** The **Edit Wire Number** drop-down*

Hiding Wire Numbers

Command: AEHIDEWIRENO

The **Hide** tool is used to hide a selected wire number. Also, this tool is used to change the attribute layer of the selected wire number to the WIRENO_HIDE layer (frozen layer). To hide the wire number, choose the **Hide** tool from the **Edit wire Number** drop-down in the **Edit Wires/Wire Numbers** panel of the **Schematic** tab; the cursor will change to a selection box on the screen and you will be prompted to select the wire number. Select the wire number; the wire number will be automatically moved to a special hidden layer and will become invisible. Next, press ENTER to exit the command.

Unhiding Wire Numbers

Command: AESHOWWIRENO

The **Unhide** tool is used to make the hidden wire numbers visible. To do so, choose the **Unhide** tool from the **Edit Wire Number** drop-down in the **Edit Wires/Wire Numbers** panel of the **Schematic** tab; the cursor will change into a selection box and you will be prompted to select the wire or the wire number to unhide the wire number. Select the wire number or the wire with which the hidden wire number is linked; the wire number will be visible automatically.

Swapping Wire Numbers

Command: AESWAPWIRENO

The **Swap** tool is used to interchange the position of wire numbers between two wire networks. To do so, choose the **Swap** tool from the **Edit Wire Number** drop-down in the **Edit Wires/Wire Numbers** panel of the **Schematic** tab; the cursor will change into a selection box and you will be prompted to select the first wire or the wire number. Next, select the first wire or the wire number; you will be prompted to select the second wire or the wire number. Select the second wire or the wire number; the wire numbers will interchange their positions. Press ENTER or ESC to exit the command.

Finding/Replacing Wire Numbers

Command: AEFINDWIRENO

The **Find/Replace** tool is used to find the required wire number and replace it with the specified wire number value. Using this tool, you can find the required wire number within the project or in the current drawing. To do so, choose the **Find/Replace** tool from the **Edit Wire Number** drop-down in the **Edit Wires/Wire Numbers** panel of the **Schematic** tab; the **Find/Replace Wire Numbers** dialog box will be displayed, as shown in Figure 3-11. This dialog box has three sets of **Find/Replace** edit boxes. As a result, you can find as well as replace three wire numbers at a time. In the **Find** edit box, enter the wire number text that you want to search and in the **Replace** edit box, enter the wire number text with which you want to replace the found text.

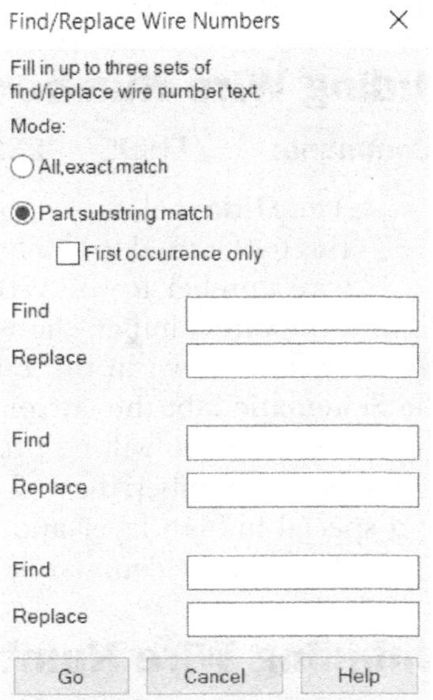

*Figure 3-11 The **Find/Replace Wire Numbers** dialog box*

Moving a Wire Number

Command: AEMOVEWIRENO

The **Move Wire Number** tool is used to move an existing wire number to a specific location on the same wire network. To do so, choose the **Move Wire Number** tool from the **Edit Wires/Wire Numbers** panel of the **Schematic** tab; the cursor will change into a selection box and you will be prompted to specify a new wire number location on the wire. Select the wire segment where you want the wire number to be moved; the wire number will automatically be moved to the specified location. Next, press ENTER or right-click on the screen to exit the command.

Scooting a Wire Number

Command: AESCOOT

The **Scoot** tool is used to slide a selected wire number on the same wire to the specified location. To do so, choose the **Scoot** tool from the **Modify Components** drop-down in the **Edit Components** panel of the **Schematic** tab, as shown in Figure 3-12; the cursor will change into a selection box and you will be prompted to select the wire number. Select the wire number; a rectangle drawn in temporary graphics will be displayed, which indicates the selected wire number. Next, specify the location where you want to place the wire number; the wire number will move on the same wire to the specified location. Next, press ENTER to exit the command. You can also scoot the entire wire, components, ladder bus, rungs, and so on.

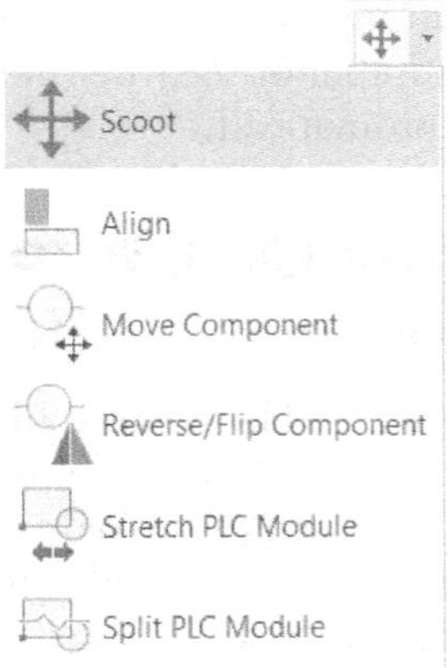

*Figure 3-12 The **Modify Components** drop-down*

Flipping a Wire Number

Command: AEFLIPWIRENO

The **Flip Wire Number** tool is used to flip or reverse the direction of the selected wire number about its respective wire. To do so, choose the **Flip Wire Number** tool from the **Edit Wires/Wire Numbers** panel of the **Schematic** tab; the cursor will change into a selection box and you will be prompted to select the wire number to mirror. Next, select the wire number; the wire number will be flipped. Now, press ENTER to exit the command.

Toggling the Wire Number Position

Command: AETOGGLEWIRENO

The **Toggle Wire Number In-line** tool is used to move the selected wire number from right, left, top, or bottom to in-line with wire and vice-versa. To do so, choose the **Toggle Wire Number In-line** tool from the **Edit Wires/Wire Numbers** panel of the **Schematic** tab; the cursor will change into a selection box and you will be prompted to select the wire number to toggle. Next, select the wire number whose position you want to change; the wire number will be toggled. Right-click or press ENTER to exit the command. You can also select the wire itself to toggle the wire number.

Repositioning the Wire Number Text with the Attached Leader

Command: AEWIRENOLEADER

The **Wire Number Leader** tool is used to reposition the wire number text with an attached leader. You can add the wire leader to every wire number or to a selected wire number, if the space for the normal positioning of the wire number is not available. To relocate the wire number, choose the **Wire Number Leader** tool from the **Wire Number Leader** drop-down in the **Insert Wires/Wire Numbers** panel of the **Schematic** tab, as shown in Figure 3-13; you will be prompted to select the wire number for the leader. Select the wire number text; you will be prompted to specify distance for the wire leader. Now, drag the cursor and click to specify a new position for the wire number for the leader. Next, right-click or press ENTER to position the wire number along with the leader. Press ENTER to exit the command.

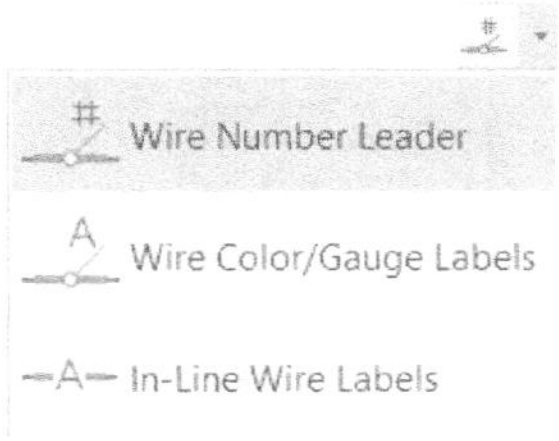

***Figure 3-13** The **Wire Number Leader** drop-down*

To remove the leader from the wire number, enter '**C**' at the Command prompt. Enter '**S**' at the Command prompt to scoot component, wire, and wire number. Press ENTER to exit the command.

Inserting In-Line Wire Markers

Command: AEINLINEWIRE

Markers are used to identify conductor color or signal name. These markers can be inserted in-line to any wire. Also, these markers are for reference and are not considered for wire numbering and reports. To insert an in-line marker, choose the **In-Line Wire Labels** tool from the **Wire Number Leader** drop-down in the **Insert Wires/Wire Numbers** panel of the **Schematic** tab; the **Insert Component** dialog box will be displayed, as shown in Figure 3-14. This dialog box displays a list of predefined in-line markers and user-defined markers.

Select a marker from the **NFPA: In-Line Wire Labels** area of the **Insert Component** dialog box; the dialog box will disappear and the cursor will change into a cross hair. Also, you will be prompted to specify the insertion point for the in-line marker. Now, specify the insertion point for the in-line marker; the marker will be inserted in between the wire. This command will continue until you press ESC. Alternatively, right-click on the screen and choose the **Cancel** option from the shortcut menu to cancel the command.

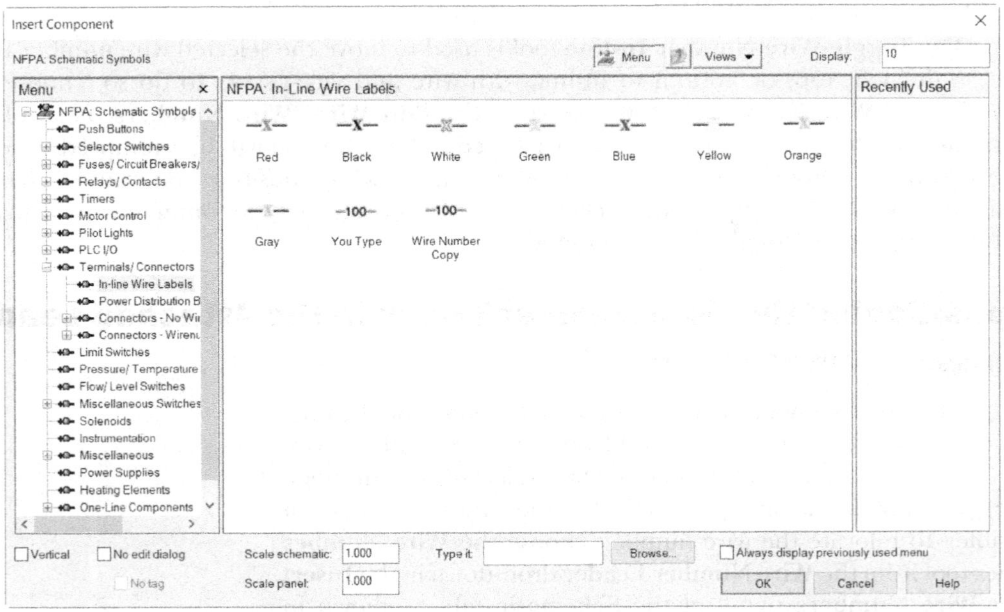

*Figure 3-14 The **Insert Component** dialog box*

You can also create and insert user-defined markers. To do so, select the **You Type** in-line wire marker from the **NFPA: In-Line Wire Labels** area; you will be prompted to specify the insertion point for the marker. Next, specify the insertion point for the marker; the **Edit Attribute - COLOR** dialog box will be displayed, as shown in Figure 3-15. In this dialog box, enter the attribute value in the edit box. You can also increase or decrease the already specified attribute value by choosing the **<** or **>** button displayed on the right of the **Edit Attribute-COLOR** dialog box. Alternatively, you can choose the **Pick** button to select a similar attribute value from the drawing that has been specified earlier. Next, choose the **OK** button; the attribute value of the in-line marker will be inserted into your drawing.

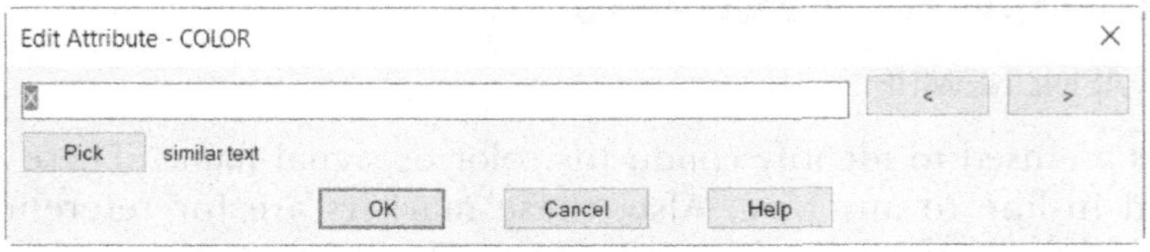

*Figure 3-15 The **Edit Attribute - COLOR** dialog box*

You can also insert the wire number as in-line marker. To do so, select the **Wire Number Copy** label from the **NFPA:In-Line Wire Labels** area; you will be prompted to specify the insertion point for the in-line marker. Specify the insertion point; the **Edit Wire Number/Attributes** dialog box will be displayed. In this dialog box, enter wire number for the wire if it is not already assigned; the wire number will be displayed as in-line wire marker. If you do not specify wire number for the wire in the **Edit Wire Number/Attributes** dialog box and if it does not have wire number, *"wn"* will get inserted into the wire.

Inserting Wire Color/Gauge Labels in a Drawing

Command: AEWIRECOLORLABEL

The **Wire Color/Gauge Labels** tool is used to insert wire color/gauge labels to a wire with or without a leader. To insert wire color/gauge labels, choose the **Wire Color/Gauge Labels** tool from the **Wire Number Leader** drop-down in the **Insert Wires/Wire Numbers** panel of the **Schematic** tab; the **Insert Wire Color/Gauge Labels** dialog box will be displayed. In this dialog box, choose the **Setup** button to change the text size, arrow size and style, and leader gap size of the label. Choose the **Auto Placement** button; you will be prompted to select objects. Next, select the wires for which you want to add wire color/gauge labels and press ENTER; the **Color/gauge text for XX** dialog box will be displayed. You can edit the text string in this dialog box and choose **OK**; the wire color/gauge labels will be added and placed automatically on selected wires. Choose the **Manual** button in the **With leader** area if wire color/gauge labels are to be placed manually with a leader. Choose the **Manual** button in the **No leader** area if wire color/gauge labels are to be placed manually without a leader.

Inserting the Special Wire Numbering in a Drawing

Command: AE3PHASEWIRENO

The **3 Phase** tool is used to insert special wire numbers into the 3 phase bus and motor circuits. Using this tool, you can speed up the work of inserting wire numbers into the 3 phase circuits. To insert special wire numbers, choose the **3 Phase** tool from the **Insert Wire Numbers** drop-down in the **Insert Wire Numbers** panel of the **Schematic** tab, refer to Figure 3-16; the **3 Phase Wire Numbering** dialog box will be displayed, as shown in Figure 3-17. The options in this dialog box are discussed next.

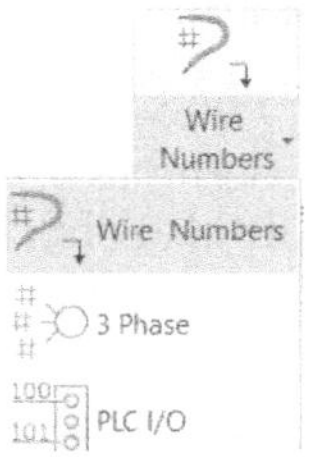

*Figure 3-16 The **Insert Wire Numbers** drop-down*

The **Prefix** area is used to specify prefix value for wire numbers. In the **Base** area, enter the starting number for the base of wire numbers in the edit box. Alternatively, choose the **Pick** button to select an existing wire number attribute value from the active drawing. Also, you can increase or decrease the base of the wire number by choosing the **>** or **<** button. In the **Suffix** area, enter the suffix value for wire numbers in the edit box. Alternatively, choose the **List** button to select a pre-defined value for the suffix value.

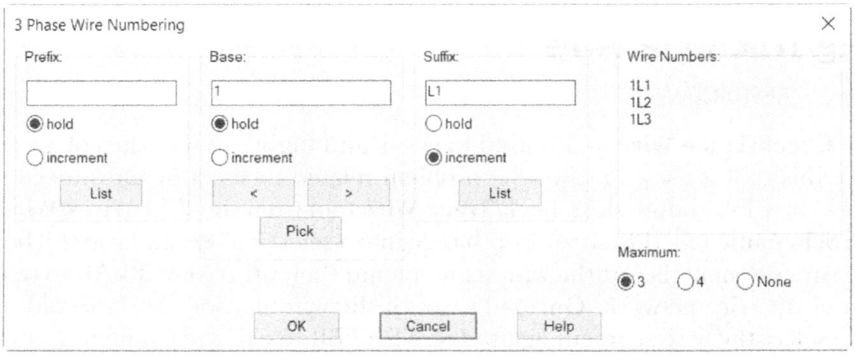

*Figure 3-17 The **3 Phase Wire Numbering** dialog box*

The **hold** and **increment** radio buttons are used to hold and increment the prefix value, base value, and suffix value for all wire numbers of a drawing. The **Wire Numbers** area displays the preview of the wire numbers specified in the edit boxes of the **Prefix**, **Base**, and **Suffix** areas. The **Maximum** area consists of **3**, **4**, and **None** radio buttons. Depending upon the radio button selected, the preview of the values specified in the **Prefix**, **Base**, and **Suffix** areas get automatically updated in the **Wire Numbers** area.

Checking Line Entities

Command: AESHOWWIRE

The **Show Wires** tool is used to highlight lines (wires) in bright red color, which are found on a valid AutoCAD Electrical wire layer. This tool is also used to trace wire numbers and find out whether these wire numbers are on a valid wire layer or not. Additionally, this tool is used to check whether an entity is a wire or not. To check whether these wires or wire numbers are on valid wire layer or not, choose the **Show Wires** tool from the **Modify Wire** drop-down in the **Edit Wires/Wire Numbers** panel of the **Schematic** tab.

Checking and Repairing Gap Pointers

Command: AEGAPPOINTER

The **Check/Repair Gap Pointers** tool is used to add or repair Xdata pointers on wire segments. This tool is also used to verify Xdata pointers on both sides of a wire gap/loop. It also checks for their validity. If the Xdata pointers on both sides of a wire gap/loop are not valid, pointers will be created accordingly.

To add or repair Xdata pointers on wire segments, choose the **Check/Repair Gap Pointers** tool from the **Modify Wire Gap** drop-down in the **Edit Wires/Wire Numbers** panel of the **Schematic** tab, as shown in Figure 3-18; the cursor will change into a selection box and you will be prompted to select the wire segment. Select the wire segment; you will be prompted to select the other wire segment. Next, select the wire segment; the gap data is added, as required in wires. The command will continue until you press ENTER.

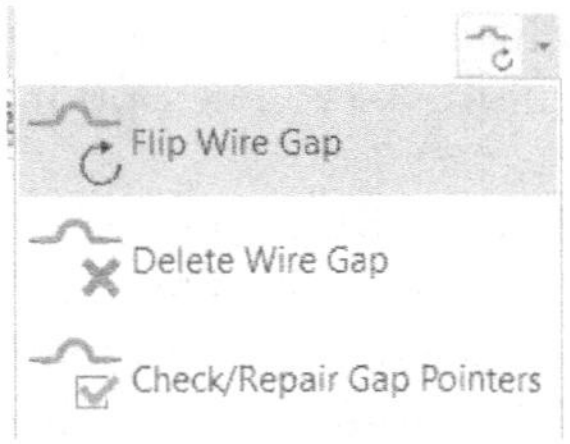

Figure 3-18 *The* ***Modify Wire Gap*** *drop-down*

Checking/Tracing a Wire

Command: AETRACEWIRE

The **Check/Trace Wire** tool is used to check and highlight the wires of a wire network. Also, this tool is used to solve the problem related to short or unconnected wires. To check the wire, choose the **Check/Trace Wire** tool from the **Edit Wires/Wire Numbers** panel of the **Schematic** tab; the cursor will change into a selection box and you will be prompted to select the wire segment. Select the wire segment and then press SPACEBAR to select the rest of the wires of the wire network. Once all wires of the wire network are selected, you will be prompted to select the wire segment again. Press ENTER to exit the command.

If you enter **A** at the **Pan/Zoom/All segments/Quit/<Space=single step>** prompt, all segments of the wire will be highlighted simultaneously. Alternatively, if you want to step through wire by wire, press SPACEBAR. If you want to pan or zoom the selected wire, enter **P** or **Z** at the Command prompt; the connected wire segment endpoints will be displayed at the Command prompt. To exit the command, enter **Q** at the Command prompt.

TUTORIALS

Tutorial 1

In this tutorial, you will insert wire numbers into the *DEMO05.dwg* drawing file, as shown in Figure 3-19. Next, you will add wire to this drawing and use the **Trim Wire** tool to trim the wire. Also, you will use the **Delete Wire Numbers** tool to delete a wire number.

(Expected time: 15 min)

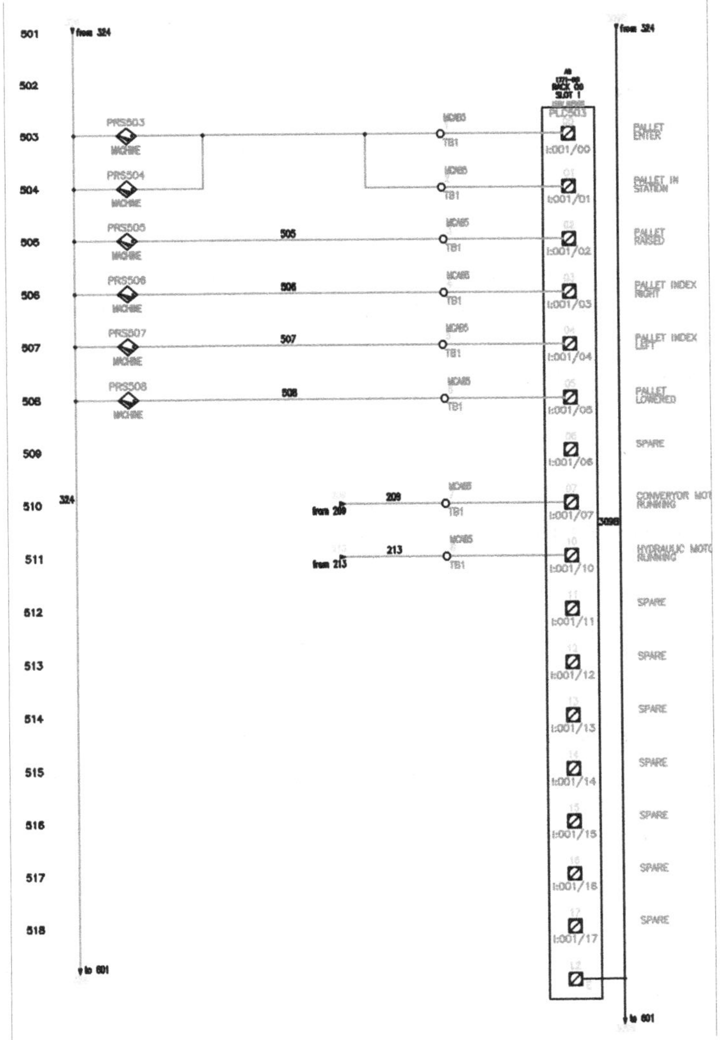

Figure 3-19 Ladder diagram for Tutorial 1

The following steps are required to complete this tutorial:

a. Open the drawing.
b. Save the drawing.
c. Add the drawing to the **CADCIM** project list.
d. Insert the wire.
e. Trim the wire.
f. Delete the wire number.
g. Save the drawing file.

Opening the Drawing

1. Make sure the **CADCIM** project is activated. If not, right-click on the **CADCIM** project; a shortcut menu is displayed. Choose the **Activate** option from it.

2. Double-click on the drawing *DEMO05.dwg* to open it.

Saving the Drawing

1. Save the drawing *DEMO05.dwg* with the name *C03_tut01.dwg*. To do so, choose **Save As > Drawing** from the **Application Menu** or choose **File > Save As** from the menu bar; the **Save Drawing As** dialog box is displayed.

Note

*If the **Save Drawing As** dialog box is not displayed, press ESC and enter **FILEDIA** at the Command prompt; you are prompted to enter a new value for FILEDIA. Next, enter **1** at the Command prompt. Now, choose **Save As > Drawing** from the **Application Menu**; the **Save Drawing As** dialog box is displayed.*

2. Next, choose **Documents > Acade 2022 > AeData > Proj > CADCIM** and then enter **C03_tut01** in the **File name** edit box.

   By default, **AutoCAD 2018 Drawing (*.dwg)** is displayed in the **Files of type** drop-down list.

3. Choose the **Save** button from the **Save Drawing As** dialog box; the *C03_tut01.dwg* file is saved.

Adding the Drawing to the CADCIM Project List

1. Right-click on the **CADCIM** project; a shortcut menu is displayed. Choose the **Add Active Drawing** option from the shortcut menu; the **Apply Project Defaults to Drawing Settings** message box is displayed. Choose the **Yes** button; the *C03_tut01.dwg* is added to the **CADCIM** project.

Note

*While adding the C03_tut01.dwg drawing file to the **CADCIM** project if the **Update Terminal Associations** message box is displayed, choose the **Yes** button in it; the C03_tut01.dwg is added to the **CADCIM** project.*

2. Drag and drop the *C03_tut01.dwg* drawing on the *TUTORIALS* subfolder; the *C03_tut01.dwg* drawing is moved to the *TUTORIALS* subfolder.

3. Next, choose **Save** from the **Application Menu** to save the drawing.

Inserting the Wire

1. Choose the **Wire** tool from **Schematic > Insert Wires/Wire Numbers > Wire** drop-down; you are prompted to specify the start point of the wire. Specify the start point of the wire on the left of wire number 503, refer to Figure 3-20; you are prompted to specify the end point of the wire. Now, specify the endpoint of the wire on the left of wire number 504; you are again prompted to specify the start point of the wire. Next, specify the start point on the right of wire number 503; you are prompted to specify the endpoint of the wire. Specify the endpoint of the wire on the right of wire number 504; the wires are inserted in the drawing, as shown in Figure 3-20.

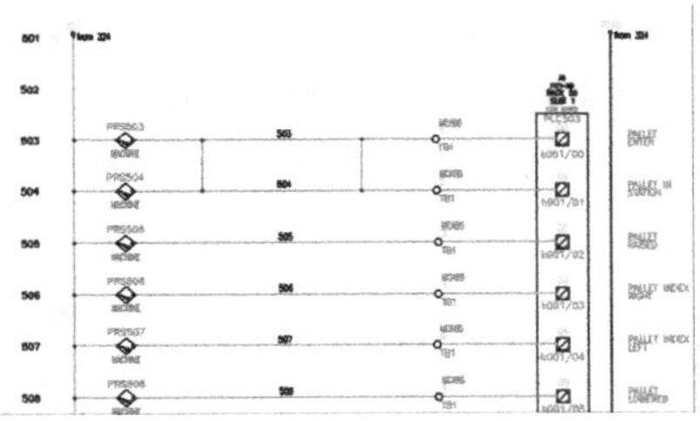

Figure 3-20 *Wires inserted in the drawing*

2. Press ENTER to exit the command.

Trimming the Wire

1. Choose the **Trim Wire** tool from the **Edit Wires/Wire Numbers** panel of the **Schematic** tab; you are prompted to select the wire. Select the middle wire of rung 504 that makes a short circuit between rung 503 and rung 504; the wire is trimmed, as shown in Figure 3-21.

2. Next, press ENTER to exit the command.

Deleting the Wire Number

1. Choose the **Delete Wire Numbers** tool from the **Edit Wires/Wire Numbers** panel of the **Schematic** tab; you are prompted to select objects. Select the wire number 503 of rung 503 and then press ENTER; the wire number 503 is deleted, as shown in Figure 3-22.

Saving the Drawing File

1. Choose **Save** from the **Application Menu** to save the *C03_tut01.dwg* drawing file.

Note

*It is recommended to save all tutorials in the forthcoming chapters in the TUTORIALS subfolder of the **CADCIM** project.*

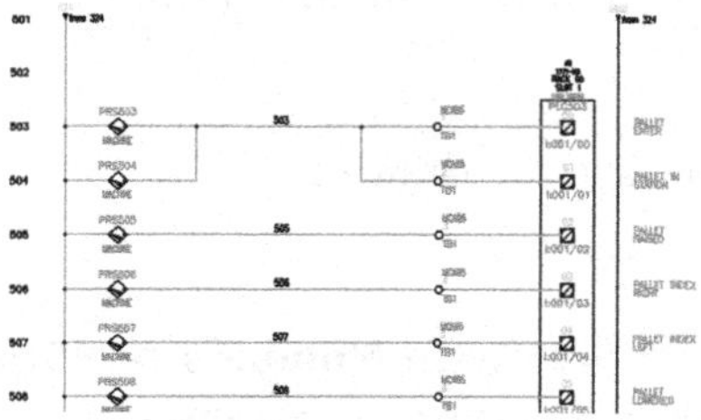

Figure 3-21 Ladder after trimming the middle portion of rung 504

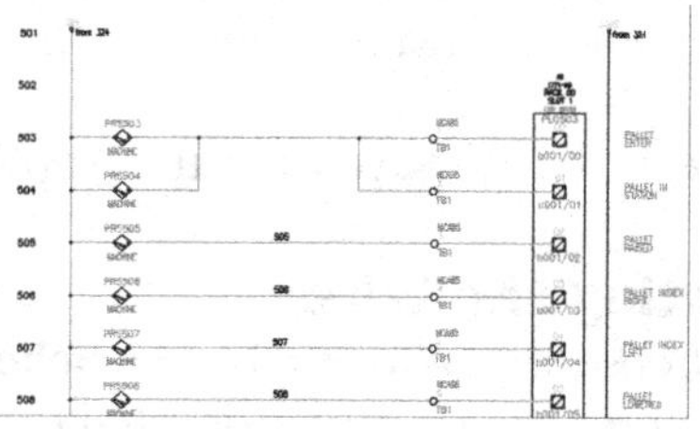

Figure 3-22 Ladder after deleting the wire number of rung 503

Tutorial 2

In this tutorial, you will use the *C03_tut01.dwg* drawing file and create a wire type using the **Create/Edit Wire Type** tool and then change the wire type of another wire to the wire type that you have created using the **Change/Convert Wire Type** tool, as shown in Figure 3-23. **(Expected time: 10 min)**

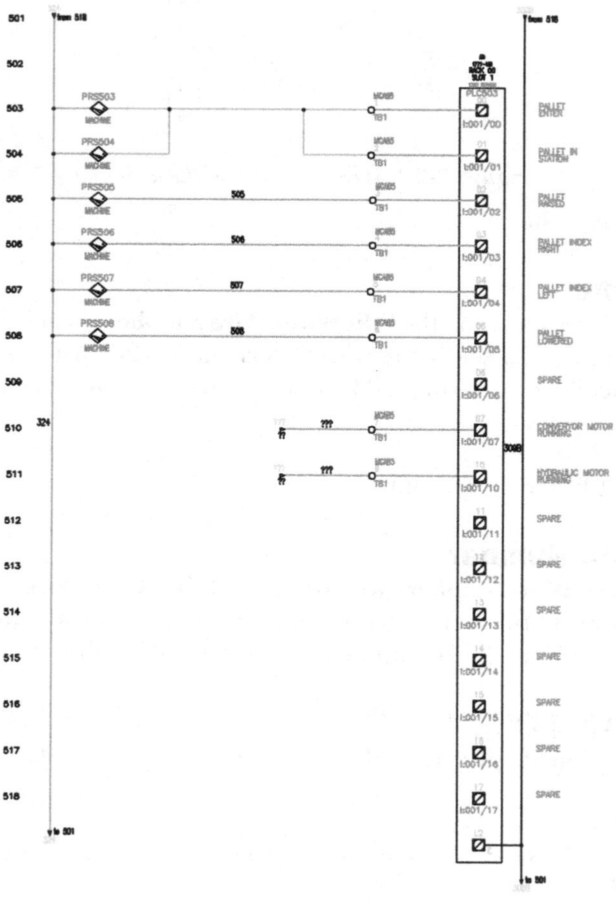

Figure 3-23 Ladder diagram for Tutorial 2

The following steps are required to complete this tutorial:

a. Open, save, and add the drawing to the **CADCIM** project list.
b. Create the wire type.
c. Change the wire type.
d. Save the drawing.

Opening, Saving, and Adding the Drawing to the CADCIM Project List

1. Make sure the **CADCIM** project is activated. Double-click on the *C03_tut01.dwg* drawing to open it. You can also download this file from *www.cadcim.com*. The path of the file is as follows:

   Textbooks > CAD/CAM > AutoCAD Electrical > AutoCAD Electrical 2022: A Tutorial Approach

2. Choose **Save As > Drawing** from the **Application Menu**; the **Save Drawing As** dialog box is displayed. Next, browse to **Documents > Acade 2022 > AeData > Proj > CADCIM** in the **Save in** drop-down list and then enter **C03_tut02** in the **File name** edit box.

   By default, **AutoCAD 2018 Drawing (*.dwg)** is displayed in the **Files of type** drop-down list.

3. Choose the **Save** button in the **Save Drawing As** dialog box; the *C03_tut02.dwg* is saved.

4. Right-click on the **CADCIM** project; a shortcut menu is displayed. Choose the **Add Active Drawing** option from the shortcut menu; the **Apply Project Defaults to Drawing Settings** message box is displayed. Choose the **Yes** button in the message box; the *C03_tut02.dwg* is added to the **CADCIM** project drawing list.

Note

*While adding the C03_tut02.dwg drawing file to the **CADCIM** project if the **Update Terminal Associations** message box is displayed, choose the **Yes** button from it; the C03_tut02.dwg is added to the **CADCIM** project.*

5. Move the drawing *C03_tut02.dwg* to the *TUTORIALS* subfolder as discussed earlier. Next, choose **Save** from the **Application Menu** to save the drawing.

Creating the Wire Type

1. Choose the **Create/Edit Wire Type** tool from **Schematic > Edit Wires/Wire Numbers > Modify Wire Type** drop-down; the **Create/Edit Wire Type** dialog box is displayed, as shown in Figure 3-24.

2. Click in the **Wire Color** column of the blank cell of the **Create/Edit Wire Type** dialog box. Next, enter **GRN** in the selected cell of the **Wire Color** column.

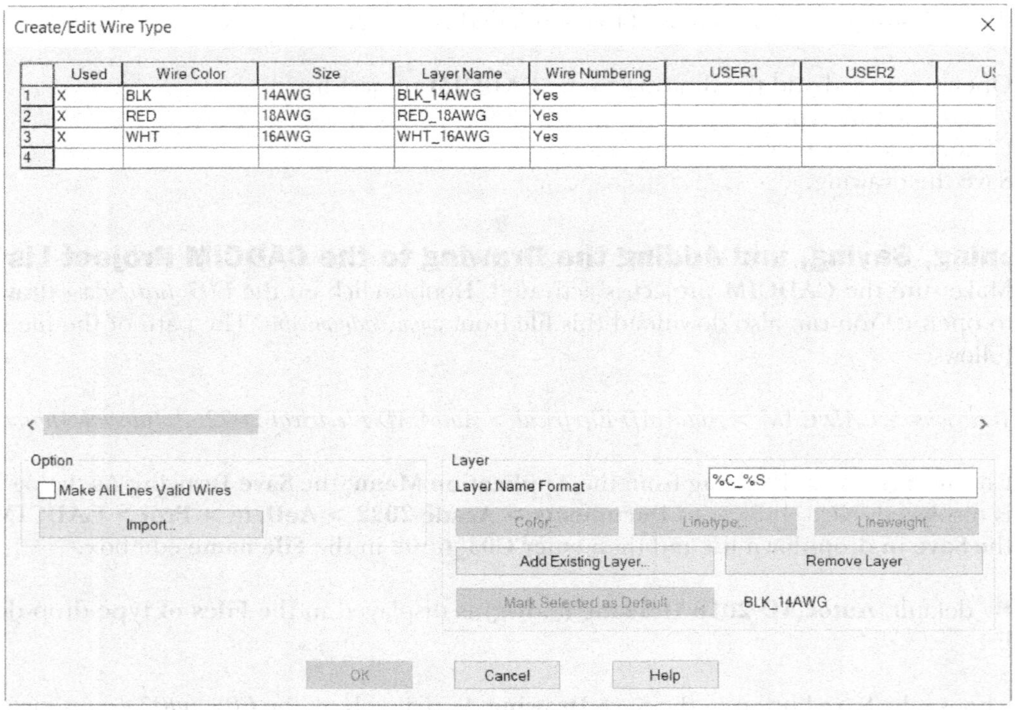

***Figure 3-24** The **Create/Edit Wire Type** dialog box*

3. Next, enter **14AWG** in the selected cell of the **Size** column and press ENTER; **GRN_14AWG** is displayed in the **Layer Name** column. Also, **Yes** is displayed in the **Wire Numbering** column, as shown in Figure 3-25.

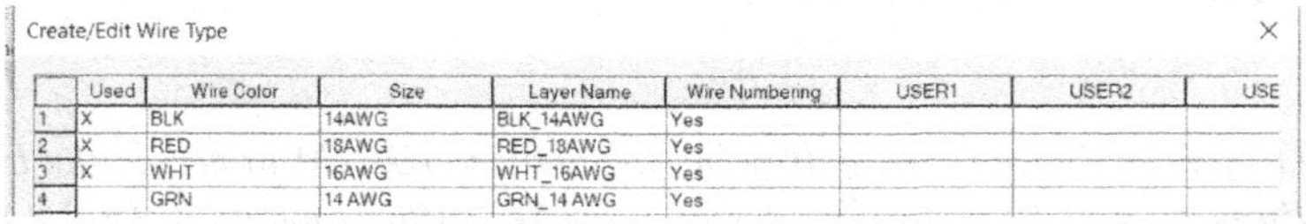

***Figure 3-25** The **Create/Edit Wire Type** dialog box showing the wire type that you have created*

4. Next, choose the **Color** button from the **Layer** area of the **Create/Edit Wire Type** dialog box; the **Select Color** dialog box is displayed. In this dialog box, select the green color and choose the **OK** button to return to the **Create/Edit Wire Type** dialog box. Next, choose the **OK** button from the **Create/Edit Wire Type** dialog box; the wire type is created.

Changing the Wire Type

Now, you need to change the wire type of the selected wire to the wire type you have created.

1. Right-click on the wire between PRS503 and TB1 of rung 503; a marking menu is displayed. Choose the **Change/Convert Wire Type** option from the marking menu; the **Change/ Convert Wire Type** dialog box is displayed. Select the row where GRN is displayed under the **Wire Color** column and choose the **OK** button; the wire type and the color of the selected wire is changed to green color (GRN_14AWG wire type). Next, press ENTER to exit the command.

Saving the Drawing

1. Choose **Save** from the **Application Menu** to save the *C03_tut02.dwg* drawing file.

Tutorial 3

In this tutorial, you will use the *C03_tut02.dwg* drawing file and insert a wire number and fix all wire numbers in it. You will also insert wire color/gauge labels, as shown in Figure 3-26.

(Expected time: 15 min)

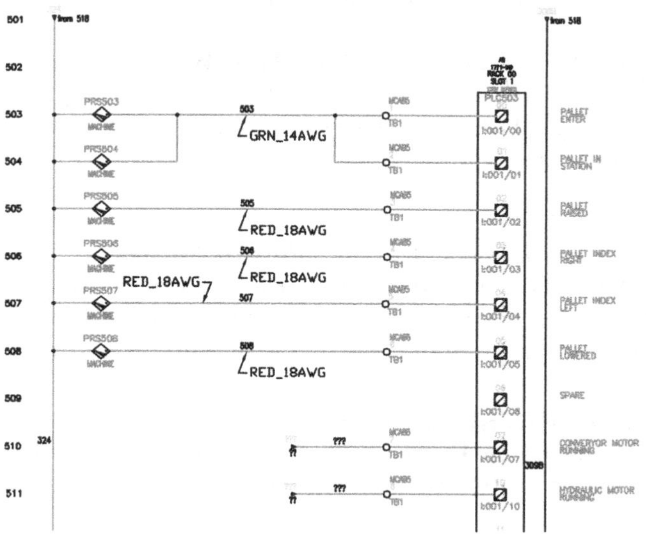

Figure 3-26 Ladder diagram for Tutorial 3

The following steps are required to complete this tutorial:

a. Open, save, and add the drawing to the **CADCIM** project list.
b. Insert a wire number and fix all wire numbers.
c. Insert wire/color gauge labels.
d. Show wires and pointers to wire numbers.
e. Save the drawing.

Opening, Saving, and Adding the Drawing to the CADCIM Project List

1. Open the *C03_tut02.dwg* file and then save it with the name *C03_tut03.dwg*.

   You can also download this file from *www.cadcim.com*. The path of the file is as follows:

   Textbooks > CAD/CAM > AutoCAD Electrical > AutoCAD Electrical 2022: A Tutorial Approach

2. Add the *c03_tut03.dwg* file to the *TUTORIALS* subfolder of the **CADCIM** project as discussed in Tutorial 2. Next, choose **Save** from the **Application Menu** to save the drawing.

Inserting a Wire Number and Fixing All Wire Numbers

1. Choose the **Wire Numbers** tool from **Schematic > Insert Wires/Wire Numbers > Wire Numbers** drop-down; the **Sheet 5 - Wire Tagging** dialog box is displayed.

2. In this dialog box, choose the **Pick Individual Wires** button; you are prompted to select objects. Select the middle wire of the rung 503 and press ENTER; the wire number 503 is inserted into the wire.

   Next, you need to fix the wire numbers in the drawing.

3. Choose the **Fix** tool from **Schematic > Edit Wires/Wire Numbers > Edit Wire Number** drop-down; you are prompted to select objects. Select the wire numbers 503, 505, 506, 507, and 508. Next, press ENTER; the selected wire numbers are fixed. Note that these wire numbers will not be updated at the time of retagging.

Note

*To unfix the fixed wire numbers, choose the **Edit Wire Number** tool from the **Schematic > Edit Wires/Wire Numbers > Edit Wire Number** drop-down; you are prompted to select the wire number. Select the wire number; the **Edit Wire Number/Attributes** dialog box is displayed. In this dialog box, clear the **Fixed** check box and choose **OK**; the selected wire number is unfixed.*

Inserting Wire Color/Gauge Labels

1. Choose the **Wire Color/Gauge Labels** tool from **Schematic > Insert Wires/Wire Numbers > Wire Number Leader** drop-down; the **Insert Wire Color/Gauge Labels** dialog box is displayed.

2. In this dialog box, choose the **Auto Placement** button in the **With leader** area; you are prompted to select objects. Select the wires of rung 503, 505, 506, 507, and 508. Next, press ENTER; the **Color/gauge text for GRN_14AWG** dialog box is displayed. Choose **OK**; the **Color/gauge text for RED_14AWG** dialog box is displayed. Again, choose **OK**; the wire color/gauge labels are inserted for the selected wires, refer to Figure 3-26.

Showing Wires and Pointers to Wire Numbers

1. Choose the **Show Wire** tool from the **Modify Wire** drop-down in the **Edit Wires/Wire Numbers** panel of the **Schematic** tab; the **Show Wires and Wire Number Pointers** dialog box is displayed.

2. Make sure the **Show wires (Lines on wire layers)** check box is selected. Next, choose **OK**; the **Drawing Audit** message box is displayed. This message box specifies that all the wires in this drawing are on wire layers and are highlighted in red color.

3. Again, choose the **Show Wire** tool from the **Modify Wire** drop-down in the **Edit Wires/Wire Numbers** panel of the **Schematic** tab; the **Show Wires and Wire Number Pointers** dialog box is displayed.

4. Clear the **Show wires (Lines on wire layers)** check box. Next, select the **Show pointers to wire numbers** check box and choose **OK**; the drawing is modified with pointers pointing to corresponding wire numbers.

Saving the Drawing

1. Choose **Save** from the **Application Menu** to save the *C03_tut03.dwg* drawing file.

Tutorial 4

In this tutorial, you will open the *C03_tut03.dwg* drawing file and use the **Show Wire Sequence** tool to trace the wire sequence of one of the wire networks in this drawing. Next, you will edit the wire sequence of this wire network using the **Edit Wire Sequence** tool.

(Expected time: 15 min)

The following steps are required to complete this tutorial:

a. Open, save, and add the drawing to the **CADCIM** project list.
b. Trace the wire sequence.
c. Edit the wire sequence.
d. Save the drawing.

Opening, Saving, and Adding the Active Drawing to the CADCIM Project List

1. Open the *C03_tut03.dwg* file and then save it with the name *C03_tut04.dwg*.

   You can also download this file from *www.cadcim.com*. The path of the file is as follows:

   Textbooks > CAD/CAM > AutoCAD Electrical > AutoCAD Electrical 2022: A Tutorial Approach

2. Add the *c03_tut04.dwg* file to the *TUTORIALS* subfolder of the **CADCIM** project as discussed in Tutorial 2. Next, choose **Save** from the **Application Menu** to save the drawing.

Tracing the Wire Sequence

1. Select the wire network, as shown in Figure 3-27.

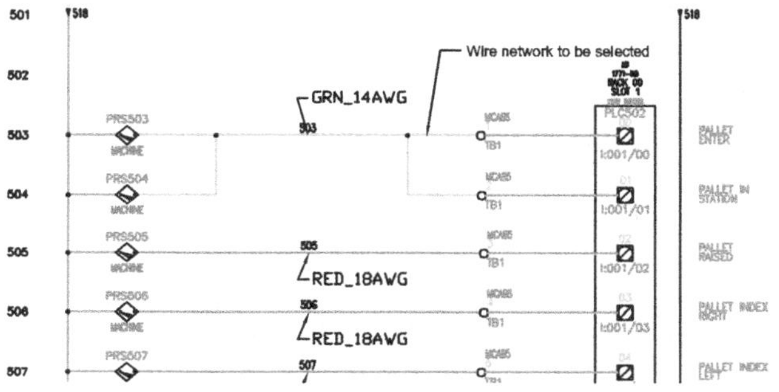

Figure 3-27 *Selected wire network*

2. Right-click on the selected wire and choose **Wire Sequence > Show Wire Sequence** from the marking menu, as shown in Figure 3-28; green arrows are displayed between **PRS503** and **PRS504** representing the start of the wire sequence, refer to Figure 3-29. Also, the Command Prompt notifies that the first wire out of the three wires in the network is displayed. Additionally, you are prompted to press SPACEBAR to see the next wire in the sequence.

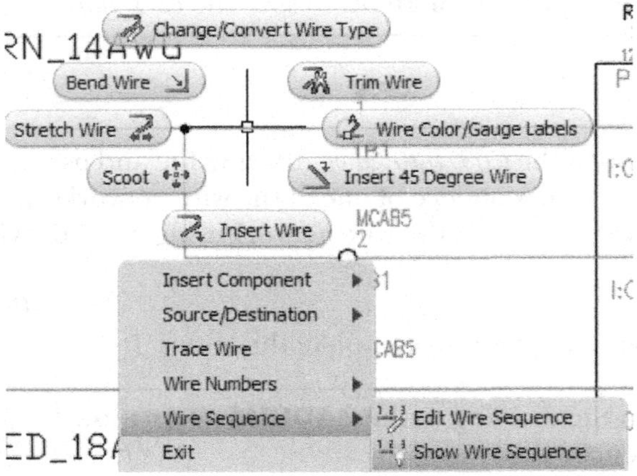

Figure 3-28 *Choosing* ***Show Wire Sequence*** *from the marking menu*

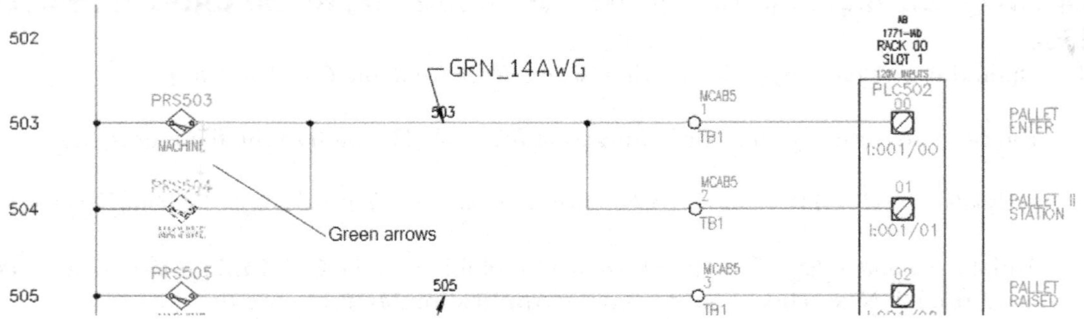

Figure 3-29 *Green arrows representing the start of the wire sequence*

3. Press SPACEBAR; the next wire sequence is shown by green arrows connecting **PRS504** with pin number 1 of **TB1** terminal strip along with the first sequence in grey arrows.

4. Press SPACEBAR again; green arrows are shown in between the pin number 1 of **TB1** and pin number 2 of **TB1** along with last two sequences in grey arrows. Figure 3-30 shows the complete wire connection sequence of the selected wire network. Next, you are prompted to press SPACEBAR again.

5. Press SPACEBAR. Now, you are prompted to press SPACEBAR to quit the command or press 1 to repeat the command. Press SPACEBAR to end the command.

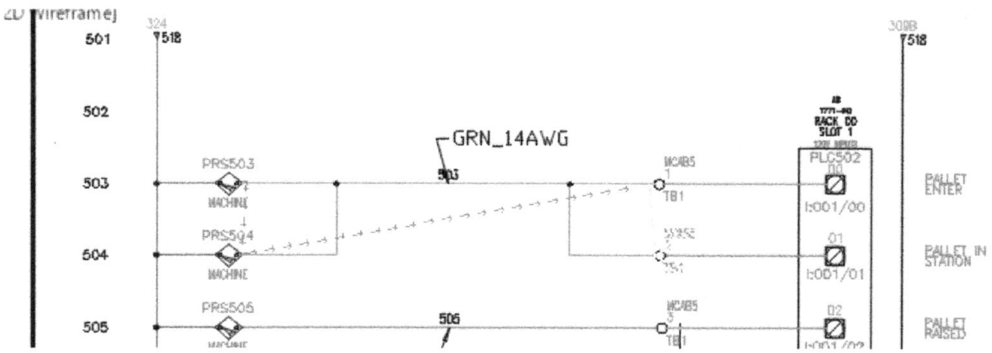

Figure 3-30 *Complete wire connection sequence of the selected wire network*

Editing the Wire Sequence

1. Choose the **Edit Wire Sequence** tool from the **Wire Sequence** drop-down in the **Edit Wires/ Wire Numbers** panel of the **Schematic** tab; you are prompted to select the wire network to be edited.

2. Select the same wire network that was selected earlier, refer to Figure 3-27; the **Edit Wire Connection Sequence** dialog box is displayed, as shown in Figure 3-31.

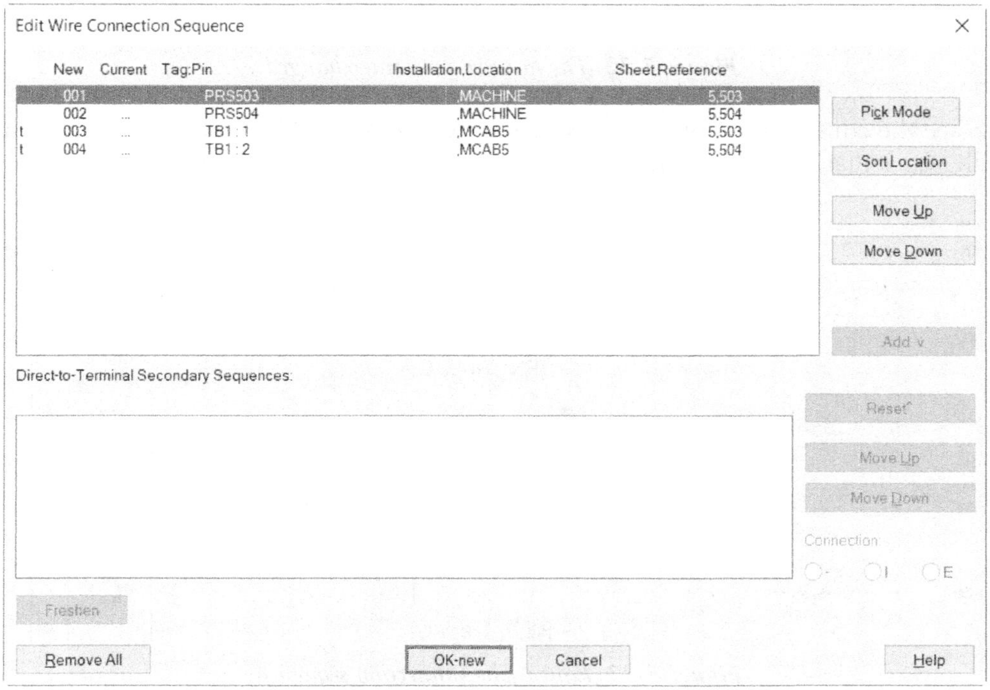

Figure 3-31 *The* ***Edit Wire Connection Sequence*** *dialog box*

3. Select the last row in the **Edit Wire Connection Sequence** dialog box and then choose **Move Up**; the last two rows get interchanged, as shown in Figure 3-32. Now, choose **OK-new**; the wire connection sequence is edited.

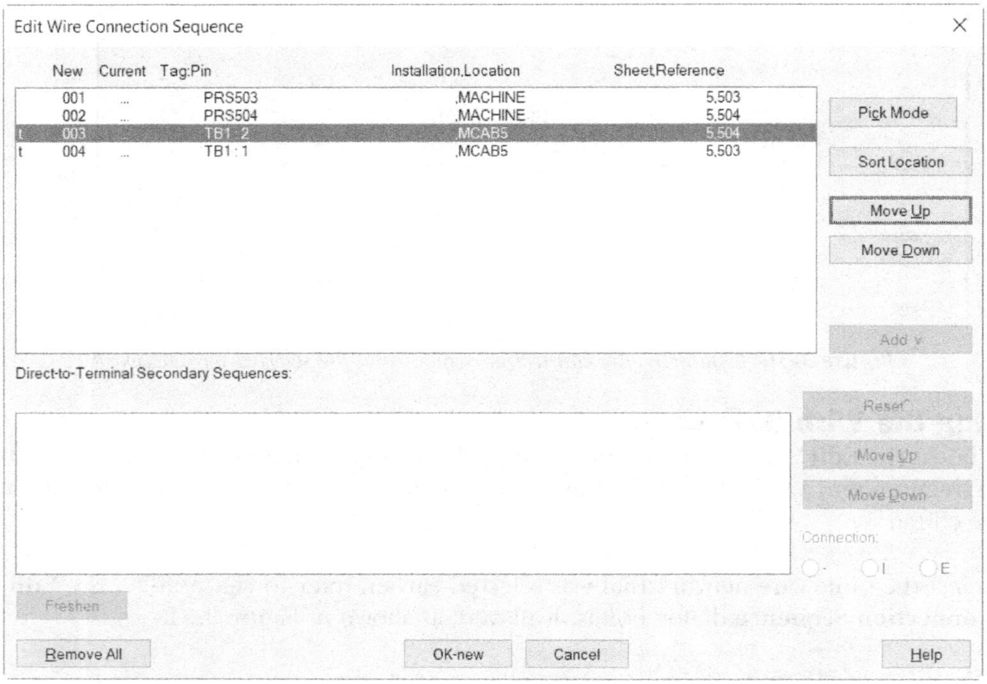

Figure 3-32 *The last two rows interchanged*

4. To see the edited wire connection sequence, use the **Show Wire Sequence** tool again. You will notice that the wire connection sequence is edited, as shown in Figure 3-33.

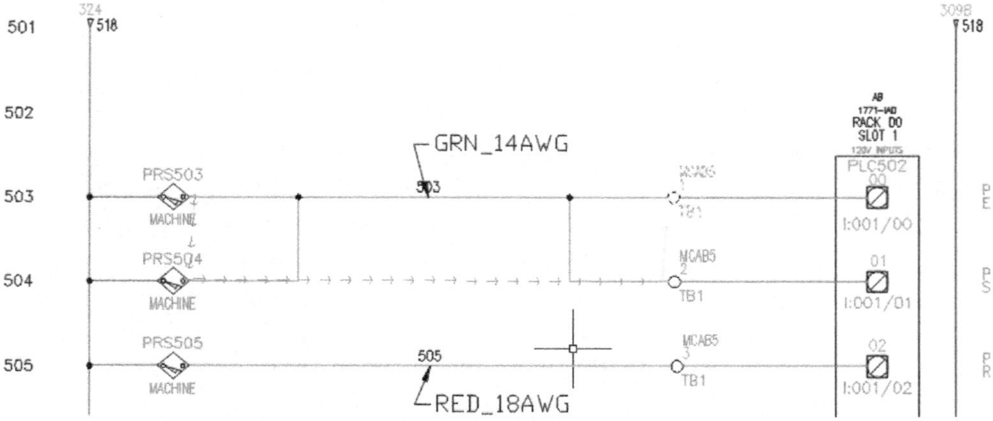

Figure 3-33 *Edited wire connection sequence*

Saving the Drawing

1. Choose **Save** from the **Application Menu** to save the *C03_tut04.dwg* drawing file.

Self-Evaluation Test

Answer the following questions and then compare them to those given at the end of this chapter:

1. Which of the following commands is used to create a wire type?

   (a) **AESOURCE** (b) **AEWIRETYPE**
   (c) **AEATTSHOW** (d) **AECONVERTWIRETYPE**

2. The __________ tool is used to stretch a wire until it meets another wire or a component.

3. The __________ tool is used to insert wire color/gauge labels into a wire.

4. The **Multiple Bus** tool is used to create multiple phase bus wiring. (T/F)

5. The **Trim Wire** tool is used to trim wires as well as the components inserted in it. (T/F)

Review Questions

Answer the following questions:

1. Which of the following tools is used to move a wire number to a different wire in the same wire network?

   (a) **Scoot** (b) **Move Wire Number**
   (c) **Copy Wire Number** (d) None of these

2. Which of the following commands is used to insert wire numbers in the drawing?

   (a) **AEWIRE** (b) **AEWIREGAP**
   (c) **AEWIRENO** (d) **AECOPYWIRENO**

3. Which of the following tools is used to move the wire number from right, left, top, or bottom to in-line with wire and vice-versa?

   (a) **Wire Number Leader** (b) **Swap Wire Numbers**
   (c) **Move/Show Attribute** (d) **Toggle Wire Number (In-line)**

4. You can automatically add wire numbers to each wire network in your project drawings. (T/F)

5. You can insert an in-line marker into any wire. (T/F)

6. The **Create/Edit Wire type** tool is used only for creating the wire type. (T/F)

7. The **Set Wire Type** dialog box is used to set a wire type for the new wires only. (T/F)

EXERCISE

Exercise 1

In this exercise, you will open the drawing *DEMO04.DWG* from the **CADCIM** project and add it to the **NEW_PROJECT** project and then you will save it with the name *C03_exer01.dwg* in this project. You will use the **Wire Numbers** tool to insert wire numbers, refer to Figure 3-34. You will create a new wire type with the name BLU_16AWG. Next, convert the wire type of the wire between CR406 and TS-B located on rung 411 to the BLU_16AWG wire type. You will also make the wire numbers of the rungs 411, 412, and 413 as fixed by using the **Edit Wire Number** tool. **(Expected time: 20 min)**

Note

It is recommended to save all exercises of the forthcoming chapters in the ***NEW_PROJECT*** *project.*

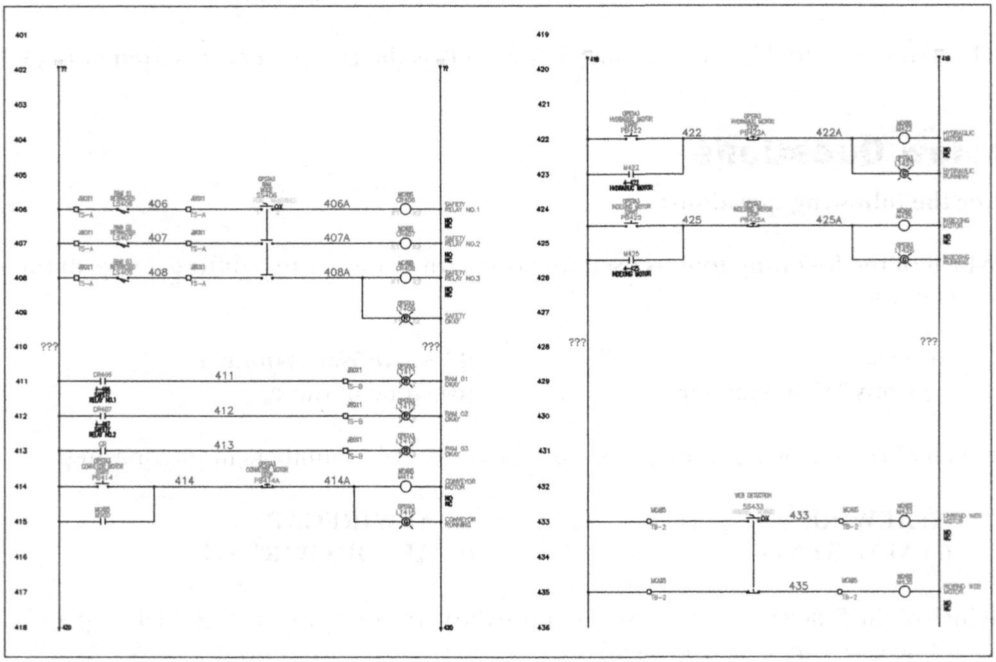

Figure 3-34 Ladder diagram for Exercise 1

Answers to Self-Evaluation Test

1. b, **2. Stretch Wire**, **3. Wire Color/Gauge Labels**, **4.** T, **5.** F

Chapter 4

Creating Ladders

Learning Objectives

After completing this chapter, you will be able to:

- *Create ladders*
- *Insert a new ladder in the drawing*
- *Modify an existing ladder*
- *Understand the format referencing of ladders*

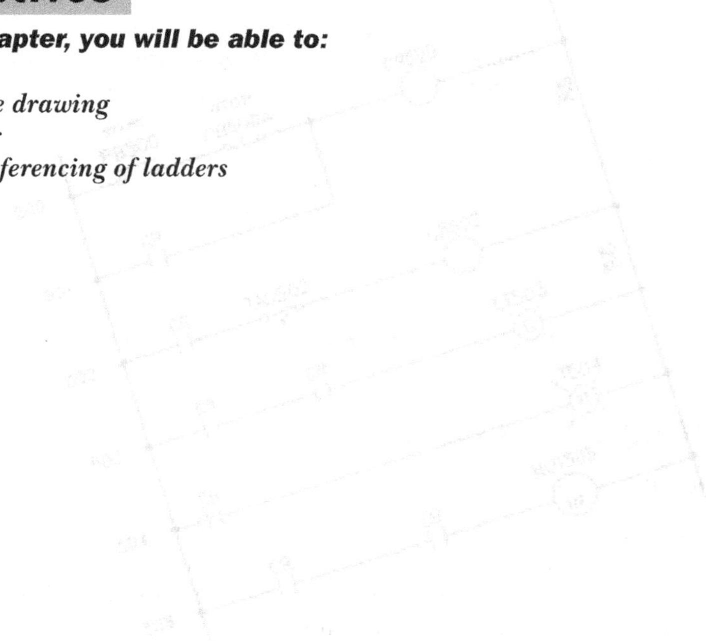

INTRODUCTION

In this chapter, you will learn about ladders and their multiple parameters such as rung spacing, number of rungs, width of ladder, ladder reference numbers, and so on. You will also learn to insert single or three phase ladders (horizontal/vertical), add and remove rungs, revise a ladder, and so on.

LADDERS

A collection of wires joined together to form a ladder-like matrix is called a ladder. The wires that are outside the ladder are called bus and the wires that are inside the ladder are called rungs. Usually, three phase ladders are created only with bus wires. Ladders are the base of schematic drawings. These schematic drawings are called ladder diagrams because they resemble the shape of a ladder. These diagrams are used to describe the logic connections of electrical control systems. Moreover, these diagrams are the specialized schematics that are commonly used to represent industrial control logic systems. The ladder diagrams have two vertical rails called bus (supply power) and several rungs (horizontal lines) that represent control circuits.

Figure 4-1 shows a ladder and its components.

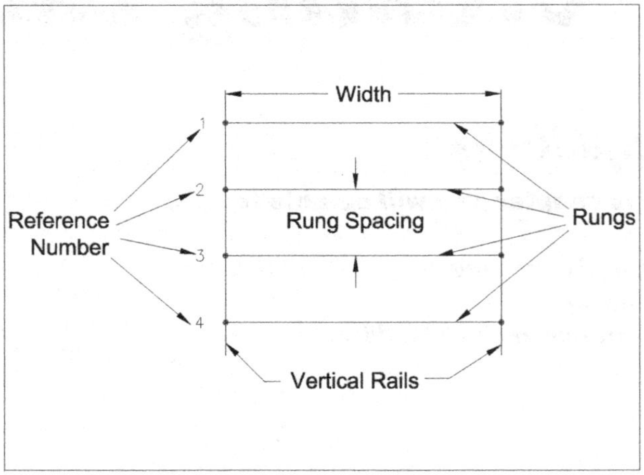

Figure 4-1 *Ladder and its components*

Inserting a New Ladder

Command: AELADDER

The **Insert Ladder** tool is used to insert a ladder along with rungs and reference numbers. You can insert a number of ladders in a drawing. To insert a ladder in a drawing, choose the **Insert Ladder** tool from the **Insert Ladder** drop-down in the **Insert Wires/ Wire Numbers** panel of the **Schematic** tab, as shown in Figure 4-2; the **Insert Ladder** dialog box will be displayed, as shown in Figure 4-3.

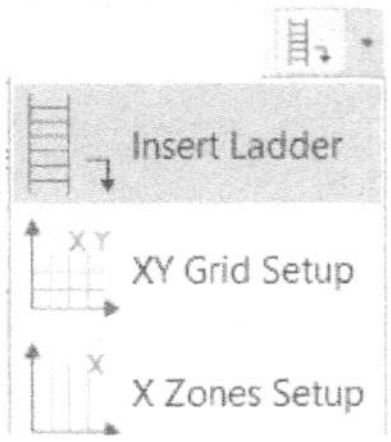

Figure 4-2 *The **Insert Ladder** drop-down*

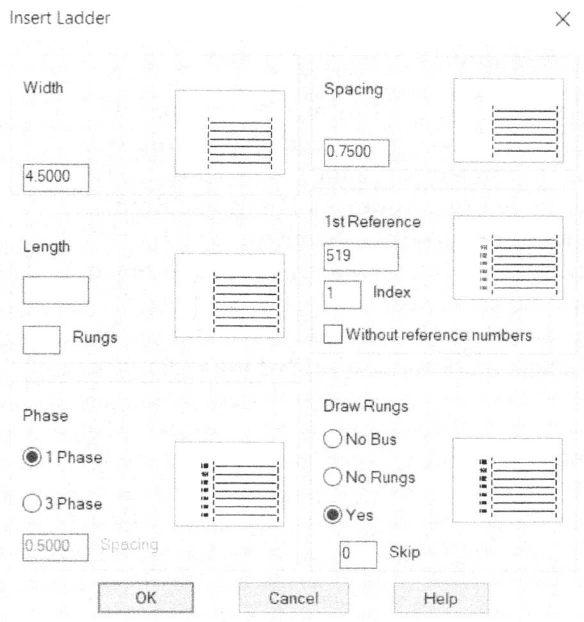

***Figure 4-3** The **Insert Ladder** dialog box*

Note

*If a drawing has sheet number specified to it and the ladder is being inserted into it, the name of the **Insert Ladder** dialog box will be **Sheet: XX - Insert Ladder** dialog box where XX stands for the sheet number.*

After specifying the desired settings in the **Insert Ladder** dialog box, choose the **OK** button; you will be prompted to specify the start position of the first rung. Specify the starting point of the first rung; the ladder will get inserted into your drawing provided you have entered the number of rungs in the **Rungs** edit box or the length of the ladder in the **Length** edit box.

Now, if you want to insert the ladder into your drawing manually, leave the **Rungs** and **Length** edit boxes blank. Choose the **OK** button; you will be prompted to specify the start position of the first rung. Specify the start position of the first rung; you will be prompted to specify the approximate position of the last reference number. Specify the approximate position of the last reference number by moving the cursor downward, and then click on the screen again; the ladder will get inserted into your drawing.

MODIFYING AN EXISTING LADDER

At times, you may need to modify an existing ladder. You can add or delete rungs, change the size of the ladder, reposition it, convert line reference numbers, renumber ladder reference, and so on. To do so, you need to use the ladder modification tools such as **Revise Ladder**, **Convert Ladder**, and so on. The existing ladder can be modified by using different methods which are discussed next.

Renumbering an Existing Ladder

Command: AEREVISELADDER

The **Revise Ladder** tool is used to modify the spacing between line reference numbers. You can also use this tool to renumber the line reference numbers of ladders. To modify a ladder, choose the **Revise Ladder** tool from the **Modify Ladder** drop-down in the **Edit Wires/Wire Numbers** panel of the **Schematic** tab, refer to Figure 4-4; the **Modify Line Reference Numbers** dialog box will be displayed, as shown in Figure 4-5. After specifying the desired values in this dialog box, choose the **OK** button to save the changes made in the **Modify Line Reference Numbers** dialog box.

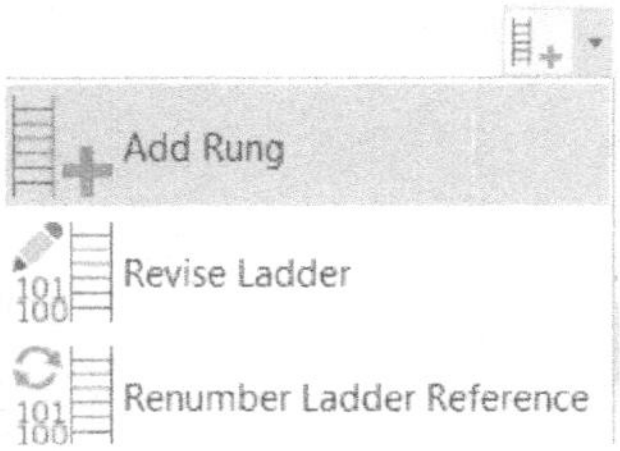

Figure 4-4 The ***Modify Ladder*** *drop-down*

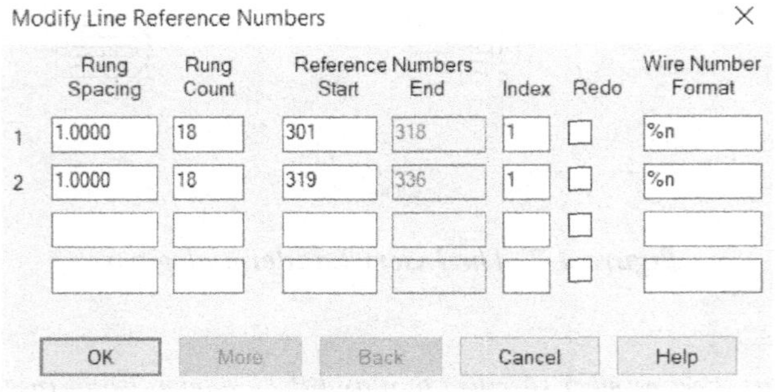

Figure 4-5 The ***Modify Line Reference Numbers*** *dialog box*

Changing the Size of a Ladder

In this section, you will learn how to shorten, lengthen, widen, or compress an existing ladder using the **Revise Ladder** tool.

Lengthening or Shortening a Ladder

Command: AEREVISELADDER

To lengthen or shorten a ladder, first you need to revise the ladder and then drag it. To do so, you need to follow two steps. These steps are discussed next.

Choose the **Revise Ladder** tool from the **Modify Ladder** drop-down in the **Edit Wires/Wire Numbers** panel of the **Schematic** tab; the **Modify Line Reference Numbers** dialog box will be displayed, refer to Figure 4-5. In this dialog box, specify the required value in the **Rung Count** edit box to match appropriate ladder length. Choose the **OK** button; the reference number of the ladder will change accordingly.

To lengthen or shorten a ladder, choose the **Stretch** tool from the **Modify** panel of the **Home** tab; the cursor will change to the selection box and you will be prompted to select the objects to stretch by using a crossing lasso. Next, select the vertical rails of a ladder, as shown in Figure 4-6 and then press ENTER; you will be prompted to specify the base point. Specify the base point or displacement and then click on the screen; the length of the ladder will change accordingly.

Note that if you move the cursor downward, the length of the ladder will increase and if you move the cursor upward, the length of the ladder will decrease.

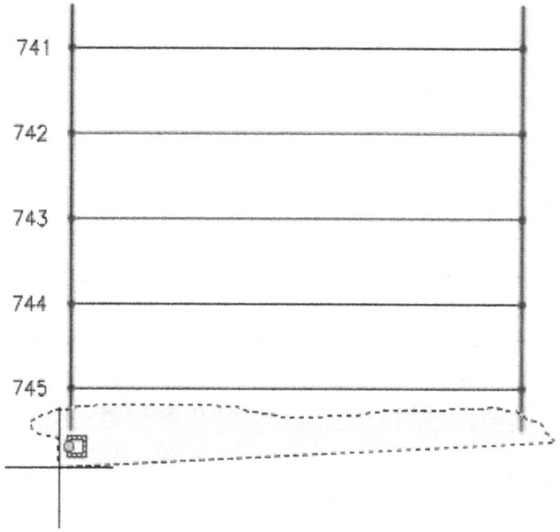

Figure 4-6 Selecting the ladder by using the crossing lasso

Widening or Compressing a Ladder

Command: AESCOOT

To widen or compress a ladder, choose the **Scoot** tool from the **Modify Components** drop-down in the **Edit Components** panel of the **Schematic** tab, as shown in Figure 4-7; the cursor will change into a selection box and you will be prompted to select the component, wire, or wire number.

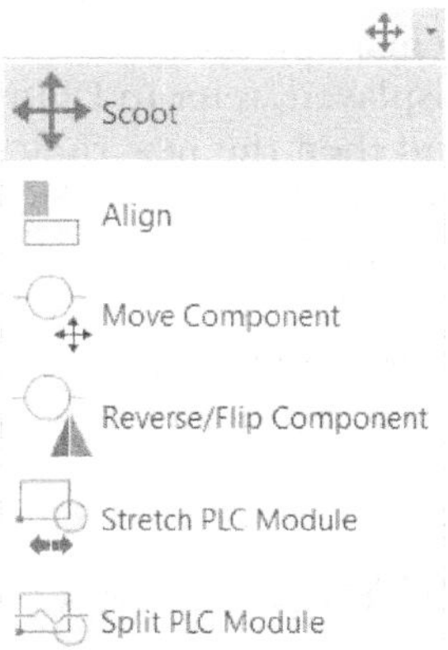

*Figure 4-7 The **Modify Components** drop-down*

Select the vertical rail of the ladder; the temporary rectangular graphics will be displayed. Move the cursor in the drawing according to your requirement and click on the screen. Next, press ENTER to exit the command. You can push in or pull out the vertical rail. If you push in the vertical rail, the ladder will be compressed and if you pull out the vertical rail, the ladder will be widened. Note that the ladder reference numbers cannot be moved when you use the **AESCOOT** command on vertical rails.

In case you want to align the components that are attached with ladders, choose the **Align** tool from the **Edit Components** panel of the **Schematic** tab; the components will be aligned. The alignment of components is discussed in detail in Chapter 6.

Note

*You can also use the **Scoot** tool to move rungs.*

Repositioning a Ladder

Command: Move

As discussed earlier, you can change the position of an existing ladder on your drawing by using the **MOVE** command of AutoCAD. But make sure you select the entire ladder while repositioning it. To do so, choose the **Move** tool from the **Modify** panel of the **Home** tab; you will be prompted to select the objects. Select the ladder using a crossing window and press ENTER; you will be prompted to specify the base point for the ladder. Specify the base point or displacement for the ladder; the ladder will be moved to the specified location.

After moving the ladder to the desired location, you need to update the new location of the ladder in AutoCAD Electrical internal ladder location list. To do so, choose the **Revise Ladder** tool from the **Edit Wires/Wire Numbers** panel of the **Schematic** tab; the **Modify Line Reference Numbers** dialog box will be displayed, refer to Figure 4-5. Choose the **Cancel** button to force AutoCAD Electrical to reread and update its internal ladder location list.

Changing the Rung Spacing

Command: AEREVISELADDER

To change the rung spacing of an existing ladder, first you need to revise the ladder and then move rungs to their new locations. To revise the ladder, choose the **Revise Ladder** tool from the **Modify Ladder** drop-down in the **Edit Wires/Wire Numbers** panel of the **Schematic** tab, refer to Figure 4-4; the **Modify Line Reference Numbers** dialog box will be displayed, refer to Figure 4-5. Specify a new rung spacing value in the **Rung Spacing** edit box and then the new rung count value in the **Rung Count** edit box to change the length of the ladder. Choose the **OK** button from this dialog box; the ladder will be modified according to the specified values.

Now, to move rungs to a new location, choose the **Scoot** tool from the **Edit Components** panel of the **Schematic** tab. Alternatively, use the **STRETCH** command of AutoCAD to move the existing rungs to a new rung location.

Adding Rungs

Command: AERUNG

The **Add Rung** tool is used to add rungs between the vertical rails of a ladder having extra line reference numbers than the number of rungs in the ladder. To add rungs to a ladder, choose the **Add Rung** tool from the **Modify Ladder** drop-down in the **Edit Wires/Wire Numbers** panel of the **Schematic** tab, refer to Figure 4-4; you will be prompted to add the rung passing through the specified location. You can specify the insertion point or enter **T** at the Command Prompt. If you specify the insertion point anywhere in the blank space between the vertical bus wires, the rung will be added between the vertical rails. If you enter **T** at the Command Prompt, the **Set Wire Type** dialog box will be displayed. This dialog has already been discussed in the previous chapter. You can select a new wire type in this dialog box and the selected wire type will become the current wire type and you can continue inserting the rungs in the ladder with the changed wire type. This command will continue till you press ESC or ENTER.

Note

While adding a new rung to the ladder, if some schematic symbols are encountered in between the rungs, then AutoCAD Electrical will break the rungs across the symbols and get connected to the vertical rails.

Converting Line Reference Numbers

Command: AE2LADDER

The **Convert Ladder** tool is used to convert normal ladder reference text or numbers to AutoCAD Electrical intelligent ladder reference number. If you have existing non-AutoCAD Electrical drawing, this tool can be used to create AutoCAD Electrical intelligent drawing. To convert ladder reference, choose the **Convert Ladder** tool from the **Tools** panel of the **Conversion Tools** tab; the cursor will change into a selection box. Select the non-intelligent line reference number and press ENTER; the **Modify Line Reference Numbers** dialog box will be displayed, refer to Figure 4-5. Next, specify the required values in this dialog box and choose the **OK** button; you will notice that the existing ladder information has been updated.

Renumbering the Ladder Line Reference

Command: AERENUMBERLADDER

The **Renumber Ladder Reference** tool is used to renumber the line reference number of ladders within a project. To do so, choose the **Renumber Ladder Reference** tool from the **Modify Ladder** drop-down in the **Edit Wires/Wire Numbers** panel of the **Schematic** tab; the **Renumber Ladders** dialog box will be displayed, as shown in Figure 4-8.

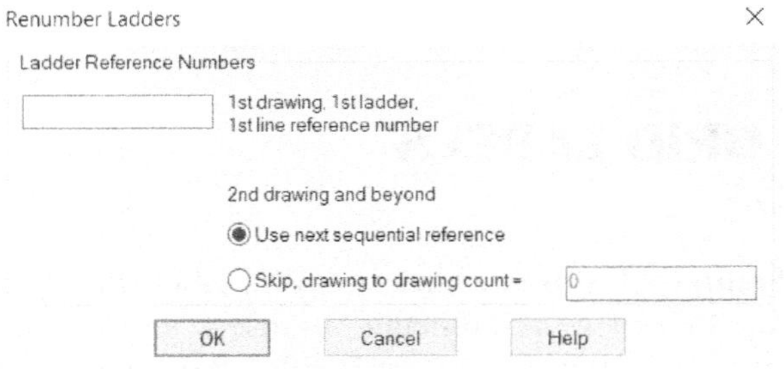

***Figure 4-8** The **Renumber Ladders** dialog box*

After entering the required parameters in the dialog box, choose the **OK** button; the **Select Drawings to Process** dialog box will be displayed. Select the drawing that you want to process and choose the **Process** button; the drawings will be transferred from the top list to the bottom list of the dialog box. Choose the **OK** button from the **Select Drawings to Process** dialog box; the line reference numbers of the ladder in the selected drawings will be changed according to the specified values.

CHANGING THE REFERENCE NUMBERING STYLE OF A LADDER

You can change the referencing style of a ladder to be inserted into the drawing. To do so, right-click on the drawing in the **PROJECT MANAGER**; a shortcut menu will be displayed. Choose **Properties > Drawing Properties** from the shortcut menu; the **Drawing Properties** dialog box will be displayed. Next, choose the **Drawing Format** tab. By default, the **Reference Numbers** radio button is selected in the **Format Referencing** area of this dialog box. On choosing the **Setup** button in this area, the **Line Reference Numbers** dialog box will be displayed, as shown in Figure 4-9.

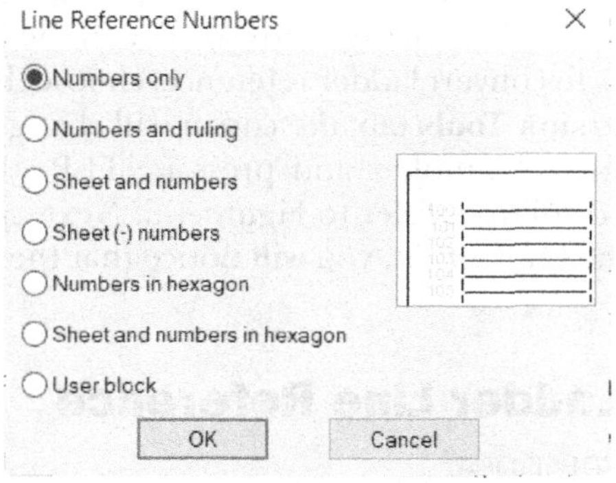

***Figure 4-9** The **Line Reference Numbers** dialog box*

By default, the **Numbers only** radio button is selected. As a result, reference numbers will be displayed along with the ladder inserted. Select the required radio button in this dialog box; the style of line reference numbers will change accordingly. Choose the **OK** button in the **Line Reference Numbers** dialog box to save the changes made and exit the dialog box. Also, choose **OK** in the **Drawing Properties** dialog box to exit the dialog box.

INSERTING X GRID LABELS

Command: AEXZONE

The **X Zones Setup** tool is used to insert X grid labels into the ladders of a drawing, which use X zone for format referencing. To insert X grid labels into ladders, choose the **X Zones Setup** tool from the **Insert Ladder** drop-down in the **Insert Wires/Wire Numbers** panel of the **Schematic** tab; the **X Zones Setup** dialog will be displayed, as shown in Figure 4-10. Note that this dialog box will be displayed only if you select the **X Zones** radio button in the **Drawing Format** tab of the **Drawing Properties** dialog box. If this radio button is not selected, the **AutoCAD Message** message box will be displayed indicating that the drawing is not configured for X Zones.

Alternatively, you can invoke the **X Zones Setup** dialog box by selecting the **X Zones** radio button in the **Drawing Format** tab of the **Drawing Properties** or the **Project Properties** dialog box and then choosing the **Setup** button. After specifying the desired settings in the **X Zones Setup** dialog box, choose the **OK** button; the X zone grid labels will be inserted into your drawing, as shown in Figure 4-11.

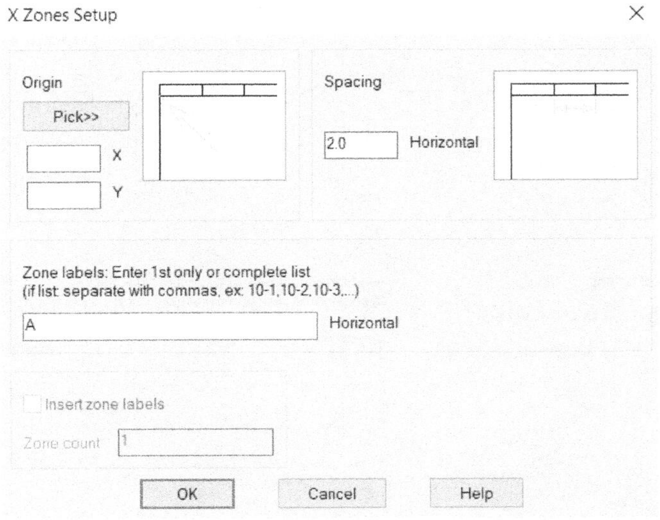

Figure 4-10 *The* ***X Zones Setup*** *dialog box*

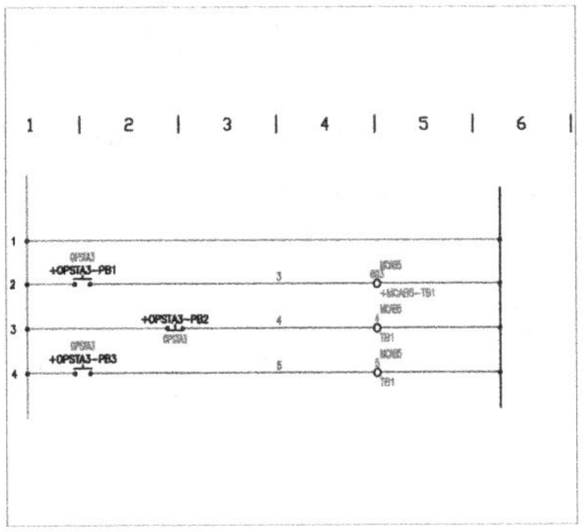

Figure 4-11 *The ladder with X Zone grid labels*

INSERTING X-Y GRID LABELS

Command: AEXYGRID

The **XY Grid Setup** tool is used to insert X-Y grid labels into the ladders of drawings, which use XY grid for format referencing. To do so, choose the **XY Grid Setup** tool from the **Insert Ladder** drop-down in the **Insert Wires/Wire Numbers** panel of the **Schematic** tab; the **X-Y Grid Setup** dialog box will be displayed, as shown in Figure 4-12. Note that this dialog box will be displayed only if you select the **X-Y Grid** radio button in the **Drawing Format** tab of the **Drawing Properties** dialog box. If this radio button is not selected, the **AutoCAD Message** message box will be displayed indicating that the drawing is not configured for X-Y Grid.

Alternatively, you can display the **X-Y Grid Setup** dialog box by selecting the **X-Y Grid** radio button in the **Drawing Format** tab of the **Drawing Properties** or the **Project Properties** dialog box and then choosing the **Setup** button.

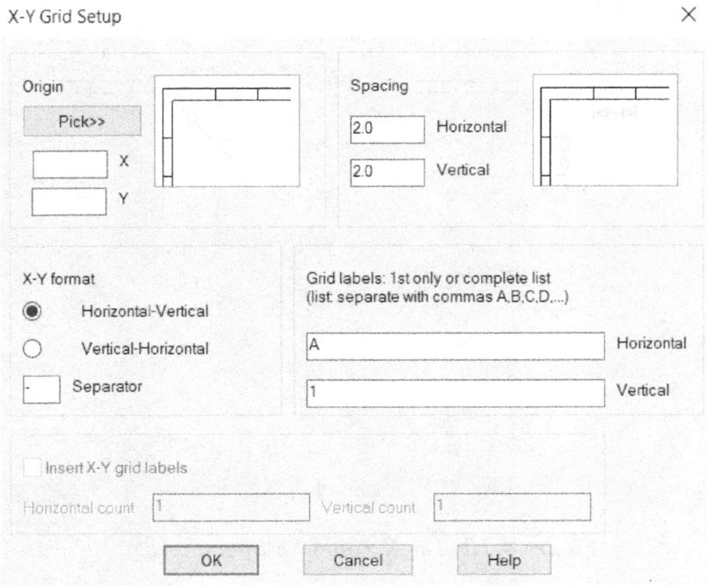

***Figure 4-12** The **X-Y Grid Setup** dialog box*

After specifying the desired settings in the **X-Y Grid Setup** dialog box, choose the **OK** button from the **X-Y Grid Setup** dialog box to insert the X-Y grid in the drawing and exit the dialog box. Figure 4-13 shows the ladder with X-Y grid setup.

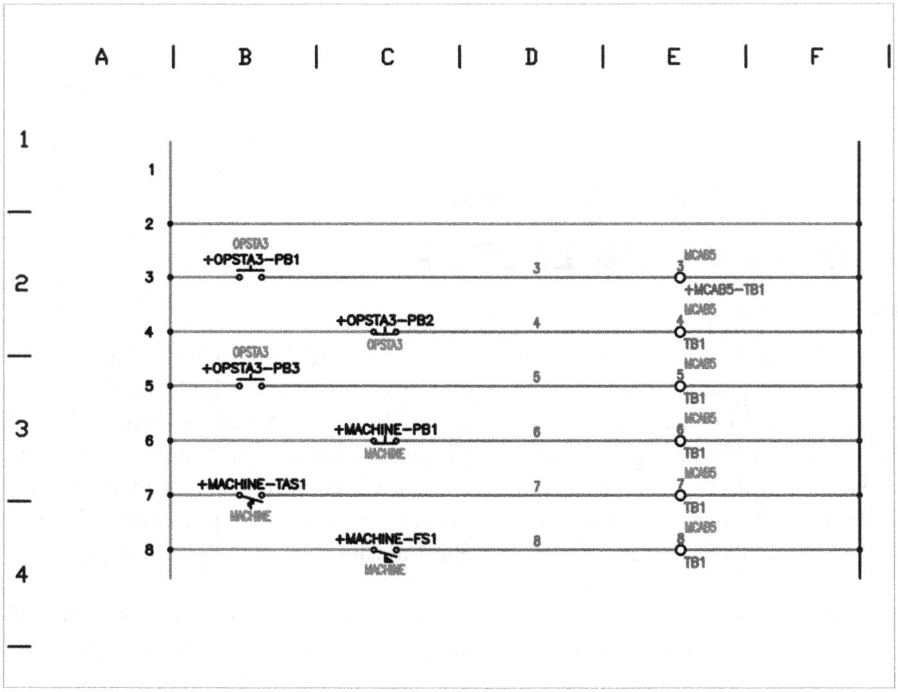

***Figure 4-13** The ladder with X-Y grid setup*

TUTORIALS

Tutorial 1

In this tutorial, you will insert a ladder into the drawing that was created in Tutorial 1 of Chapter 2 and then modify it by trimming rungs. You will also add wires between rungs, as shown in Figure 4-14. **(Expected time: 15 min)**

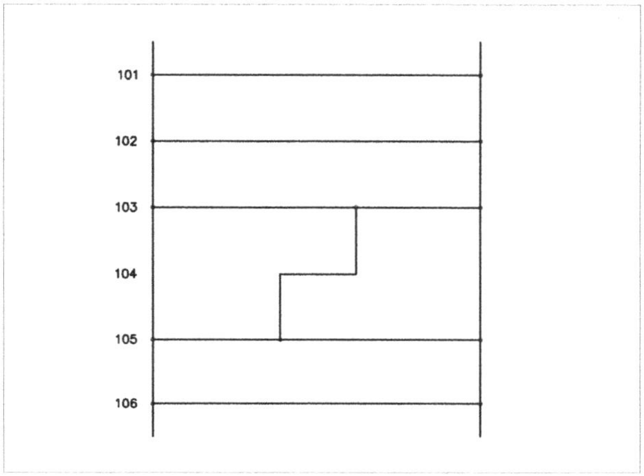

Figure 4-14 *The ladder diagram for Tutorial 1*

The following steps are required to complete this tutorial:

a. Open, save, and add the drawing to the **CADCIM** project drawing list.
b. Insert the ladder into the drawing.
c. Trim ladder rungs.
d. Add wires to the ladder.
e. Save the drawing file.

Opening, Saving, and Adding the Drawing to the CADCIM Project list

1. Open the *C02_tut01.dwg* file and then save it with the name *C04_tut01.dwg*.

   You can also download this file from *www.cadcim.com*. The path of the file is as follows:

   Textbooks > CAD/CAM > AutoCAD Electrical > AutoCAD Electrical 2022: A Tutorial Approach

2. Right-click on the *TUTORIALS* subfolder of the **CADCIM** project; a shortcut menu is displayed. Next, choose the **Add Active Drawing** option from the shortcut menu; the **Apply Project Defaults to Drawing Settings** message box is displayed. Choose the **Yes** button from this message box; *C04_tut01.dwg* is added in the *TUTORIALS* subfolder of the **CADCIM** project at the bottom of the drawing list. Next, choose **Save** from the **Application Menu** to save the drawing.

Inserting the Ladder into the Drawing

1. To insert the ladder into the drawing, choose the **Insert Ladder** tool from **Schematic> Insert Wires/Wire Numbers > Insert Ladder** drop-down; the **Insert Ladder** dialog box is displayed, refer to Figure 4-15. By default, the **1 Phase** radio button is selected in the **Phase** area. If it is not selected, then select it.

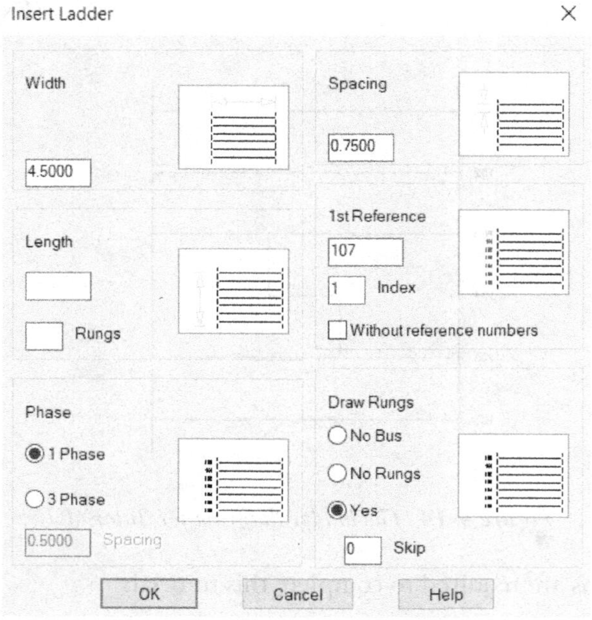

***Figure 4-15** The **Insert Ladder** dialog box*

2. Enter **5.0** in the **Width** edit box of the **Width** area.

3. Enter **1.0** in the **Spacing** edit box of the **Spacing** area.

4. Enter **101** in the **1st Reference** edit box. Also, make sure 1 is entered in the **Index** edit box.

   Make sure the **Without reference numbers** check box is clear.

5. Enter **6** in the **Rungs** edit box and click in the **Length** edit box of the **Insert Ladder** dialog box; the length of the ladder is automatically calculated and the value **5.5000** is displayed in the **Length** edit box. The **Yes** radio button is selected by default in the **Draw Rungs** area.

6. Make sure **0** is entered in the **Skip** edit box of the **Draw Rungs** area.

7. Choose the **OK** button; you are prompted to specify the start position of the first rung. Specify the start position as **7,14** at the Command prompt and press ENTER; the ladder is inserted into the drawing, as shown in Figure 4-16.

Trimming Ladder Rung

1. Choose the **Trim Wire** tool from the **Edit Wires/Wire Numbers** panel of the **Schematic** tab; a selection box is displayed and you are prompted to select a wire to trim. Select the rung **104**; the rung is trimmed from the ladder, as shown in Figure 4-17.

2. Press ENTER to exit the command.

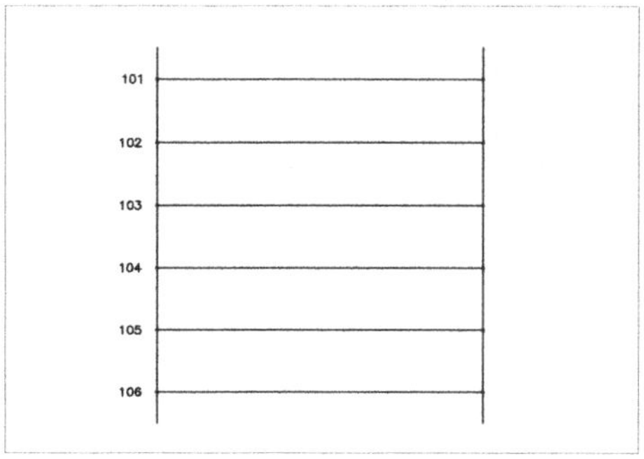

Figure 4-16 *The ladder inserted into the drawing*

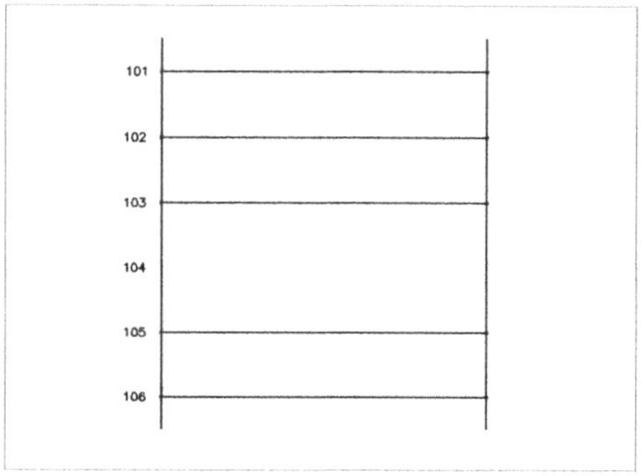

Figure 4-17 *The ladder with trimmed rung*

Adding Wires to the Ladder

1. Choose the **Wire** tool from the **Insert Wires/Wire Numbers** panel of the **Schematic** tab; you are prompted to specify the start point of the wire. Add the wire to the ladder at the locations shown in Figure 4-18. Next, press ENTER to exit the command.

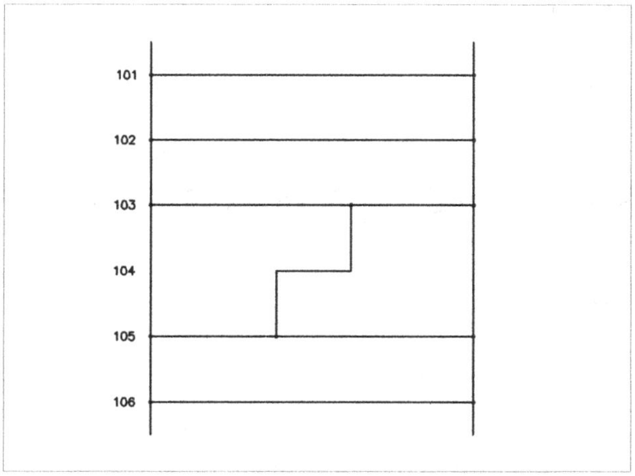

Figure 4-18 *The ladder with the wire added*

Saving the Drawing File

1. Choose **Save** from the **Application Menu** to save the *C04_tut01.dwg* drawing file.

Note

It is recommended that you save all tutorials given in the forthcoming chapters in the TUTORIALS subfolder of the ***CADCIM*** *project.*

Tutorial 2

In this tutorial, you will stretch and revise the ladder that you have created in Tutorial 1 of this chapter. Also, you will add rungs to the ladder, as shown in Figure 4-19.

(Expected time: 10 min)

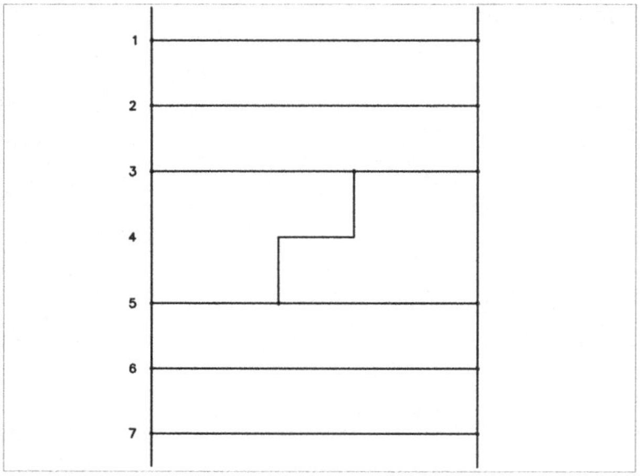

Figure 4-19 *The ladder diagram for Tutorial 2*

The following steps are required to complete this tutorial:

a. Open, save, and add the drawing to the **CADCIM** project drawing list.
b. Stretch the ladder.
c. Revise the ladder.
d. Add rungs to the ladder.
e. Save the drawing file.

Opening, Saving, and Adding the Drawing to the CADCIM Project list

1. Open the *C04_tut01.dwg* file and then save it with the name *C04_tut02.dwg*.

   You can also download this file from *www.cadcim.com*. The path of the file is as follows:

   Textbooks > CAD/CAM > AutoCAD Electrical > AutoCAD Electrical 2022: A Tutorial Approach

2. Right-click on the *TUTORIALS* subfolder of the **CADCIM** project; a shortcut menu is displayed. Next, choose the **Add Active Drawing** option from the shortcut menu; the **Apply Project Defaults to Drawing Settings** message box is displayed. Choose the **Yes** button from this message box; *C04_tut02.dwg* is added in the *TUTORIALS* subfolder of the **CADCIM** project at the bottom of the drawing list. Next, choose **Save** from the **Application Menu** to save the drawing.

Stretching the Ladder

1. Choose the **Stretch** tool from the **Modify** panel of the **Home** tab. Select the ladder by using the crossing lasso from right to left, as shown in Figure 4-20 and press ENTER. Make sure you select only the vertical bus of the ladder.

2. Specify the base point by selecting the lower end of the vertical bus. Now, move the cursor vertically down and click to stretch the vertical bus, refer to Figure 4-21.

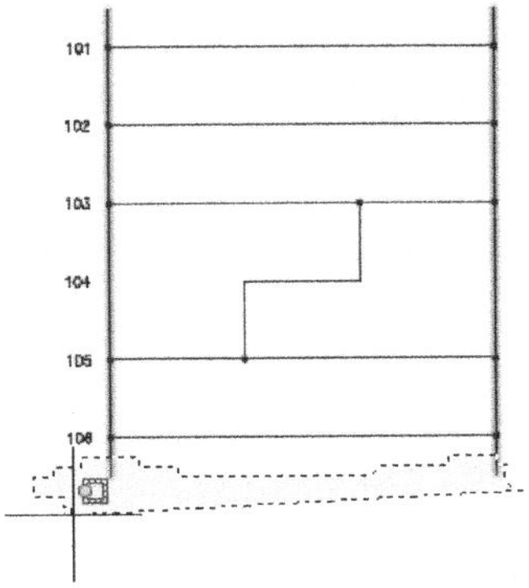

Figure 4-20 *Selecting the ladder by using the crossing lasso*

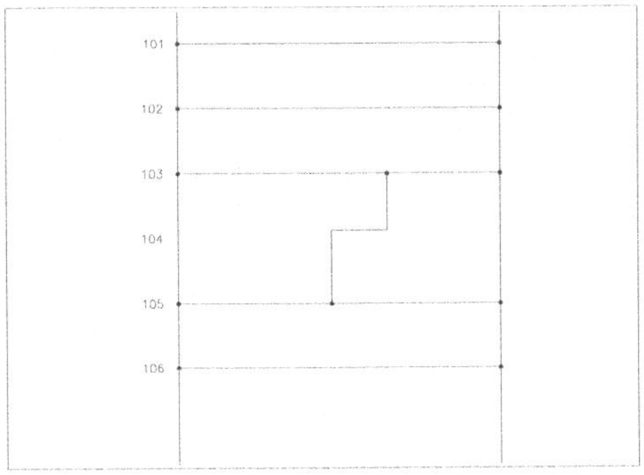

Figure 4-21 *The stretched ladder*

Revising the Ladder

1. After stretching the ladder, you need to revise it. To do so, choose the **Revise Ladder** tool from **Schematic > Edit Wires/Wire Numbers > Modify Ladder** drop-down; the **Modify Line Reference Numbers** dialog box is displayed, as shown in Figure 4-22.

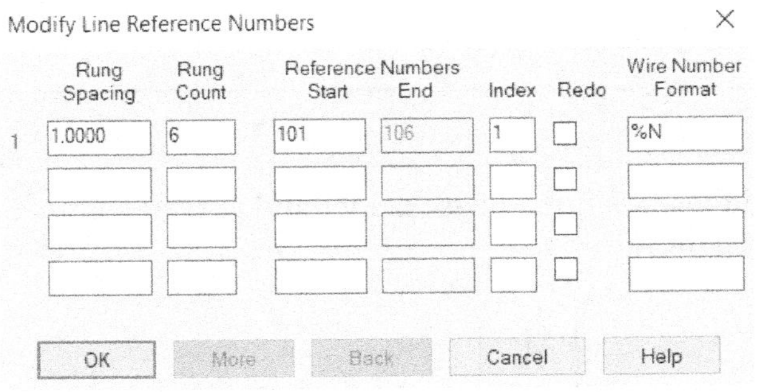

Figure 4-22 *The* ***Modify Line Reference Numbers*** *dialog box*

Since you entered **1.0000** in the **Spacing** edit box of the **Insert Ladder** dialog box while creating the ladder, the space between rungs in the **Rung Spacing** edit box is displayed as 1.0000.

2. Enter **7** and **1** in the **Rung Count** and **Reference Numbers Start** edit boxes, respectively. Also, make sure 1 is displayed in the **Index** edit box.

3. Once you enter the values in the **Rung Spacing**, **Rung Count**, and **Reference Numbers Start** edit boxes, the **Redo** check box is selected automatically. If it is not selected, select it manually. By default, %N is displayed in the **Wire Number Format** edit box. Do not change it.

4. Choose the **OK** button; the new reference numbers are assigned to the ladder, refer to Figure 4-23.

Adding Rungs to the Ladder

1. Choose the **Add Rung** tool from **Schematic > Edit Wires/Wire Numbers > Modify Ladder** drop-down; you are prompted to specify the location for the rung.

2. Click in the area between the vertical buses, under the rung **6**; a rung is added below the rung 6, as shown in Figure 4-24.

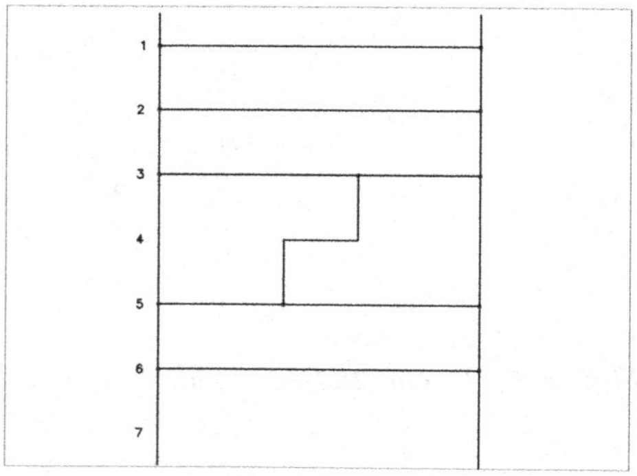

Figure 4-23 *The modified ladder*

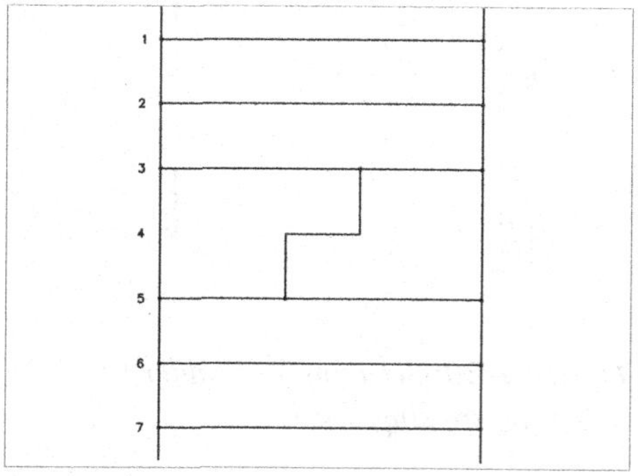

Figure 4-24 *A rung added to the ladder*

3. Press ENTER to exit the command.

Saving the Drawing

1. Choose **Save** from the **Application Menu** to save the *C04_tut02.dwg* drawing file.

Tutorial 3

In this tutorial, you will insert a three-phase ladder and also add a three-phase wire bus to the drawing, as shown in Figure 4-25. **(Expected time: 15 min)**

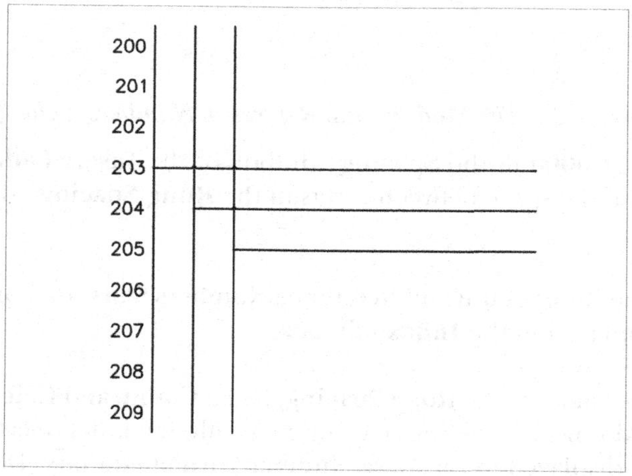

Figure 4-25 *Three-phase ladder and three-phase wire bus*

The following steps are required to complete this tutorial:

a. Create a new drawing *C04_tut03.dwg*.
b. Insert a three-phase ladder into the drawing.
c. Insert a three-phase wire bus.
d. Save the drawing file.

Creating a New Drawing

1. Create a new drawing with the name *C04_tut03.dwg* in the *TUTORIALS* subfolder of the **CADCIM** project. Refer to Tutorial 1 of Chapter 2 for the method of creating the drawing.

Inserting the Three-Phase Ladder into the Drawing

1. Choose the **Insert Ladder** tool from **Schematic > Insert Wires/Wire Numbers > Insert Ladder** drop-down; the **Insert Ladder** dialog box is displayed.

2. In this dialog box, select the **3 Phase** radio button from the **Phase** area.

3. Enter **1** in the **Spacing** edit box of the **Phase** area.

4. Enter **1** in the **Spacing** edit box of the **Spacing** area.

5. Enter **200** in the **1st Reference** edit box to specify the beginning of the line reference for the ladder.

6. Enter **1** in the **Index** edit box and enter **10** in the **Rungs** edit box.

7. Choose **OK**; you are prompted to specify the start position of the first rung. Enter **8,15** at the Command prompt and press ENTER; the ladder is inserted into your drawing, as shown in Figure 4-26.

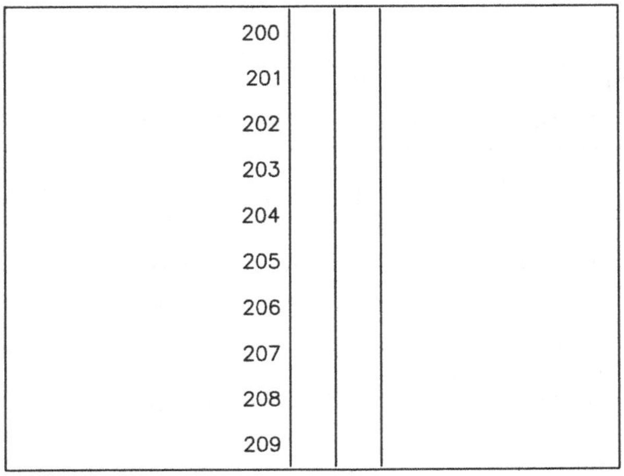

Figure 4-26 *Three-phase ladder*

Inserting the Three-Phase Wire Bus into the Drawing

1. Choose the **Multiple Bus** tool from the **Insert Wires/Wire Numbers** panel of the **Schematic** tab; the **Multiple Wire Bus** dialog box is displayed, as shown in Figure 4-27.

2. Enter **1** in the **Spacing** edit box of the **Horizontal** and **Vertical** areas.

   In this dialog box, by default, the **Another Bus (Multiple Wires)** radio button is selected in the **Starting at** area and the value 3 is displayed in the **Number of Wires** edit box.

3. Choose the **OK** button; a selection box appears on the screen. Select the start point on the first (extreme left) vertical bus at reference 203. Move the cursor to the right up to the desired length and specify the endpoint in the drawing area. Next, click on the screen.

4. Right-click on the screen or press ENTER to exit the command. The ladder with three-phase wire bus is shown in Figure 4-28.

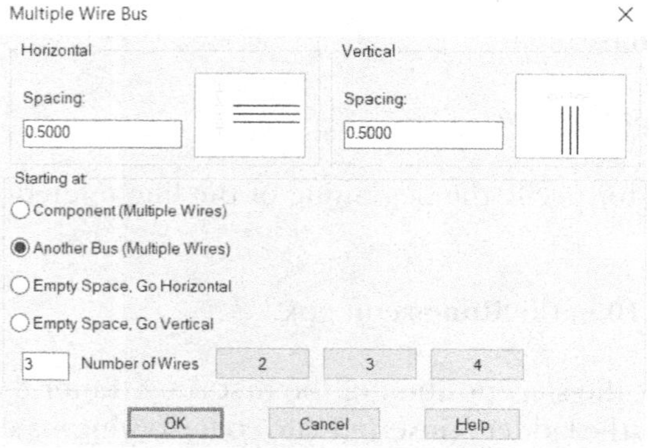

*Figure 4-27 The **Multiple Wire Bus** dialog box*

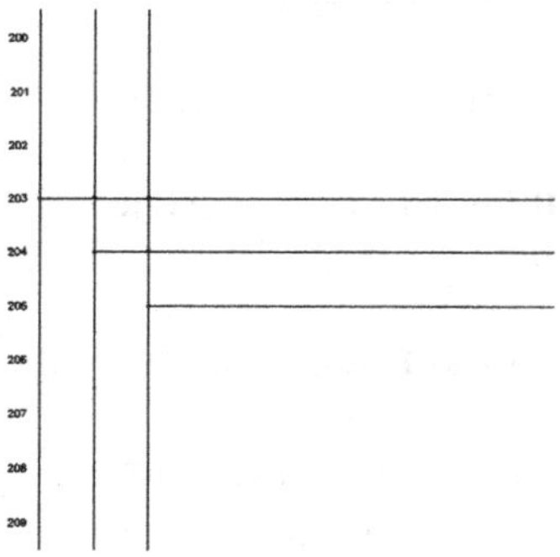

Figure 4-28 The ladder with three-phase wire bus

Saving the Drawing

1. Choose **Save** from the **Application Menu** to save the *C04_tut03.dwg* drawing file.

Tutorial 4

In this tutorial, you will set the format for reference numbers of a ladder and then insert a ladder into the drawing. Next, you will insert another ladder and insert X-Y grid labels into the drawing, refer to Figure 4-29. **(Expected time: 15 min)**

The following steps are required to complete this tutorial:

a. Create a new drawing.
b. Set format for reference numbers.
c. Insert a ladder into the drawing.
d. Insert X-Y grid labels into the drawing.

e. Renumber ladder line reference numbers.
f. Save the drawing file.

***Figure 4-29** The ladder diagram for Tutorial 4*

Creating a New Drawing

1. Create a new drawing with the name *C04_tut04.dwg* in the *TUTORIALS* subfolder of the **CADCIM** project. Refer to Tutorial 1 of Chapter 2 for the method of creating the drawing.

Setting the Format for Reference Numbers

1. Right-click on *C04_tut04.dwg* in the **PROJECT MANAGER**; a shortcut menu is displayed. Choose **Properties > Drawing Properties** from it, as shown in Figure 4-30; the **Drawing Properties** dialog box is displayed.

2. Make sure the **Drawing Settings** tab is chosen and then enter **A1** in the **Sheet** edit box of the **Sheet Values** area. Now, choose the **Drawing Format** tab.

3. Make sure the **Reference Numbers** radio button is selected in the **Format Referencing** area and then choose the **Setup** button, refer to Figure 4-31; the **Line Reference Numbers** dialog box is displayed, as shown in Figure 4-32.

4. In this dialog box, select the **Sheet and numbers** check box and choose the **OK** button to close the dialog box. Next, close the **Drawing Properties** dialog box. The format for the reference numbers of the ladder is set.

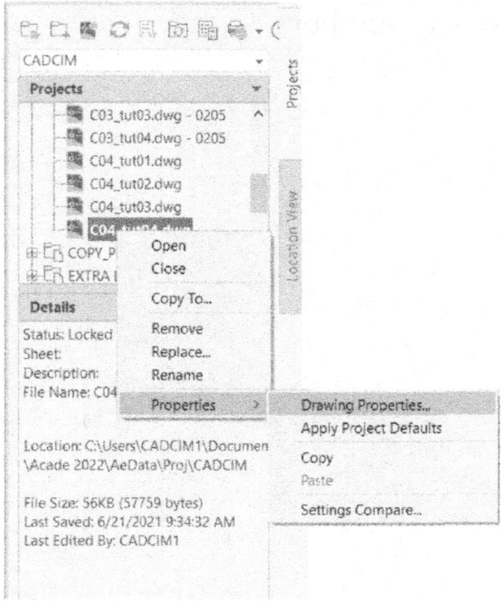

Figure 4-30 *Choosing* ***Drawing Properties*** *from the shortcut menu*

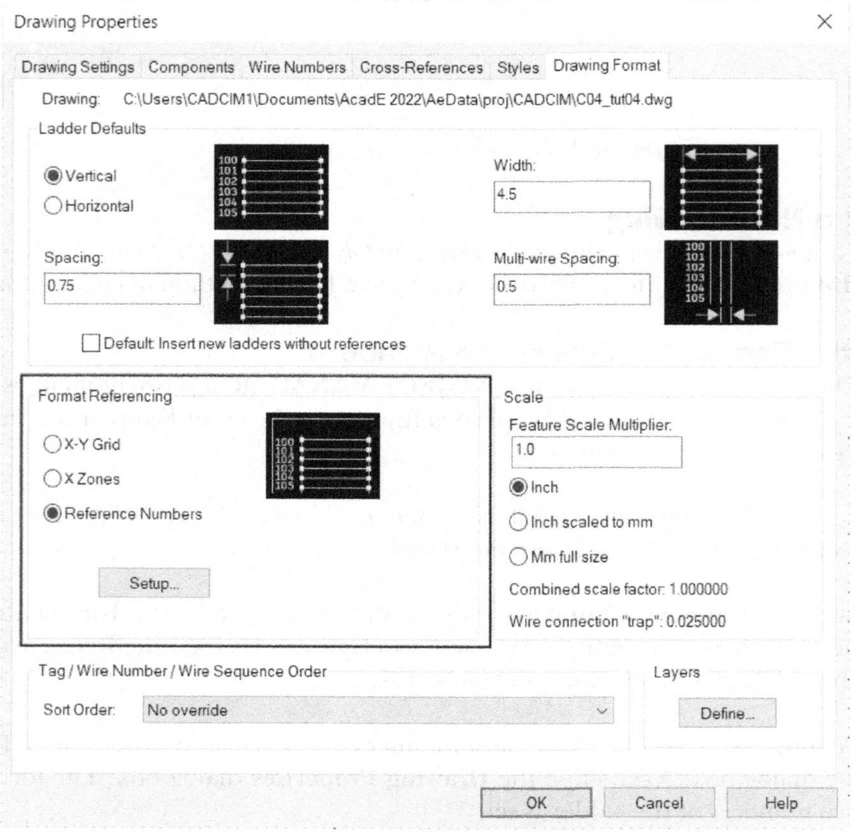

Figure 4-31 *The* ***Drawing Properties*** *dialog box*

Inserting a Ladder into the Drawing

1. Choose the **Insert Ladder** tool from **Schematic > Insert Wires/Wire Numbers > Insert Ladder** drop-down; the **Sheet: A1 - Insert Ladder** dialog box is displayed. Make sure the **1 Phase** radio button is selected in the **Phase** area.

2. Enter **10** in the **Width** edit box of the **Width** area. Next, enter **1.5** in the **Spacing** edit box of the **Spacing** area.

3. Enter **201** in the **1st Reference** edit box. Make sure 1 is displayed in the **Index** edit box and the **Without reference numbers** check box is clear.

4. Enter **10** in the **Rungs** edit box and click in the **Length** edit box of the **Sheet: A-Insert Ladder** dialog box; the length of the ladder is automatically calculated and is displayed as **14.25**. Make sure the **Yes** radio button is selected in the **Draw Rungs** area.

5. Enter **0** in the **Skip** edit box of the **Draw Rungs** area. Next, choose the **OK** button; you are prompted to specify the start position of the first rung. Specify the start position as **5,17.5** at the Command prompt and press ENTER; the ladder is inserted into the drawing, as shown in Figure 4-33.

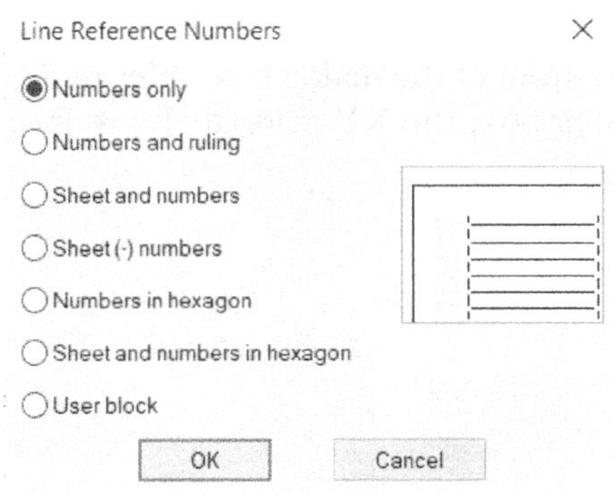

*Figure 4-32 The **Line Reference Numbers** dialog box*

Figure 4-33 The inserted ladder

Inserting X-Y Grid Labels

1. Right-click on *C04_tut04.dwg* in the **PROJECT MANAGER**; a shortcut menu is displayed. Choose **Properties > Drawing Properties** from it; the **Drawing Properties** dialog box is displayed.

2. In this dialog box, choose the **Drawing Format** tab. Next, select the **X-Y Grid** radio button in the **Format Referencing** area and then choose the **OK** button to close the dialog box.

3. Choose the **Insert Ladder** tool from **Schematic > Insert Wires/Wire Numbers > Insert Ladder** drop-down; the **Sheet: A1 - Insert Ladder** dialog box is displayed.

Make sure the **1 Phase** radio button is selected in the **Phase** area.

4. Enter **10** in the **Width** edit box of the **Width** area. Next, enter **1.5** in the **Spacing** edit box of the **Spacing** area.

5. Enter **10** in the **Rungs** edit box and click in the **Length** edit box of the **Sheet: A1 - Insert Ladder** dialog box; the length of the ladder is automatically calculated and is displayed as **14.25**. Make sure the **Yes** radio button is selected in the **Draw Rungs** area.

6. Make sure **0** is displayed in the **Skip** edit box of the **Draw Rungs** area. Next, choose the **OK** button; you are prompted to specify the start position of the first rung. Specify the start position as **19,17.5** at the Command prompt and press ENTER; the ladder is inserted into the drawing.

7. Choose the **XY Grid Setup** tool from **Schematic > Insert Wires/Wire Numbers > Insert Ladder** drop-down; the **X-Y Grid Setup** dialog box is displayed.

8. In this dialog box, enter **1.9** in the **Horizontal** edit box and **2.5** in the **Vertical** edit box of the **Spacing** area.

9. Enter **A1,A2,A3,A4,A5,A6** in the **Horizontal** edit box and **10,20,30,40,50,60** in the **Vertical** edit box of the **Grid Labels** area.

10. Enter **17.625** in the **X** edit box and **19** in the **Y** edit box of the **Origin** area. Next, select the **Insert X-Y Grid labels** check box at the bottom of the dialog box, refer to Figure 4-34. Now, choose the **OK** button to close the dialog box; the X-Y grid labels are inserted into the drawing, as shown in Figure 4-35.

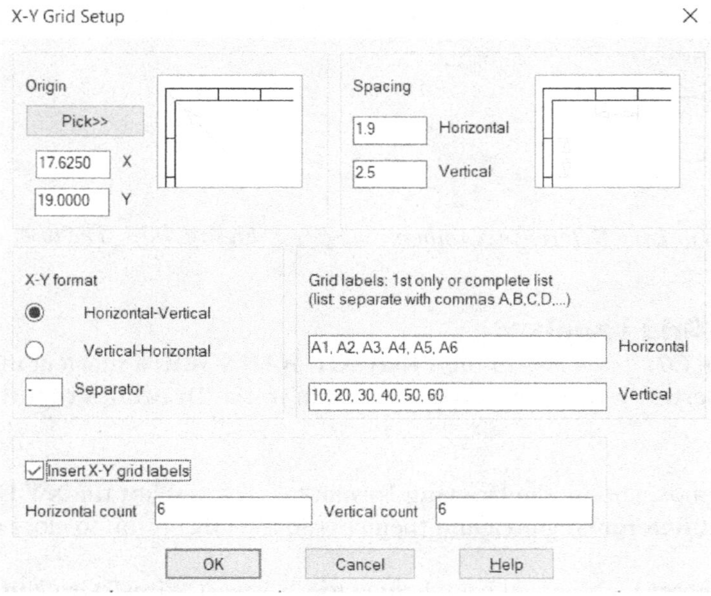

*Figure 4-34 The **X-Y Grid Setup** dialog box*

Figure 4-35 *The X-Y grid labels*

Renumbering Ladder Line Reference Numbers

1. Choose the **Renumber Ladder Reference** tool from **Schematic > Edit Wires/Wire Numbers > Modify Ladder** drop-down; the **Renumber Ladders** dialog box is displayed.

2. In this dialog box, enter **401** in the **1st drawing, 1st ladder, 1st line reference number** edit box. Also, make sure the **Use next sequential reference** radio button is selected. Next, choose **OK**; the **Select Drawings to Process** dialog box is displayed.

3. In this dialog box, select all the drawings of Chapter 4 (*C04_tut01.dwg, C04_tut02.dwg, C04_tut03.dwg,* and *C04_tut04.dwg)* and choose the **Process** button; the selected drawings are transferred from the top list to the bottom list. Next, choose **OK**; the line reference numbers for all the ladders are changed.

   You may open the above mentioned drawings one by one to see the change in line reference numbers.

Saving the Drawing

1. Choose **Save** from the **Application Menu** to save the drawing file, *C04_tut04.dwg*.

Self-Evaluation Test

Answer the following questions and then compare them to those given at the end of this chapter:

1. Which of the following tools is used to move the rungs of a ladder?

   (a) **Add Rung** (b) **Scoot**
   (c) **Move** (d) **Revise Ladder**

2. The __________ tool is used to add rungs to a ladder.

3. Rungs can be added only if you choose the **Add Rung** tool from the **ACE:Main Electrical** toolbar and click in the blank space between the __________ rails.

4. Ladder diagrams are used to describe the logic of electrical control systems. (T/F)

5. If you update a ladder reference, the existing components and the wire numbers will be automatically updated. (T/F)

Review Questions

Answer the following questions:

1. Which of the following tools is used to adjust the line reference number along the side of the ladder?

   (a) **Convert Ladder** (b) **Renumber Ladder Reference**
   (c) **Revise Ladder** (d) **Insert Ladder**

2. Which of the following tools is used to insert wires between the rails of a ladder?

   (a) **Revise Ladder** (b) **Add Rung**
   (c) **Insert Ladder** (d) None of these

3. The **XY Grid Setup** tool is used to insert the X-Y grid labels into the drawing. (T/F)

4. The **STRETCH** command of AutoCAD is used only to stretch the ladder and not for shortening it. (T/F)

5. The **Revise Ladder** tool is used to shorten, lengthen, compress, and widen a ladder. (T/F)

EXERCISES

Exercise 1

Create a new drawing with the name *C04_exer01.dwg* in the **NEW_PROJECT** project and insert a ladder with 10 rungs starting with reference number 100, spacing between rungs = 0.5, and width = 6. You also need to add a wire between rungs 100 and 101, as shown in Figure 4-36. Trim the rung 104 and the middle portion of the rung 101 using the **Trim Wire** tool.

(Expected time: 10 min)

Figure 4-36 *Ladder diagram for Exercise 1*

Exercise 2

Create a new drawing with the name *C04_exer02.dwg* in the **NEW_PROJECT** project and insert a ladder with 12 rungs starting with reference number 500, spacing between rungs = 1, and width = 6. You also need to add a rung below the rung 12, as shown in Figure 4-37.

(Expected time: 10 min)

Note

It is recommended to save all exercises of the forthcoming chapters in the ***NEW_PROJECT*** *project.*

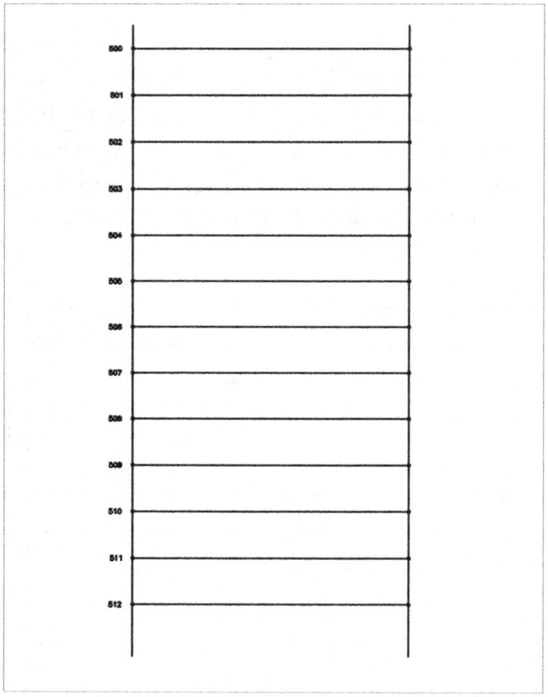

Figure 4-37 *Ladder diagram for Exercise 2*

Answers to Self-Evaluation Test

1. b, **2. Add Rung**, **3.** vertical, **4.** T, **5.** F

Chapter 5

Schematic Components

Learning Objectives

After completing this chapter, you will be able to:

- *Insert schematic components*
- *Annotate and edit schematic symbols*
- *Assign catalog part number to components*
- *Swap or update blocks*
- *Insert components from the equipment and user defined lists*

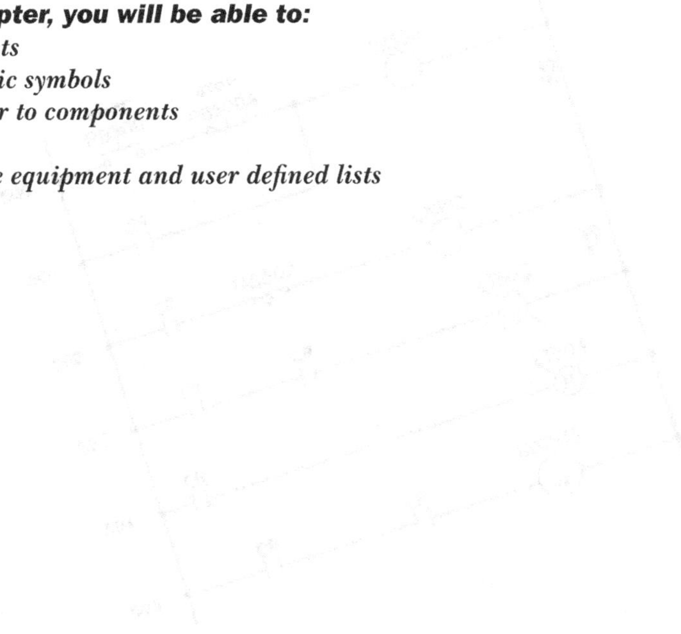

INTRODUCTION

In AutoCAD Electrical, schematic components are AutoCAD blocks with attributes such as TAG1, TAG2, DESC, RATING, LOC, INST, and so on. These components consist of visible attributes such as component tags, descriptions, pin numbers, and so on and invisible attributes such as manufacturer information, wire connection points, and so on. In this chapter, you will learn how to insert and edit schematic components such as push buttons, selector switches, and relays. You will also learn how to insert components from the equipment, panel, and catalog lists. Also, you will learn how to use AutoCAD Electrical tools to break wires, assign component tags, and cross reference to the related components while inserting components. Finally, this chapter describes how these tools can be used to specify the catalog information, component descriptions, and location codes.

INSERTING SCHEMATIC COMPONENTS USING ICON MENU

Command: AECOMPONENT

You can insert components into a drawing by selecting a component from the icon menu. To insert a component, choose the **Icon Menu** tool from the **Icon Menu** drop-down in the **Insert Components** panel of the **Schematic** tab, refer to Figure 5-1. On doing so, the **Insert Component** dialog box will be displayed, as shown in Figure 5-2.

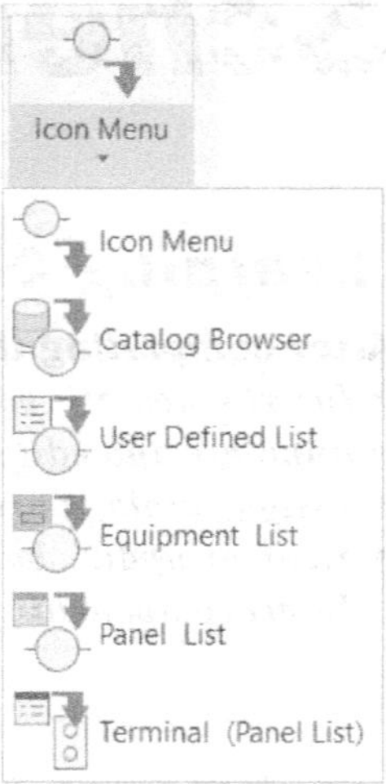

*Figure 5-1 The **Icon Menu** drop-down*

Note

*In order to use the **Insert Component** tool, the invisible WD_M block should be inserted into the drawing. The process of inserting a block into a drawing has been discussed in Chapter 1.*

To insert a component into your drawing, select a component from the list displayed in the **Menu** area; the **NFPA: Schematic Symbols** area will be changed according to the component selected and the area will be populated with various symbols of the selected icon. For example, if you select **Push Buttons** from the **Menu** area or the **Push Buttons** icon (first row, first column) from the **NFPA: Schematic Symbols** area, you will notice that the **NFPA: Schematic Symbols** area is changed to the **NFPA: Push Buttons** area, displaying various symbols of push buttons.

Select the symbol to be inserted; the symbol will appear on the cursor as a reference. The horizontal symbol will appear on the cursor by default and you will be prompted to specify the insertion point for the component. Next, specify the insertion point for the component; the **Insert / Edit Component** dialog box will be displayed. Specify the details of the component in this dialog box and choose the **OK** button; the component will be inserted into your drawing. Note that when you insert a symbol, the underlying wire breaks and then reconnects again. It happens only when you place the component directly on the wire or close to it within a trap distance.

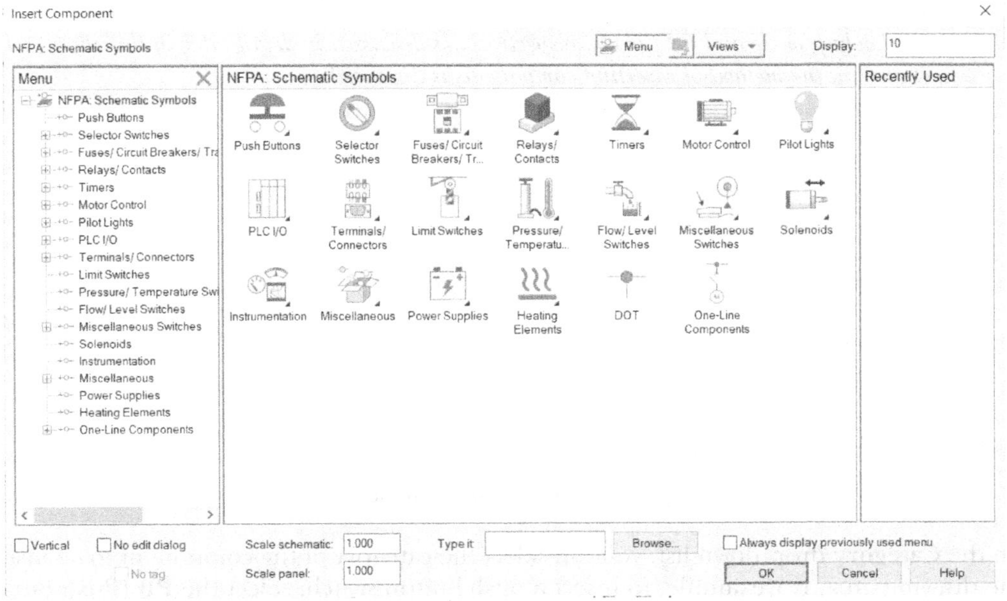

Figure 5-2 *The* ***Insert Component*** *dialog box*

Note

1. If you select a vertical wire, AutoCAD Electrical will automatically insert a vertically oriented symbol.

2. You can insert multiple components at a time. To do so, choose the ***Multiple Insert (Icon Menu)*** *tool from the* ***Multiple Insert*** *drop-down in the* ***Insert Components*** *panel of the* ***Schematic*** *tab. You can also insert the copies of a selected component. To do so, choose the* ***Multiple Insert (Pick Master)*** *tool from the* ***Multiple Insert*** *drop-down in the* ***Insert Components*** *panel of the* ***Schematic*** *tab.*

INSERTING COMPONENTS USING Catalog Browser

Command: AECATALOGOPEN

You can insert components into a drawing by using the **Catalog Browser** dialog box. To do so, choose the **Catalog Browser** tool from the **Icon Menu** drop-down in the **Insert Components** panel of the **Schematic** tab, refer to Figure 5-1. On doing so, the **Catalog Browser** dialog box will be displayed, refer to Figure 5-3.

Note

The ***Catalog Browser*** *dialog box is used in three modes as follows:*

1. Insertion mode - In this mode, you can insert schematic components into a drawing.

2. Lookup mode - In this mode, you can assign a catalog value to a component.

3. Edit mode - In this mode, you can edit the catalog database.

The method of assigning the catalog information to a component and editing the catalog database in the Lookup mode and Edit mode, respectively, is discussed in detail later in this chapter. In the next section, the method of inserting components is discussed.

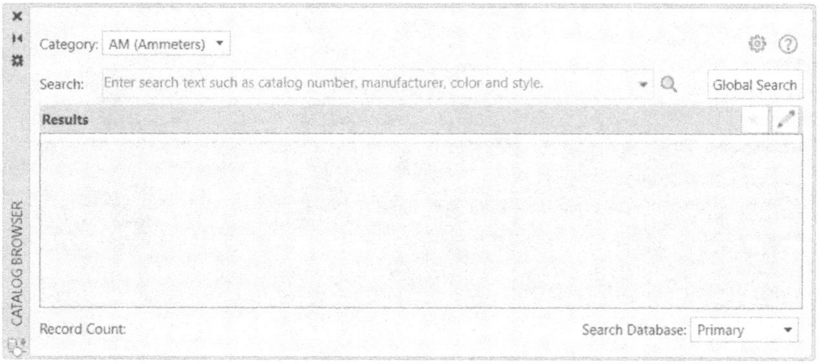

Figure 5-3 *The* ***Catalog Browser*** *dialog box*

From the **Category** drop-down list, you can select the category of the component to be inserted in the drawing area. For example, to insert a push button switch, select the **PB (Push buttons)** option from the drop-down list. The **Search** field is used to search the required entries from the database based on the keywords mentioned in it.

The **Global Search** button appears next to the **Search** button. This button is used to search a keyword mentioned in the **Search** field from all the category tables of the catalog database. When you enter the keywords in the **Search** field and choose the **Global Search** button, the **Catalog Global Search** dialog box will be displayed. In this dialog box, a table list is displayed at the left side. In this table list, two columns are there. The first column specifies the number of entries and the second column specifies the name of the respective category. At the right of the table list, catalog record list is there. In this list, details of each catalog record are specified in various columns. When you select any cell from the table list, catalog record list at the right gets updated accordingly.

The Database grid displays the resulting entries as per the keywords mentioned in the **Search** field. There are a number of columns in this grid. If you click on any of these columns headers, the entries will be sorted in ascending or descending order. If you right-click on any of these column titles, a shortcut menu will be displayed.

All the columns displayed in the Database grid are available as options in the shortcut menu. If you click on the tick mark corresponding to any of the columns in the shortcut menu, that column will be removed from the Database grid. When you choose the **More** option, the **Columns to display** dialog box will be displayed. This dialog box displays the names of the columns that can be displayed in the Database grid. To display a column in the Database grid, you need to select the corresponding check box.

You can choose the **Freeze column** option from the shortcut menu to freeze the selected column. Once the column is frozen, it will be visible even if you scroll the Database grid to the extreme left or extreme right. Choose the **Restore all columns to defaults** option to undo the changes made to the column(s) such as size change, display order change, or visibility mode.

When you click (or right-click) on an entry in any of the columns of the Database grid, a flyout will be displayed. The options in this flyout depend on the type of entry selected. The **Favorites List** button is used to display the Favorites list for the selected category. This button is activated only if one or more entries from the selected category are added to the Favorites list.

The **Edits catalog database** button enables the **Catalog Browser** dialog box to switch from the Insertion mode to the Edit mode. The **Search Database** drop-down list is available at the bottom in the **Catalog Browser** dialog box. It provides two options, **Primary** and **Secondary** for selecting a catalog database. The **Primary** option is selected by default and displays the default catalog database. The **Secondary** option will be activated if you have specified the secondary file for the catalog database. To activate this option, choose the **Other File** button in the **Project Setting** tab of the **Project Properties** dialog box and then specify the secondary file.

The **Configure your Database** button is available at the upper right corner of the **Catalog Browser** dialog box. It is used to configure the catalog database. To configure the database, choose the **Configure your Database** button; the **Configure Database** dialog box will be displayed. In this dialog box, you need to select the **Microsoft Access** or **Microsoft SQL Server** radio button to select the data source for configuring the database.

ANNOTATING AND EDITING THE SYMBOLS

Command: AEEDITCOMPONENT

You can enter the details while inserting components. However, you can also change or edit these details after inserting a component. To do so, choose the **Edit** tool from the **Edit Components** drop-down in the **Edit Components** panel of the **Schematic** tab, as shown in Figure 5-4; you will be prompted to select the component. Select the component; the **Insert / Edit Component** dialog box will be displayed.

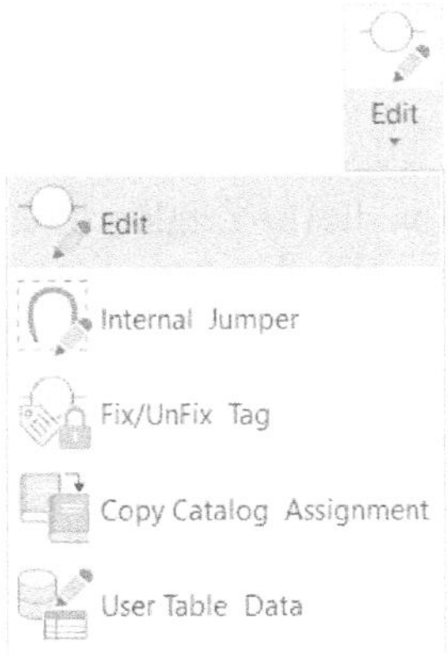

***Figure 5-4** The **Edit Components** drop-down*

The **OK-Repeat** button will be activated in the **Insert / Edit Component** dialog box only when a new component is inserted in the drawing. This button will not be activated while you edit a component. Change the desired parameters and choose the **OK** button from the **Insert / Edit Component** dialog box to complete the editing or insertion of the selected component. The **OK** button will be activated while editing or inserting the component in the drawing.

ASSIGNING CATALOG INFORMATION AND EDITING THE CATALOG DATABASE

You can assign catalog information to a component using the **Catalog Browser** dialog box. The **Lookup** button in the **Catalog Data** area of the **Insert / Edit Component** dialog box is used to assign the catalog information from the **Catalog Browser** dialog box to a component. To assign catalog information, choose the **Edit** tool from the **Edit Components** panel of the **Schematic** tab; you will be prompted to select the component. Next, select the component; the **Insert / Edit Component** dialog box will be displayed. Choose the **Lookup** button from the **Catalog Data** area of this dialog box; the **Catalog Browser** dialog box will be displayed in the Lookup mode, refer to Figure 5-5. This dialog box displays the catalog database of the components from where you can select the manufacturer or catalog values. The **Category** drop-down list displays the category of the selected component. Select the entry from the Database grid of the dialog box. You can also search for the required catalog information by using the **Search** field. After selecting the required entry, choose the **OK** button from the **Catalog Browser** dialog box; the selected catalog information will be displayed in the **Manufacturer** and the **Catalog** edit boxes of the **Insert/Edit Component** dialog box.

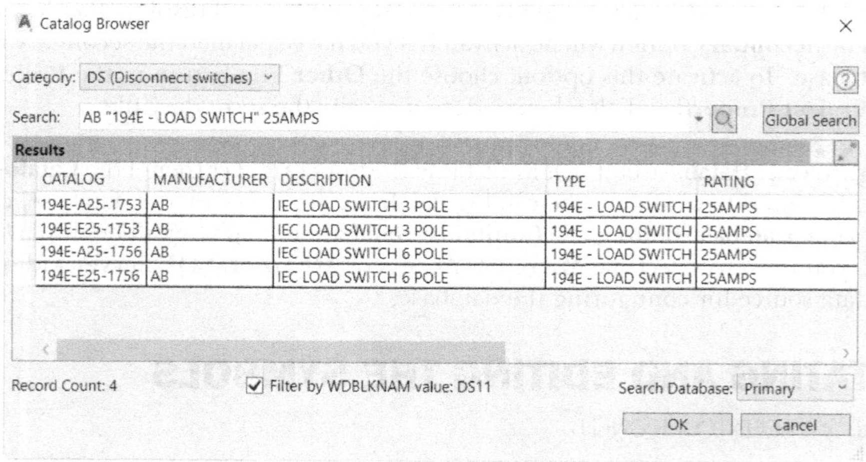

*Figure 5-5 The **Catalog Browser** dialog box*

To edit the catalog database available in the **Catalog Browser** dialog box, choose the **Edit catalog database** button; the **Catalog Browser** dialog box will switch to the Edit mode. Also, the name of the dialog box will change to the **Catalog Browser - Edit Mode** dialog box and the color of the Database grid will change to yellow, refer to Figure 5-6.

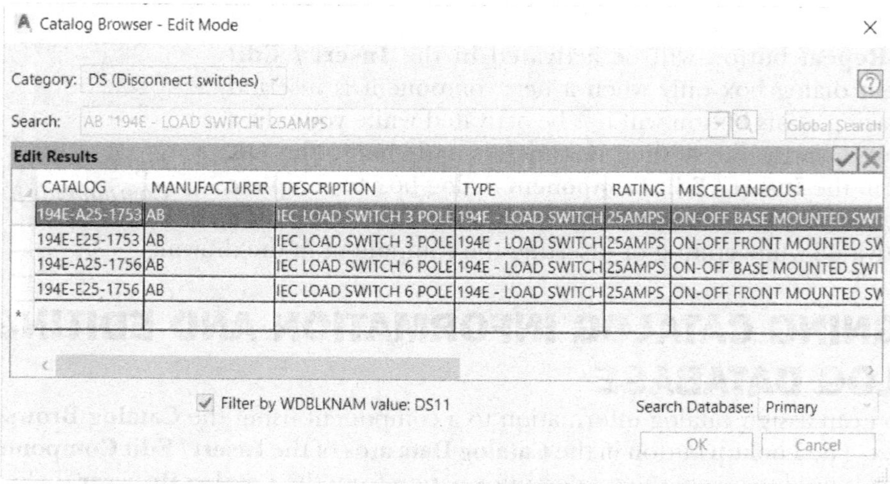

*Figure 5-6 The **Catalog Browser - Edit Mode** dialog box*

You can also click in any cell of the Database grid and change the content by simply entering the desired text. After editing the Database as explained above, if you choose the **Accept changes** button, the changes will be reflected in the **Catalog Browser - Edit Mode** dialog box and it will also exit from the Edit mode. If you want to cancel any changes made in the database, choose the **Cancel changes** button.

CREATING A PROJECT SPECIFIC CATALOG DATABASE

Command: AECREATEPROJCATALOG

When you install AutoCAD Electrical, the catalog information in the **Catalog Browser** dialog box is stored in a default Microsoft Access Database file. The extension of the file is *.mdb*. This file contains a huge sample vendor data. When you create a new project, a part of this sample data is used in the project. You can create a project specific catalog database which contains entries only for the components used in the project. You can send this project specific catalog database file to the client. This file is much smaller in size than the default catalog database file.

To create a project specific catalog database, activate the project for which you want to create it. Next, choose the **Create Project-Specific Catalog Database** tool from the **Other Tools** panel of the **Project** tab; the **Create Project-Specific Catalog Database** dialog box will be displayed. Specify the path and other options in this dialog box. Then, choose the **OK** button to create a project specific catalog database at the specified path.

INSERTING COMPONENTS FROM THE EQUIPMENT LIST

Command: AECOMPONENTE

The **Equipment List** tool is used to insert schematic component from the equipment list. Using this tool, you can find and insert appropriate schematic symbol from *schematic_lookup.mdb*. To insert a schematic component from the equipment list, choose the **Equipment List** tool from the **Icon Menu** drop-down in the **Insert Components** panel of the **Schematic** tab; the **Select Equipment List Spreadsheet File** dialog box will be displayed.

Note

You can insert either a single or multiple schematic components from the equipment list.

Select the spreadsheet file from the **Select Equipment List Spreadsheet File** dialog box and choose the **Open** button; the **Table Edit** dialog box will be displayed. This dialog box will be displayed only if multiple sheets/tables of the equipment list are found in the data file. Next, select the table to be edited from the **Table Edit** dialog box and choose **OK**; the **Settings** dialog box will be displayed. When you choose the **Default settings** button from the **Settings** dialog box, the options in the **View/Edit Settings** area will be activated. Using this button, you will be able to use the AutoCAD Electrical default settings for reading the format of the database file.

The **Spreadsheet/Table columns** button will be activated when you choose the **Default settings** button from the **Settings** dialog box. This button is used to define the order of data in the selected equipment list file. Also, you can modify the settings of a selected equipment list file. To define the order of data, choose the **Spreadsheet/Table columns** button from the **View/Edit Settings** area; the **Equipment List Spreadsheet Settings** dialog box will be displayed.

In this dialog box, you can assign column numbers to data categories such as **Manufacturer**, **Catalog**, **Assembly Code**, and so on. After specifying the options in this dialog box, choose the **OK** button to return to the **Settings** dialog box.

The **Save Settings** button will be activated only if you choose the **Default settings** button. This button is used to save the settings to the file so that it can be used later. To save the settings to a file, choose the **Save Settings** button; the **Save Settings** dialog box will be displayed. Enter the name of the file in the **File name** edit box. Next, choose the **Save** button; the column information will be saved in *.wde* format.

Now, choose the **OK** button in the **Settings** dialog box; the **Schematic equipment in** dialog box will be displayed, refer to Figure 5-7.

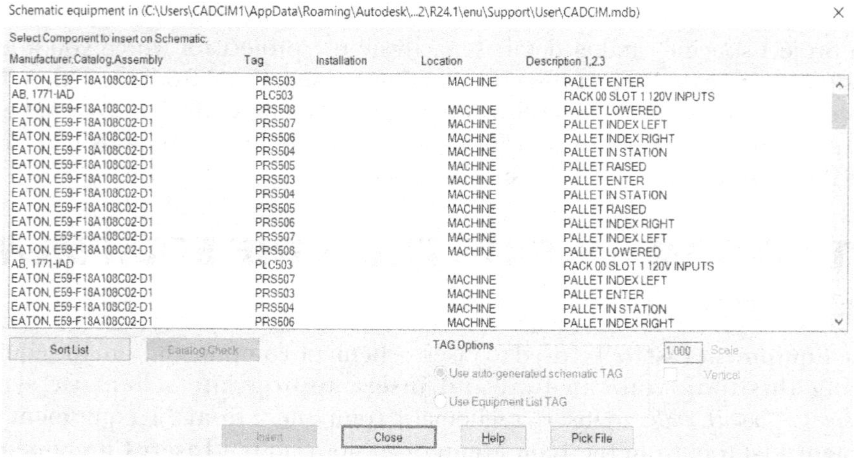

*Figure 5-7 The **Schematic equipment in** dialog box*

The **Insert** button will be activated only after you have selected a component from the list displayed. This button is used to find and insert a selected schematic component. To insert the selected component, choose the **Insert** button; the **Insert** dialog box will be displayed. This dialog box displays the block name of the selected component with a short description of each block name.

The **OK** button of the **Insert** dialog box will be activated only if you select the block name from the list displayed. Choose the **OK** button; you will be prompted to specify the insertion point for the block. Specify the insertion point; the **Insert/Edit Component** dialog box will be displayed. Enter the required information in this dialog box and choose the **OK** button from the **Insert/Edit Component** dialog box; the **Schematic equipment in** dialog box will be displayed again. Choose the **Close** button to exit from this dialog box.

INSERTING COMPONENTS FROM THE USER DEFINED LIST

Command: AECOMPONENTCAT

The **User Defined List** tool is used to insert a schematic symbol from a user-defined pick list. The data displayed in the pick list is stored in a database in *.mdb* format with the file name as *wd_picklist.mdb*. The location of this file is *C:\users\user name\documents\acade 2022\aedata\en-us\catalogs*. From the pick list, you can select the catalog number or the component description to be inserted into the drawing. You can edit, add, or delete the pick list data any time. To insert a schematic component from a user defined list, choose the **User Defined List** tool from the **Icon Menu** drop-down in the **Insert Components** panel of the **Schematic** tab; the **Schematic Component or Circuit** dialog box will be displayed, as shown in Figure 5-8.

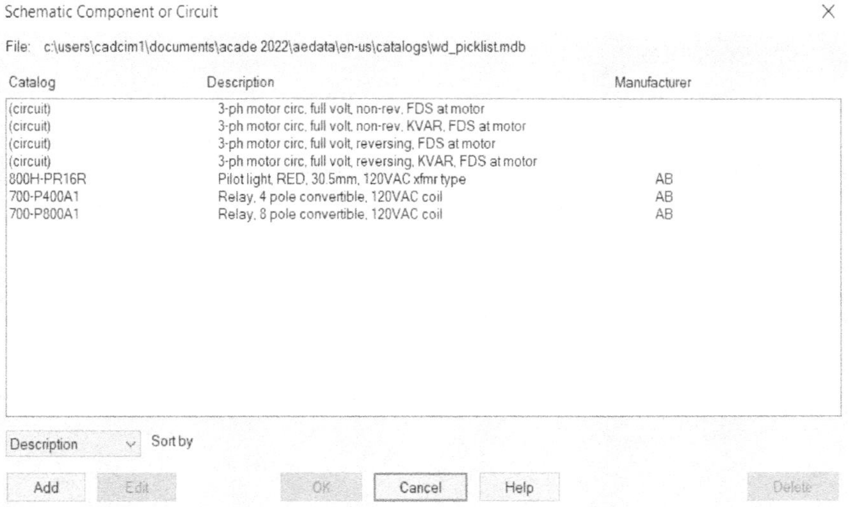

Figure 5-8 The ***Schematic Component or Circuit*** *dialog box*

The **OK** button in the **Schematic Component or Circuit** dialog box will be activated only if you select a component or a circuit from the list displayed. Now, select the component in the **Schematic Component or Circuit** dialog box and then choose the **OK** button; the component along with the cursor will be displayed and you will be prompted to specify the insertion point for the component. Next, specify the insertion point for the component; the **Insert / Edit Component** dialog box will be displayed. Enter the required information in this dialog box and choose the **OK** button; the component will be inserted into the drawing.

You can add a record using the **Schematic Component or Circuit** dialog box. To invoke this dialog box, choose the **User Defined List** tool from the **Icon Menu** drop-down in the **Insert Components** panel of the **Schematic** tab; the **Schematic Component or Circuit** dialog box will be displayed, refer to Figure 5-8. To add a new record in this dialog box, choose the **Add** button in this dialog box; the **Add record** dialog box will be displayed.

After specifying all values, choose the **OK** button in the **Add record** dialog box; the component or circuit will automatically be added to the list displayed in the **Schematic Component or Circuit** dialog box.

You can edit an existing record in the **Schematic Component or Circuit** dialog box. To edit an existing record of the schematic component in the component list, select a component or a circuit from this dialog box; the **Edit** button will be activated. Choose this button; the **Edit Record** dialog box will be displayed.

Make necessary changes in the **Edit Record** dialog box and then choose the **OK** button; you will notice that the modified component or the circuit name is displayed in the **Schematic Component or Circuit** dialog box.

SWAPPING AND UPDATING BLOCKS

Command: AESWAPBLOCK

The **Swap/Update Block** tool is used to replace or update a block in a drawing. Also, this tool is used to update drawings if the library has been changed. To replace or update a block, choose the **Swap/Update Block** tool from the **Edit Components** panel of the **Schematic** tab; the **Swap Block / Update Block / Library Swap** dialog box will be displayed, as shown in Figure 5-9.

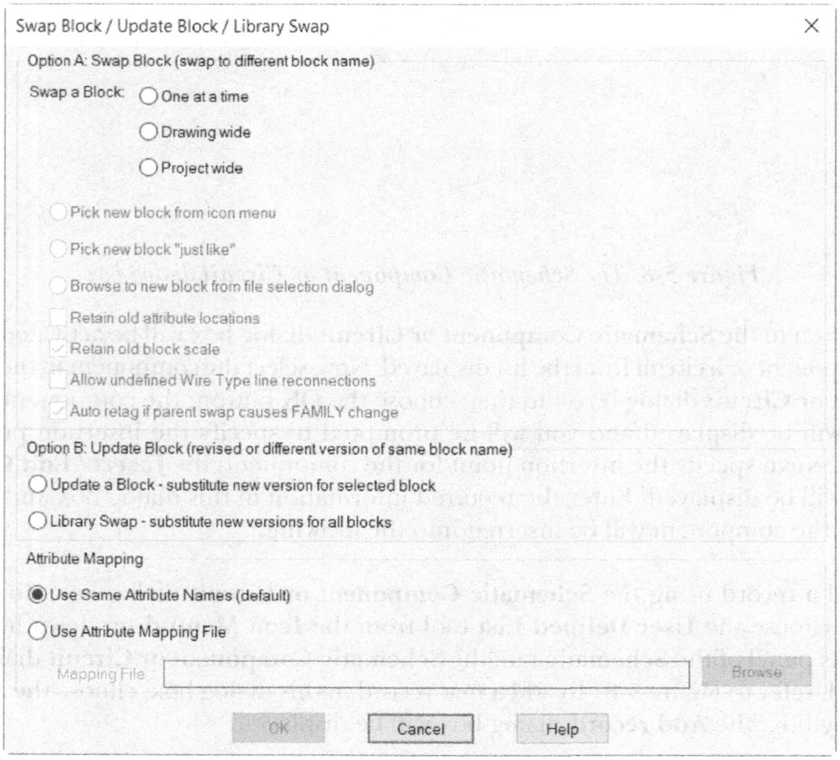

*Figure 5-9 The **Swap Block / Update Block / Library Swap** dialog box*

In this dialog box, only one area out of the **Option A: Swap Block (swap to different block name)** and **Option B: Update Block (revised or different version of same block name)** areas can be used at a time.

The **OK** button in the **Swap Block / Update Block / Library Swap** dialog box will be activated only if you select any one of the options from the **Option A: Swap Block (swap to different block name)** or **Option B: Update Block (revised or different version of same block name)** areas. Choose this button after specifying the settings.

TUTORIALS

Tutorial 1

In this tutorial, you will insert a ladder with four rungs into the drawing and then insert push buttons, relays, and pilot lights from the **Insert Component** dialog box. Also, you will add description to components, as shown in Figure 5-10. **(Expected time: 20 min)**

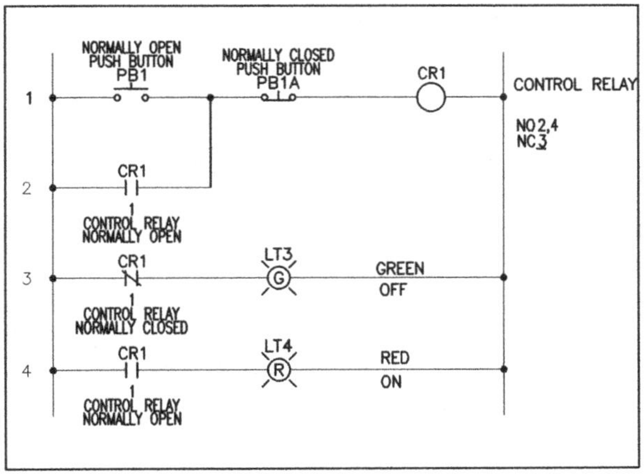

Figure 5-10 *Ladder and components inserted in it*

The following steps are required to complete this tutorial:

a. Create a new drawing.
b. Insert ladder into the drawing.
c. Insert components and add descriptions to components.
d. Add the wire.
e. Trim the rung of the ladder.
f. Save the drawing file.

Creating a New Drawing

1. Open the **PROJECT MANAGER**, if it is not already displayed.

2. Activate the **CADCIM** Project.

3. Choose the **New Drawing** button in the **PROJECT MANAGER**; the **Create New Drawing** dialog box is displayed. Enter **C05_tut01** in the **Name** edit box of the **Drawing File** area.

4. Next, choose the **Browse** button on the right of the **Template** edit box; the **Select template** dialog box is displayed. Select the **ACAD_ELECTRICAL** template from this dialog box and choose the **Open** button; the path and location of the template file is displayed in the **Template** edit box.

5. Enter **Schematic Components** in the **Description 1** edit box.

6. Choose the **OK** button; the *C05_tut01.dwg* is created in the **CADCIM** project and displayed at the bottom of the drawing list in it.

7. Move the *C05_tut01.dwg* to the *TUTORIALS* subfolder as discussed earlier.

Inserting Ladder into the Drawing

1. Choose the **Insert Ladder** tool from **Schematic > Insert Wires/Wire Numbers > Insert Ladder** drop-down; the **Insert Ladder** dialog box is displayed.

2. Set the following parameters in the **Insert Ladder** dialog box:

   Width: **5.0**
   1st Reference: **1**
   1 Phase: Select this radio button
   Spacing: **1.0**
   Rungs: **4**
   Yes: Select this radio button

   After setting these parameters, click in the **Length** edit box; the length of the ladder is automatically calculated and is displayed in this edit box.

3. Choose the **OK** button; you are prompted to specify the start position of the first rung. Enter **10, 15** at the Command prompt and press ENTER; the ladder is inserted in the drawing, as shown in Figure 5-11.

4. Zoom in the drawing.

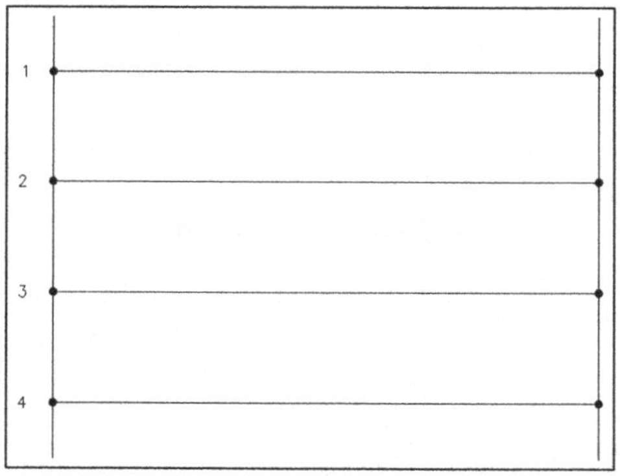

Figure 5-11 *The ladder inserted in the drawing*

Inserting Components and Adding Description to Components

1. Choose the **Icon Menu** tool from **Schematic > Insert Components > Icon Menu** drop-down; the **Insert Component** dialog box is displayed, as shown in Figure 5-12.

2. Select the **Push Buttons** icon from the **NFPA: Schematic Symbols** area displayed on the right in the **Insert Component** dialog box; the **NFPA: Schematic Symbols** area is replaced by the **NFPA: Push Buttons** area.

3. Select **Push Button NO** from the **NFPA: Push Buttons** area. You will notice that the component attached with the cursor is displayed. Also, you are prompted to specify the insertion point for the push button. Next, place the push button at the extreme left of the rung 1; the **Insert / Edit Component** dialog box is displayed.

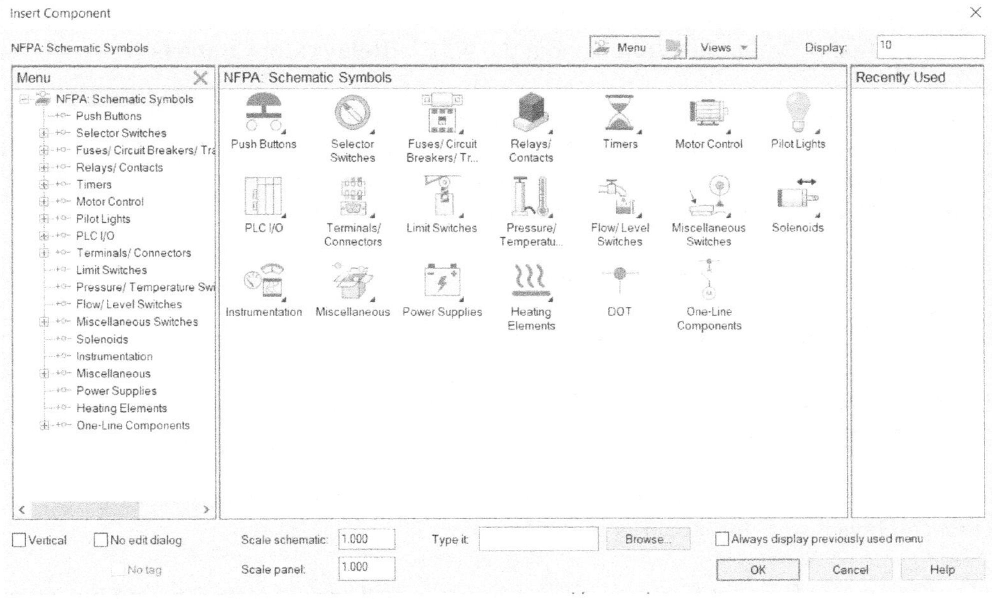

*Figure 5-12 The **Insert Component** dialog box*

The PB1 is displayed in the edit box of the **Component Tag** area. Do not change the value.

4. Enter **NORMALLY OPEN** and **PUSH BUTTON** in the **Line 1** and **Line 2** edit boxes of the **Description** area, respectively.

5. Choose the **OK** button from the **Insert / Edit Component** dialog box; the PB1 component is inserted in the drawing.

6. Again, choose the **Icon Menu** tool from **Schematic > Insert Components > Icon Menu** drop-down; the **Insert Component** dialog box is displayed.

7. Select the **Push Buttons** icon from the **NFPA: Schematic Symbols** area displayed on the right in the **Insert Component** dialog box; the **NFPA: Schematic Symbols** area is modified to the **NFPA: Push Buttons** area, displaying various types of push buttons.

8. Select the **Push Button NC** component; you will notice that the component attached with the cursor is displayed. Also, you are prompted to specify the insertion point for the push button. Next, place the push button to the right of PB1 on the rung 1, refer to Figure 5-13;

the **Insert / Edit Component** dialog box is displayed. PB1A is displayed in the edit box of the **Component Tag** area. Do not change this value.

9. Enter **NORMALLY CLOSED** and **PUSH BUTTON** in the **Line 1** and **Line 2** edit boxes in the **Description** area, respectively.

10. Choose the **OK** button from the **Insert / Edit Component** dialog box; the PB1A symbol is inserted in the drawing.

11. Repeat Step 1. Select the **Relays / Contacts** icon displayed in the **NFPA: Schematic Symbols** area of the **Insert Component** dialog box; the **NFPA: Relays and Contacts** area is displayed.

12. Select **Relay Coil** that is located in the first row and the first column in the **NFPA: Relays and Contacts** area of the **Insert Component** dialog box; you are prompted to specify the insertion point. Next, place the relay to the right of the PB1A push button on rung1; the **Insert / Edit Component** dialog box is displayed.

    CR1 is displayed in the edit box of the **Component Tag** area. Do not change this value.

13. Enter **CONTROL RELAY** in the **Line 1** edit box of the **Description** area.

14. Choose the **OK** button in the **Insert / Edit Component** dialog box; the component is inserted into rung 1 of the ladder.

15. Repeat Step1. Select the **Relays/Contacts** icon from the **Insert Component** dialog box; the **NFPA: Relays and Contacts** area is displayed.

16. Select **Relay NO Contact** from the **NFPA: Relays and Contacts** area; you are prompted to specify the insertion point for the symbol. Place **Relay NO Contact** at the extreme left of rung 2; the **Insert / Edit Child Component** dialog box is displayed.

17. In this dialog box, choose the **Drawing** button from the **Component Tag** area; the **Active Drawing list for FAMILY = "CR"** dialog box is displayed. Select **CR1** from this dialog box and choose the **OK** button; **CR1** is displayed in the **Tag** edit box.

18. Enter **CONTROL RELAY** in the **Line 1** edit box of the **Description** area, if it is not displayed already.

19. Enter **NORMALLY OPEN** in the **Line 2** edit box of the **Description** area.

20. Choose the **OK** button from the **Insert / Edit Child Component** dialog box; CR1 is inserted in the drawing in rung 2.

21. Repeat Step 1. Select the **Relays/Contacts** icon from the **NFPA: Schematic Symbols** area in the **Insert Component** dialog box; the **NFPA: Relays and Contacts** area is displayed.

22. Select the **Relay NC Contact** symbol; you are prompted to specify the insertion point. Place the symbol on the left of rung 3; the **Insert / Edit Child Component** dialog box is displayed.

23. In this dialog box, choose the **Drawing** button from the **Component Tag** area; the **Active Drawing list for FAMILY = "CR"** dialog box is displayed. In this dialog box, select the first tag entry, CR1, whose description is CONTROL RELAY and then choose the **OK** button; CR1 is displayed in the **Tag** edit box and CONTROL RELAY is displayed in the **Line 1** edit box of the **Description** area in the **Insert / Edit Child Component** dialog box.

24. Next, enter **NORMALLY CLOSED** in the **Line 2** edit box of the **Description** area.

25. Choose the **OK** button in the **Insert / Edit Child Component** dialog box; CR1 is inserted on rung 3 in the drawing.

26. Repeat Step 1. Next, select the **Pilot Lights** icon from the **NFPA: Schematic Symbols** area; the **NFPA: Schematic Symbols** area is modified to the **NFPA: Pilot Lights** area. Next, select the **Green Standard** symbol that is located on the first row and the second column; you are prompted to specify the insertion point. Place the symbol on the right of the **Relay NC Contact** symbol on the rung 3; the **Insert / Edit Component** dialog box is displayed. By default, LT3 is displayed in the edit box of the **Component Tag** area. Do not change this value.

27. Enter **GREEN** and **OFF** in the **Line 1** and **Line 2** edit boxes of the **Description** area, respectively.

28. Choose **OK** from the **Insert / Edit Component** dialog box; the LT3 symbol is inserted on rung 3.

29. Similarly, place CR1 (Relay NO Contact) with the description CONTROL RELAY and NORMALLY OPEN at extreme left of rung 4.

30. Insert the Pilot light (**Red Standard)** from the **Insert Component** dialog box with the description RED and ON to the right of CR1 on rung 4. Figure 5-13 shows the ladder with the following components inserted in it:

    (a) PB1, (b) PB1A, (c) CR1 and its contacts, (d) LT3 and (e) LT4

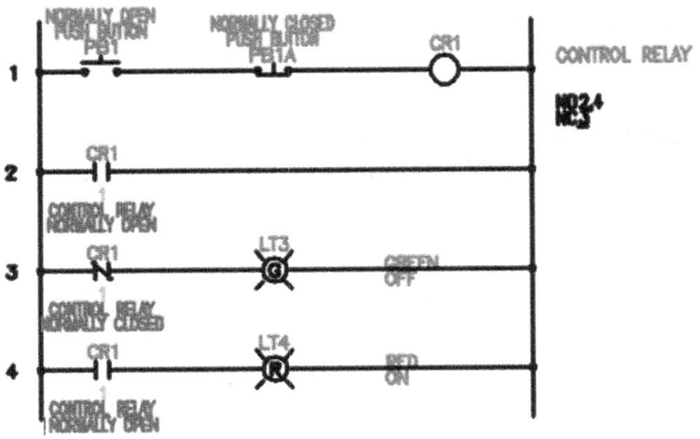

Figure 5-13 *The ladder and components*

Note

In order to check the cross-reference fields of the relay coil (CR1) inserted on rung 1, choose the ***Edit*** *tool from the* ***Edit Components*** *panel of the* ***Schematic*** *tab; you are prompted to select component. Next, select CR1 that is located on rung 1; the* ***Insert / Edit Component*** *dialog box is displayed. In this dialog box, the* ***Cross-Reference*** *area displays the number of NO and NC contacts used along with this relay coil.*

Adding the Wire

1. Choose the **Wire** tool from **Schematic > Insert Wires/Wire Numbers > Wire** drop-down; you are prompted to specify the start point of the wire.

2. Click in the middle of PB1 and PB1A of rung 1 and drag the cursor downward and click on rung 2; a wire is drawn between rung 1 and rung 2 and these rungs are joined, as shown in Figure 5-14. Press ESC to exit the command.

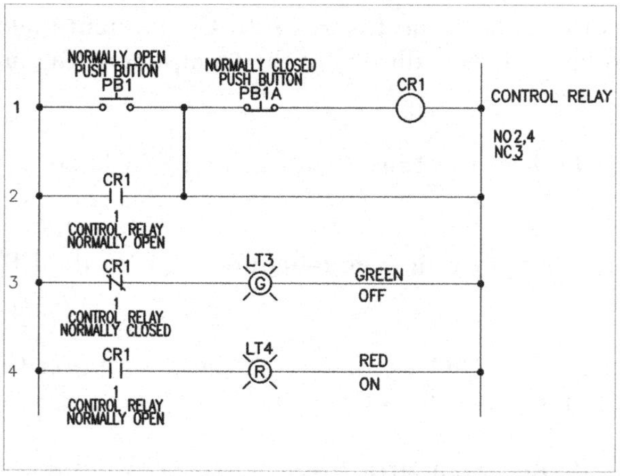

Figure 5-14 Wire inserted between rung 1 and rung 2

Trimming the Rung

1. Choose the **Trim Wire** tool from the **Edit Wires/Wire Numbers** panel of the **Schematic** tab; you are prompted to select the wire. Select the right portion of rung 2; the selected portion is removed, as shown in Figure 5-15. Next, press ESC to exit the command.

Saving the Drawing File

1. Choose **Save** from the **Application Menu** to save the *C05_tut01.dwg* drawing file.

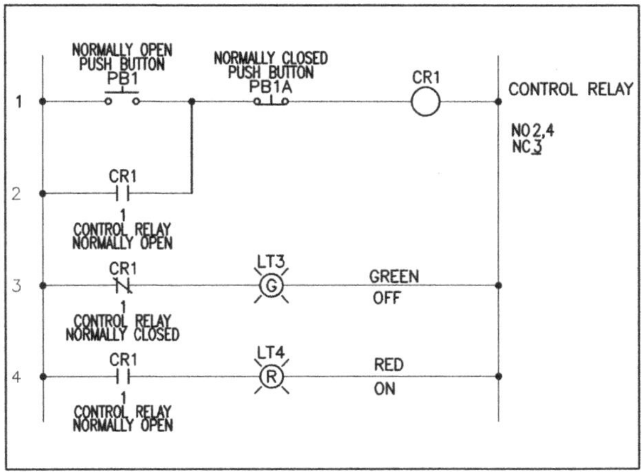

Figure 5-15 *The ladder with the half portion of rung 2 trimmed*

Tutorial 2

In this tutorial, you will open the *C05_tut01.dwg* file of Tutorial 1 of this chapter. Next, you will edit the component PB1 and add the catalog information to this component. **(Expected time: 15 min)**

The following steps are required to complete this tutorial:

a. Open, save, and add the drawing to the **CADCIM** project.
b. Edit the component.
c. Save the drawing file.

Opening, Saving, and Adding the Drawing to the CADCIM Project

1. Open *C05_tut01.dwg* drawing file. Save it with the name *C05_tut02.dwg*. Add it to the **CADCIM** project list, as discussed in the earlier chapters.

Editing the Component

1. Choose the **Edit** tool from **Schematic > Edit Components > Edit Components** drop-down; you are prompted to select the component to be edited. Select the PB1 push button; the **Insert/Edit Component** dialog box is displayed.

2. Choose the **Lookup** button in the **Catalog Data** area; the **Catalog Browser** dialog box is displayed, as shown in Figure 5-16.

3. Select **800H-BR6A** from the **Catalog Browser** dialog box.

   Make sure **AB**, **30.5mm EXTENDED**, and **RED** are selected in the **MANUFACTURER**, **TYPE**, and **STYLE** columns, respectively.

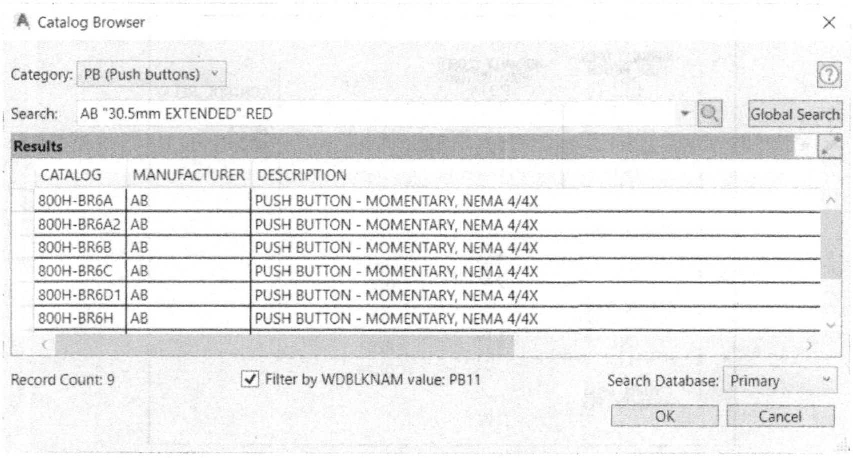

***Figure 5-16** The **Catalog Browser** dialog box*

4. Choose the **OK** button in the **Catalog Browser** dialog box; AB and 800H-BR6A are displayed in the **Manufacturer** and **Catalog** edit boxes in the **Catalog Data** area of the **Insert / Edit Component** dialog box.

5. Choose the **OK** button in the **Insert / Edit Component** dialog box to add the catalog data to the component PB1; the **Update other drawings?** message box is displayed. Choose the **OK** button in this message box; the **QSAVE** message box is displayed. Choose the **OK** button in this message box to save the changes in the drawing.

Saving the Drawing File

1. Choose **File > Save** from the menu bar to save the *C05_tut02.dwg* drawing file.

Tutorial 3

In this tutorial, you will use the **Swap/Update Block** tool to swap a PB1A block with a limit switch LS1 that you have inserted in the drawing of Tutorial 1 of this chapter, as shown in Figure 5-17. You will also create a project specific catalog database for the **CADCIM** project.

(Expected time: 15 min)

The following steps are required to complete this tutorial:

a. Open, save, and add the drawing to the **CADCIM** project.
b. Swap the block.
c. Create project specific catalog database.
d. Save the drawing file.

Opening, Saving, and Adding the Drawing to the CADCIM Project

1. Open *C05_tut01.dwg* drawing file. Save it with the name *C05_tut03.dwg*. Add it to the **CADCIM** project list, as discussed in the earlier chapters.

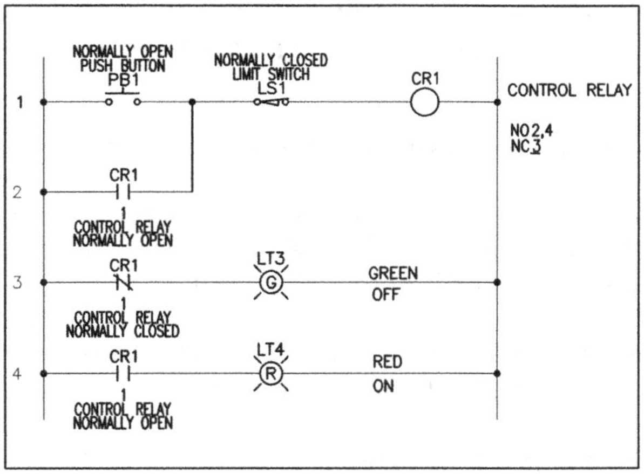

Figure 5-17 *The push button PB1A swapped with the limit switch LS1*

Swapping the Block

1. Choose the **Swap/Update Block** tool from the **Edit Components** panel of the **Schematic** tab; the **Swap Block / Update Block / Library Swap** dialog box is displayed, as shown in Figure 5-18.

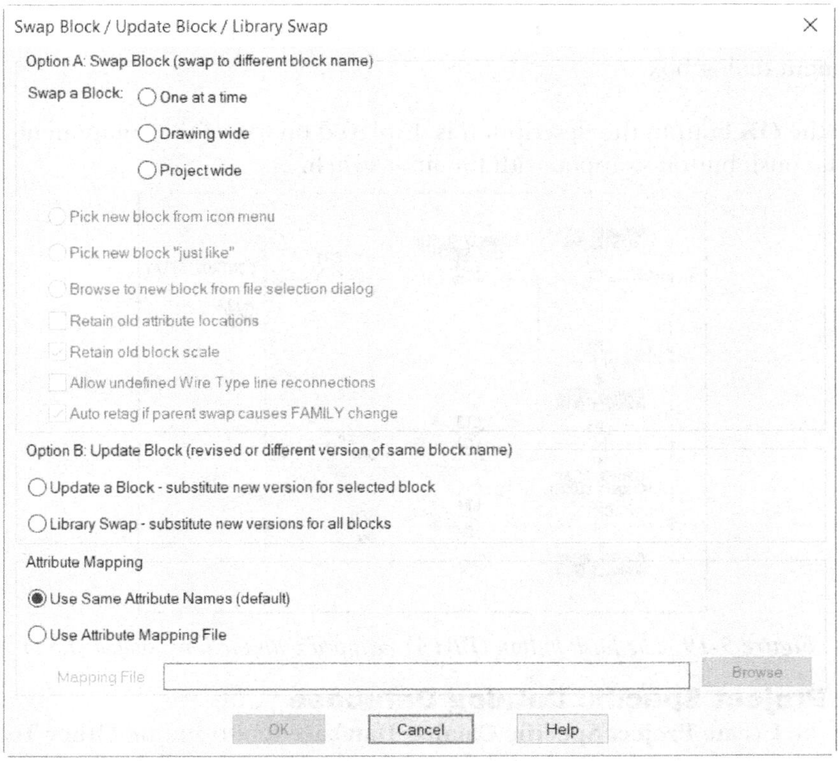

Figure 5-18 *The* ***Swap Block / Update Block / Library Swap*** *dialog box*

2. Select the **One at a time** radio button in the **Option A: Swap Block (swap to different block name)** area; the other options in this area are activated.

3. Select the **Retain old block scale** and **Auto retag if parent swap causes FAMILY change** check boxes in the **Option A: Swap Block (swap to different block name)** area, if they are not selected.

4. Clear the **Allow undefined Wire Type line reconnections** check box, if it is selected.

5. Select the **Pick new block from icon menu** radio button from the **Option A: Swap Block (swap to different block name)** area; the **OK** button is activated. Next, choose the **OK** button; the **Insert Component** dialog box is displayed.

6. Next, select the **Limit Switches** icon from the **NFPA: Schematic Symbols** area; a list of limit switches is displayed in the **NFPA: Limit Switches** area.

7. Select **Limit Switch, NC**; you are prompted to select the component to swap out. Next, select the PB1A push button that you have placed on rung 1; the push button is swapped with LS1. Now, press ENTER to exit the command.

8. To enter description for the limit switch, choose the **Edit** tool from **Schematic > Edit Components > Edit Components** drop-down; you are prompted to select the component. Select LS1; the **Insert / Edit Component** dialog box is displayed.

9. Enter **LIMIT SWITCH** in the **Line 2** edit box of the **Description** area in the **Insert / Edit Component** dialog box.

10. Choose the **OK** button; the description is displayed on top of the component. Figure 5-19 shows the push button swapped with the limit switch.

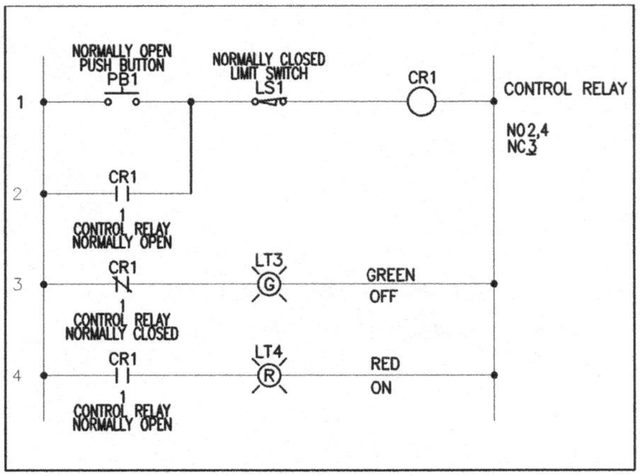

Figure 5-19 *The push button (PB1A) swapped with the limit switch (LS1)*

Creating Project Specific Catalog Database

1. Choose the **Create Project Specific Catalog Database** tool from the **Other Tools** panel of the **Project** tab; the **Create Project-Specific Catalog Database** dialog box is displayed. In this dialog box, select the **No - keep it separate** radio button; the edit box and the browse button below this radio button get activated.

Note that if you select the **Yes - make it the active project default catalog file** radio button, the project specific catalog database file will be the default project catalog file.

2. Choose the browse button; the **Save** dialog box is displayed. In this dialog box, specify the location in the **Look in** drop-down list. Note that the location you specify should be other than the default location specified in the edit box, refer to Figure 5-20. Next, enter **CADCIM** in the **File name** edit box and then choose the **Open** button to close the **Save** dialog box; the specified location is displayed in the edit box located below the **No - keep it separate** radio button. Now, choose the **OK** button in the **Create Project-Specific Catalog Database** dialog box; the *CADCIM.mdb* file is created at the specified location.

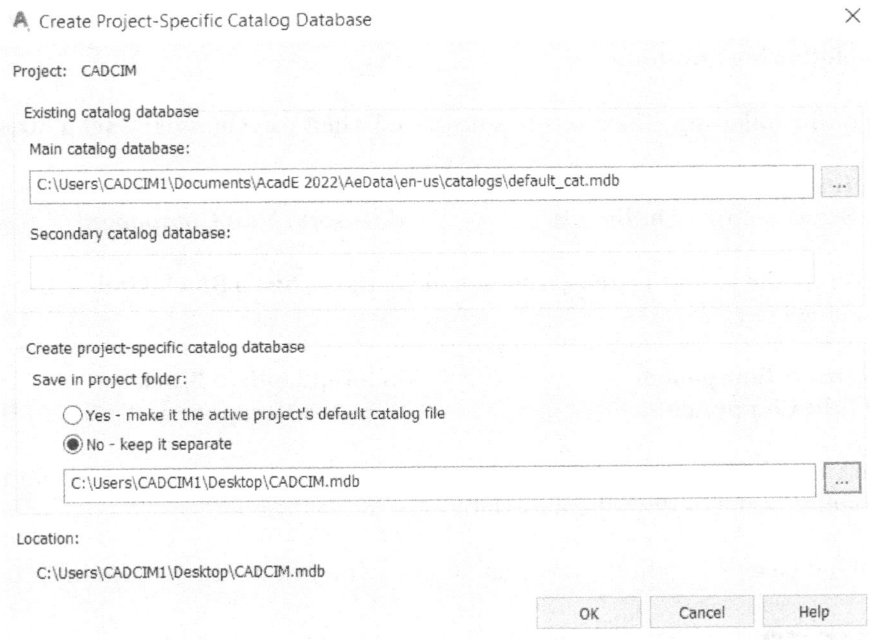

***Figure 5-20** The **Create Project-Specific Catalog Database** dialog box*

Saving the Drawing File

1. Choose **Save** from the **Application Menu** to save the *C05_tut03.dwg* drawing file.

Self-Evaluation Test

Answer the following questions and then compare them to those given at the end of this chapter:

1. Which of the following commands is used to insert a component into a drawing?

(a) **AECOMPONENT** (b) **AEEDITCOMPONENT**
(c) **AEPLCP** (d) **AECOPYCOMP**

2. The __________ tool is used to replace one block instance with another block.

3. When you choose the **Add** button in the **Schematic Component or Circuit** dialog box, the __________ dialog box is displayed.

4. The **OK-Repeat** button in the **Insert / Edit Component** dialog box is used to insert another component similar to the one inserted into the drawing. (T/F)

5. The **Catalog Browser** dialog box, in its Lookup mode, displays the database table that matches the family type of the inserted component. (T/F)

Review Questions

Answer the following questions:

1. Which of the following dialog boxes is displayed when you choose the **Icon Menu** tool?

   (a) **Insert Component** (b) **Schematic Component or Circuit**
   (c) **Select component list file** (d) **Insert / Edit Component**

2. Which one of the following tools, if chosen, displays the **Swap Block/Update Block/Library Swap** dialog box?

   (a) **Insert Component** (b) **Swap/Update Block**
   (c) **Edit Component** (d) **Insert Component (Catalog List)**

3. To insert multiple components at a time, choose the __________ tool from the **Insert Components** panel of the **Schematic** tab.

4. You will not be able to edit the component, once you insert it into a drawing. (T/F)

EXERCISES

Exercise 1

Create a new drawing with the name *C05_exer01.dwg* in the **NEW_PROJECT** project and insert a ladder with the following specifications: Width = 5; Spacing = 1.000; Rungs = 12, and 1st Reference = 1. Also, you will insert components into the ladder, as shown in Figure 5-21. Use the **Wire** tool to add wires and the **Trim Wire** tool to trim wires. Figure 5-21 shows the complete ladder diagram for Exercise 1. **(Expected time: 25 min)**

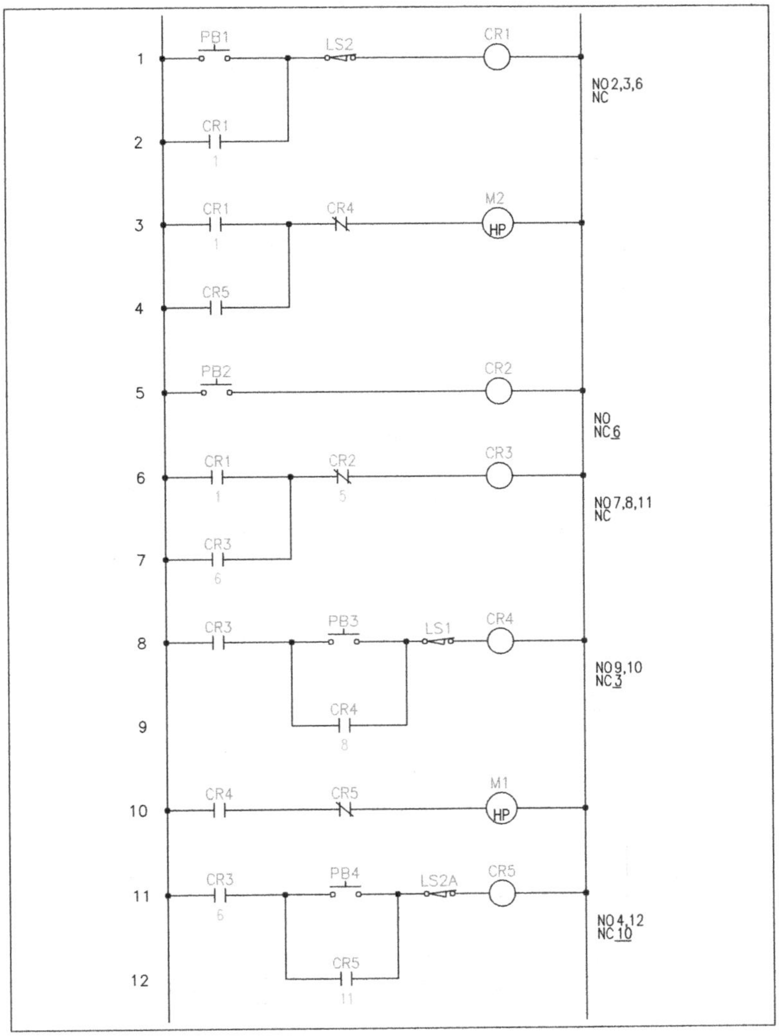

Figure 5-21 Ladder with components inserted for Exercise 1

Exercise 2

In this exercise, you will change the symbol library setting from NFPA to JIC 125. Next, you will swap a limit switch LS1 of the **NFPA** library with a toggle switch NO TG8 of the **JIC 125** library and switch LS2A with a proximity switch PRS11, as shown in Figure 5-22. Also, you will create a project specific catalog database for the NEW_PROJECT project. **(Expected time: 20 min)**

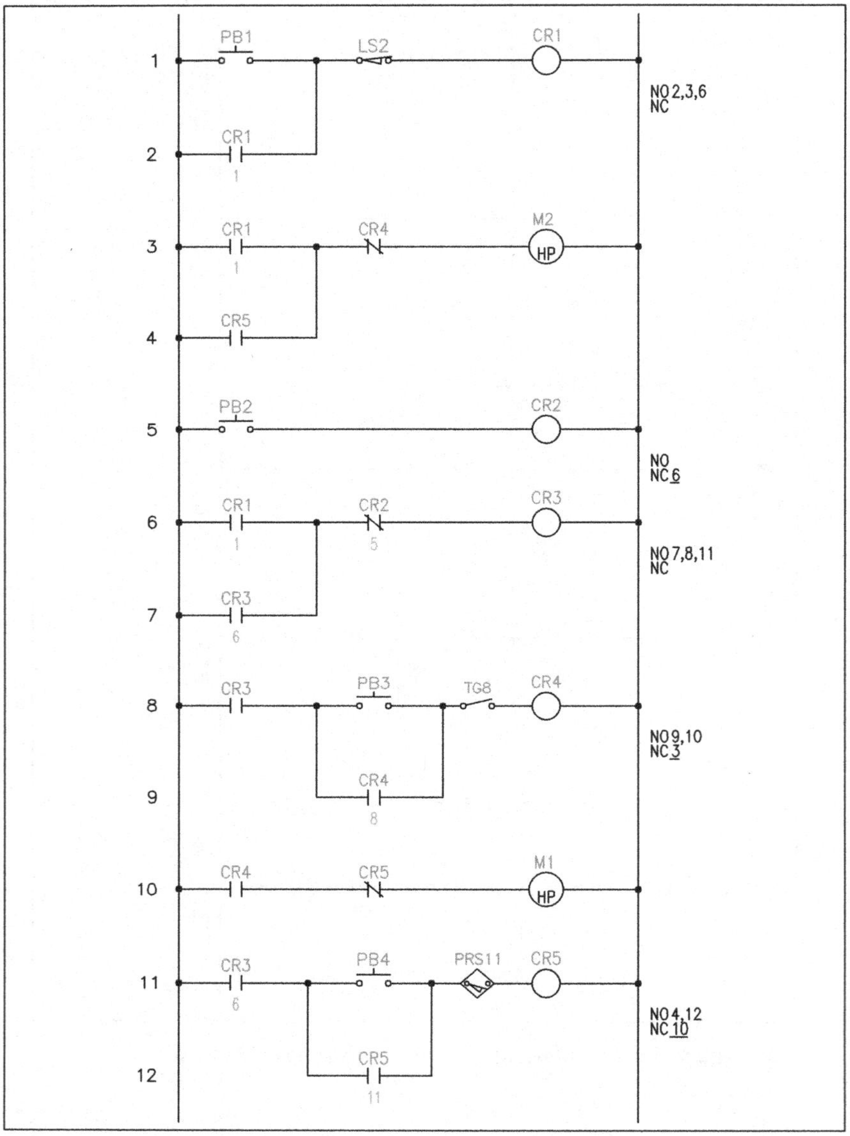

Figure 5-22 *Ladder diagram for Exercise 2*

Answers to Self-Evaluation Test

1. a, **2. Swap/Update Block**, **3. Add record**, **4.** T, **5.** T

Chapter 6

Schematic Editing

Learning Objectives

After completing this chapter, you will be able to:

- *Use basic schematic editing commands*
- *Update a schematic component from a one-line component*
- *Copy the catalog assignment and location values from one component to multiple components*
- *Use auditing tools to find out errors in the project and rectify them*
- *Retag the drawings*
- *Use the tools for editing attributes*

INTRODUCTION

In this chapter, you will learn about different tools that are used to create or modify electrical schematic. You can edit, move, copy, and align components in your drawing. These tools are very important as they provide extra functionalities to electrical schematic drawings. Also, later in this chapter, you will learn about auditing tools, editing tools of attributes, and retagging of drawings.

SCOOT TOOL

Command: AESCOOT

The **Scoot** tool is used to move objects such as components, terminals, PLC I/O modules, signal arrows, wire segments, wires with wire crossing-loops, ladder rungs, ladder buses, and so on in the drawing. The **Scoot** tool is similar to the AutoCAD **Move** tool. However, the **Scoot** tool has intelligence about electrical objects, which enables you to reposition electrical objects. Also, this tool is used to reconnect wires after repositioning the components. To scoot components, choose the **Scoot** tool from the **Modify Components** drop-down in the **Edit Components** panel of the **Schematic** tab, as shown in Figure 6-1; you will be prompted to select component, wire, or wire number for scoot. Select the component to scoot along with its connected wire; a temporary graphic indicating that you have selected the component will be displayed. Next, move the cursor to a place where you want to locate the component; the selected object will move in the orthogonal direction to the specified location.

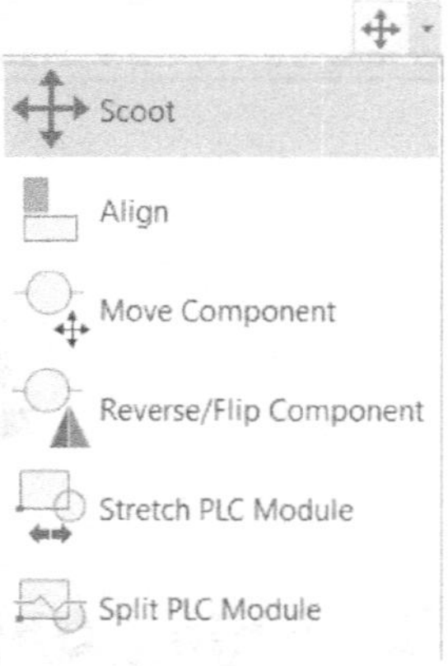

***Figure 6-1** The **Modify Components** drop-down*

Also, the wires will be reconnected after you scoot the components, the components will be updated, and the existing wire numbers will be recentered. Next, press ESC or click on the screen to exit the command. If you move components to a location that requires component updates, the **Component(s) Moved** dialog box will be displayed. Choose the **OK to Retag** button to retag the components automatically. Now, if the moved component is linked to other components in the current drawing, the **AutoCAD** message box will be displayed. Choose the **OK** button to update the related component in the drawing.

Also, if the component that you have moved consists of child components or related panel components in the other drawings of the active project, the **Update other drawings?** message box will be displayed. Choose the **OK** button from the **Update other drawings?** message box to batch process or update the other drawings of a project. Choose the **Task** button to save all modifications in the task list to be run later. If you do not want to retag a component, choose the **No Retag** button. Choose the **Update child cross-references only** button to update only the child cross-references.

MOVE COMPONENT TOOL

Command: AEMOVE

The **Move Component** tool is used to move the selected component from its current location or wire location to the specified location. To do so, choose the **Move**

Component tool from the **Modify Components** drop-down in the **Edit Components** panel of the **Schematic** tab, refer to Figure 6-1; you will be prompted to select a component to move. Select the component; you will be prompted to specify insertion point. Specify the insertion point for the component; the selected component will be moved to the specified location, the underlying wires will be reconnected, and the component tags will be updated automatically.

COPYING A COMPONENT

Command: AECOPYCOMP

The **Copy Component** tool is used to copy the selected component and insert the copied component to the specified location in the drawing. To do so, choose the **Copy Component** tool from the **Edit Components** panel of the **Schematic** tab; you will be prompted to select the component to copy. Select the component to be copied; you will be prompted to specify the insertion point for the copied component. Specify the insertion point; the **Insert / Edit Component** dialog box will be displayed.

Next, enter the required values in this dialog box and choose the **OK** button; the values along with the copied component will be displayed in your drawing. If you do not change the values displayed in the **Insert / Edit Component** dialog box, the values of the original component will be transferred to the copied component. Also, the copied component will be retagged automatically and wire numbers will get updated accordingly.

Note

1. If the selected component that you want to copy is a child component, the ***Insert / Edit Child Component*** *dialog box will be displayed instead of the* ***Insert / Edit Component*** *dialog box.*

2. You can also copy the components using the ***Copy*** *command of AutoCAD, but in this case, the components will not get retagged and the wires will not get reconnected automatically.*

ALIGNING COMPONENTS

Command: AEALIGN

The **Align** tool is used to line up components vertically or horizontally. To do so, choose the **Align** tool from the **Modify Components** drop-down in the **Edit Components** panel of the **Schematic** tab; you will be prompted to select the components to be aligned. Select the components; a temporary line passing through the center of the component will be displayed. This component is treated as the reference or master component with which you need to align the rest of the components that you select. Next, select the components to align with the master component and then press ENTER; all the selected components will be aligned to the master component.

DELETING COMPONENTS

Command: AEERASECOMP

The **Delete Component** tool is used to delete a component from the drawing. Also, after deleting the component, the wires get reconnected and the wire numbers get updated. To delete a component, choose the **Delete Component** tool from the **Edit Components** panel of the **Schematic** tab from the menu bar; you will be prompted to select components. Select the component(s) and press the ENTER key or right-click on the screen; the component(s) will be deleted. If you select a parent component that has child contacts, the **Search for / Surf to Children?** dialog box will be displayed. Choose the **No** button in the **Search for / Surf to Children?** dialog box to delete the component.

UPDATING A SCHEMATIC COMPONENT FROM A ONE-LINE COMPONENT

You can update a schematic component from a one-line component. In other words, you can copy information such as descriptions, catalog values, installation/location codes, and so on from a one-line component to a schematic component. To do so, choose the **Edit** tool from the **Edit Components** drop-down in the **Edit Components** panel of the **Schematic** tab and select the schematic component to be updated from the active drawing; the **Insert/Edit Component** dialog box will be displayed. In this dialog box, choose the **Schematic** button from the **Component Tag** area; the **XX Tags in Use** dialog box will be displayed. Note that **XX** stands for the schematic component to be updated. In this dialog box, select the **Show one-line components (1-*)** check box from the **Show** area; one-line components will be displayed along with the other components, refer to Figure 6-2.

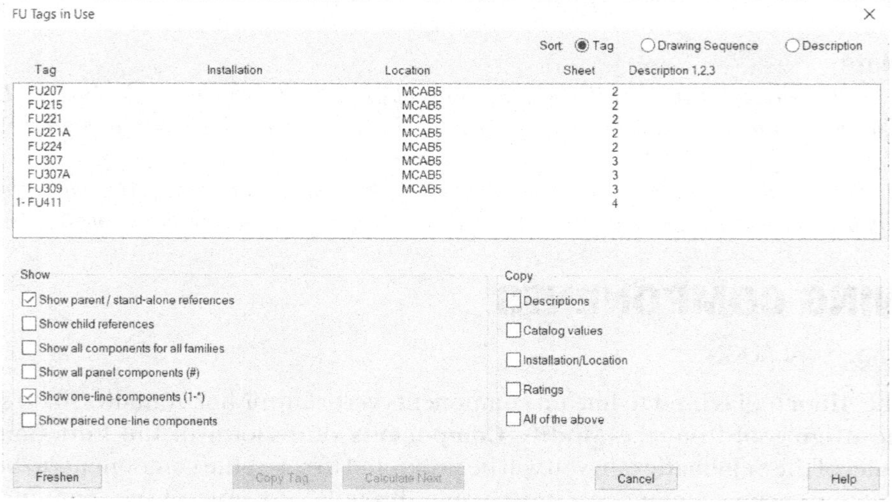

***Figure 6-2** The **M Tags in Use** dialog box*

Next, select the desired one-line component from the list. Also, select the check boxes in the **Copy** area based on the requirement and choose the **Copy Tag** button; the **Copy Tag** dialog box will be displayed. Choose the **OK** button from this dialog box; the **Copy Tag** dialog box will disappear and the **Insert/Edit Component** dialog box will be displayed with the updated values of the schematic component. Choose the **OK** button in this dialog box to update the schematic component from a one-line component.

The **Show paired one-line components** check box is available in the **Show** area of the **XX Tags in Use** dialog box. You need to select this check box along with the **Show one-line components**

(1-*) check box and the **Show parent / stand-alone references** check box to show pair of matching one-line components with schematic or panel component.

COPYING THE CATALOG ASSIGNMENT

Command: AECOPYCAT

The **Copy Catalog Assignment** tool is used to copy the catalog part numbers from one component to another. Also, this tool is used to insert or edit the catalog data of the selected component or footprint. This tool helps in assigning the same catalog information to multiple components. To copy manufacturer, catalog, assembly, and multiple catalog attributes from one component to other selected components, choose the **Copy Catalog Assignment** tool from the **Edit Components** drop-down in the **Edit Components** panel of the **Schematic** tab, as shown in Figure 6-3; you will be prompted to select the master component. Select the component to be treated as master component; the **Copy Catalog Assignment** dialog box will be displayed, as shown in Figure 6-4. This dialog box displays the catalog data of the selected component.

*Figure 6-3 The **Edit Components** drop-down*

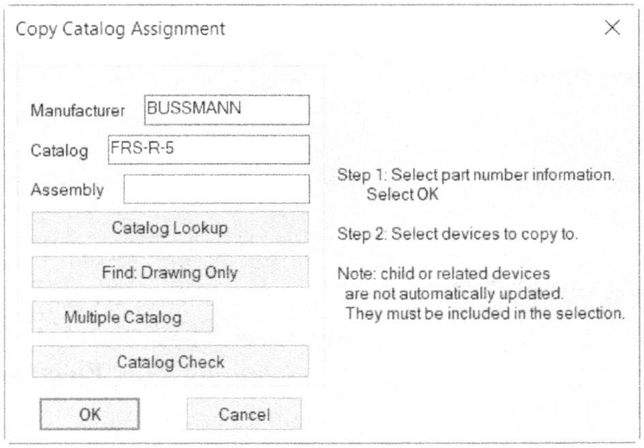

*Figure 6-4 The **Copy Catalog Assignment** dialog box*

Note

*If the catalog assignment is not assigned to the master component, the **Copy Catalog Assignment** dialog box will not display any catalog data. Also, if you want to modify the catalog assignment both for the master and child components, choose the **Catalog Lookup** button from the **Copy Catalog Assignment** dialog box. On doing so, the **Catalog Browser** dialog box will be displayed. Next, select the desired catalog information and then choose the **OK** button from this dialog box; the catalog data will be assigned to the selected component.*

After specifying the options in the **Copy Catalog Assignment** dialog box, choose the **OK** button; you will be prompted to pick the target component(s). Select the target component(s) and press ENTER; the catalog information will be copied to the target component(s). If the target component consists of catalog data different from the master component, the **Different symbol block names** dialog box will be displayed.

If you choose the **OK** button, the **Caution: Existing Data on Target** dialog box will be displayed. To overwrite the catalog information of the master component to the target component(s), choose the **Overwrite** button; the catalog information of the master component will be applied to the target component(s). But if the target component has its child contacts or references, then after choosing the **Overwrite** button, the **Update Related Components?** message box will be displayed. In this message box, choose the **Yes-Update** button; the related components will be updated. Choose the **Skip** button to skip the update of the related components.

COPYING INSTALLATION/LOCATION CODE VALUES

Command: AECOPYINSTLOC

The **Copy Installation/Location Code Values** tool is used to copy the installation and location code assignments from one component to another. To do so, choose the **Copy Installation/Location Code Values** tool from the **Edit Components** panel of the **Schematic** tab; the **Copy Installation/Location to Components** dialog box will be displayed.

After specifying the required options in the **Copy Installation/Location to Components** dialog box, choose the **OK** button; you will be prompted to select components. Select the component(s) to copy the installation/location values and press ENTER; the installation/location values will be copied to the selected component.

Note

AutoCAD Electrical does not show any warning if you overwrite existing installation and location codes.

AUDITING DRAWINGS

The auditing tools are used to check errors in the project. These tools help in troubleshooting and improving the accuracy of a drawing. There are two auditing tools: **Electrical Audit** and **Drawing Audit**. These tools are discussed next.

Electrical Auditing

Command: AEAUDIT

The **Electrical Audit** tool is used to find out the problems that affect the drawings of a project. This tool can be used to correct some of the errors in drawings. To find errors, choose the **Electrical Audit** tool from the **Schematic** panel of the **Reports** tab; the **Electrical Audit** dialog box will be displayed, refer to Figure 6-5. The progress of the electrical audit will be displayed in the edit box. Once the audit is finished, the edit box will display the total number of errors that occurred in the active project. In this dialog box, the **Project** radio button is selected by default. As a result, the total number of errors found in the active project will be displayed in the edit box of the **Electrical Audit** dialog box. If you select the **Active Drawing** radio button, then the total number of errors found in the active drawing will be displayed in the edit box. This dialog box also displays the date and time of the electrical audit report.

The **Details** button is used to view errors as well as to expand or collapse the **Electrical Audit** dialog box. To view errors, choose the **Details** button; the **Electrical Audit** dialog box will expand and display detailed information about the errors found in the project. If you choose this button again, the **Electrical Audit** dialog box will collapse.

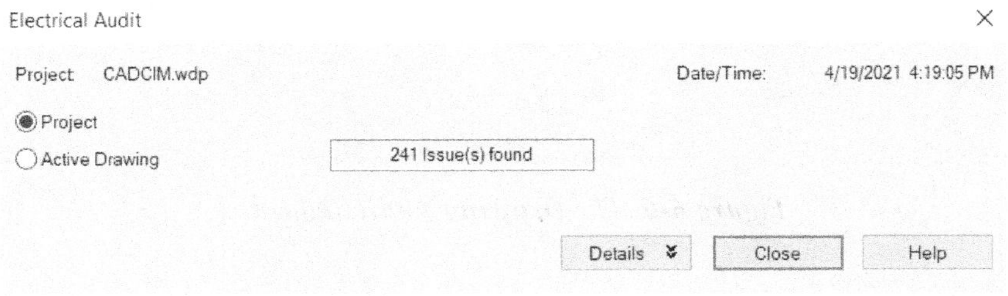

Figure 6-5 *The **Electrical Audit** dialog box*

Note

*If the active drawing is not a part of the active project, then the **Active Drawing** radio button will not be activated in the **Electrical Audit** dialog box.*

*If errors are not found in a project or active drawing, then the **Details** button in the **Electrical Audit** dialog box will not be activated.*

The expanded **Electrical Audit** dialog box has ten different tabs. If the tab has a red circle and a white 'x', it means an error is present in that category. To view the errors present in a tab, choose the tab; the information about the errors will be displayed in the lower part of the **Electrical Audit** dialog box.

Auditing a Drawing

Command: AEAUDITDWG

The **DWG Audit** tool is used to find out the problems in wiring that affect the wire connectivity of a design. Using this tool, you can audit a single drawing or multiple drawings in a project. The auditing of a drawing is performed to check for wire gaps, wire number, color, zero length wires, gauge labels, and wire number floaters for errors, and so on. To find out errors in wires and to rectify them, choose the **DWG Audit** tool from the **Schematic** panel of the **Reports** tab; the **Drawing Audit** dialog box will be displayed, as shown in Figure 6-6.

The **Audit drawing or project** area consists of two radio buttons: **Active drawing** and **Project**. These radio buttons are discussed next.

The **Active drawing** radio button is selected by default and is used to audit an active drawing only. Choose the **OK** button from the **Drawing Audit** dialog box; the modified **Drawing Audit** dialog box will be displayed, as shown in Figure 6-7. After specifying the options in the modified **Drawing Audit** dialog box, choose the **OK** button; the **Drawing Audit** message box will be displayed, as shown in Figure 6-8. Choose the **OK** button in this message box; the **Report: Audit**

for this drawing dialog box will be displayed. You can save and print a report by choosing the **Save As** button and the **Print** button, respectively from this dialog box.

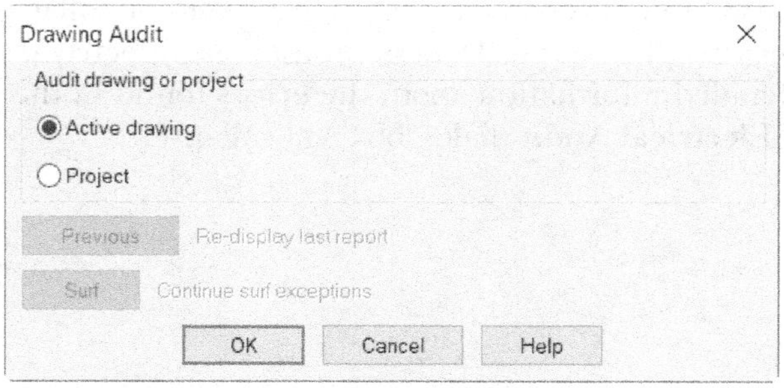

***Figure 6-6** The **Drawing Audit** dialog box*

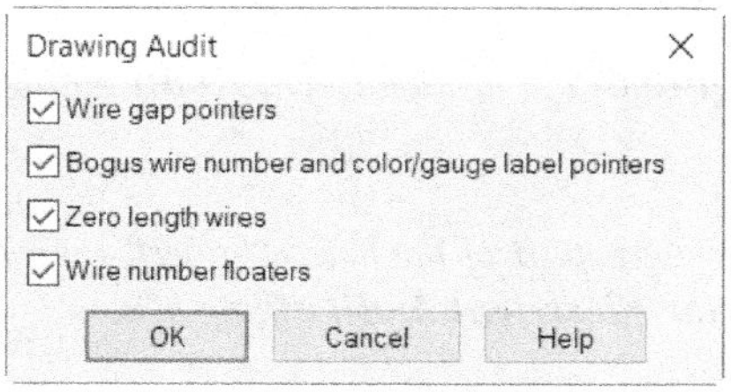

***Figure 6-7** The modified **Drawing Audit** dialog box*

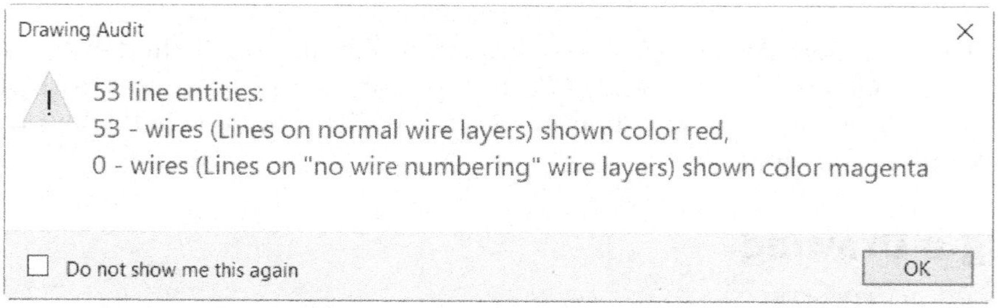

***Figure 6-8** The **Drawing Audit** message box*

The **Previous** button of the **Drawing Audit** dialog box is activated if you have already generated the audio report, and is used to display the previous audit report. To do so, choose the **Previous** button; the **Report: Audit** dialog box will be displayed. This dialog box re-displays the last run audit report. The **Project** radio button is used to audit an active project.

RETAGGING DRAWINGS

Command: AEPROJUPDATE

The **Project-Wide Update/Retag** tool is used to update the selected drawings in a project. Also, this tool is used to retag components, update the cross-reference of component, retag wire numbers and signal references, and so on. To update or retag the drawings of a project, choose the **Update/Retag** tool from the **Project Tools** panel of the **Project** tab or choose **Projects > Project-Wide Update/Retag** from the menu bar; the **Project-Wide Update or Retag** dialog box will be displayed, as shown in Figure 6-9.

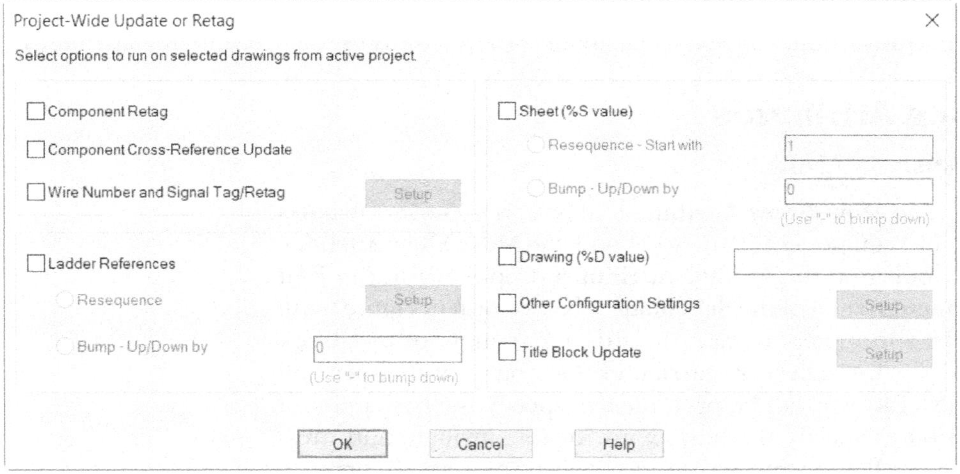

Figure 6-9 *The* ***Project-Wide Update or Retag*** *dialog box*

The **Component Retag** check box is used to retag all non-fixed components of the selected drawings. The **Component Cross-Reference Update** check box is used to update the cross-reference of components of the selected drawings.

The **Wire Number and Signal Tag/Retag** check box is used to update signal symbols and the wire numbers that are not fixed. Select the **Wire Number and Signal Tag/Retag** check box; the **Setup** button will be activated. Using this button, you can insert or update wire numbers that are linked to wire networks.

The **Ladder References** check box is used to renumber the ladders sequentially. To renumber each ladder of a drawing sequentially, select this check box; the options below the **Ladder References** check box will be activated. These options are discussed next.

The **Resequence** radio button is selected automatically when you select the **Ladder References** check box and is used to define the starting reference number for a ladder. Also, it is used to define the sequence of ladders in different drawings. The **Setup** button located on the right of the **Resequence** radio button is used to renumber ladder reference. To do so, choose the **Setup** button; the **Renumber Ladders** dialog box will be displayed. Using this dialog box, you can renumber the reference numbers of a ladder of the selected drawings in an active project. Specify the required options in this dialog box and choose the **OK** button to return to the **Project-Wide Update or Retag** dialog box.

You can use any one of the options in the **Project-Wide Update or Retag** dialog box or use all of them simultaneously by selecting the options displayed in this dialog box. After selecting the options, choose the **OK** button from the **Project-Wide Update or Retag** dialog box; the **Select Drawings to Process** dialog box will be displayed. Select the required drawings and choose the **Process** button; the selected drawings will be displayed in the bottom list. Next, choose the **OK** button from this dialog box; the selected drawings will be processed and updated accordingly.

USING TOOLS FOR EDITING ATTRIBUTES

In AutoCAD Electrical, there are a number of tools that can be used to modify attributes. These tools are used to modify the attributes of selected symbols but they are not used to modify the block. In this section, you will learn about different tools used for modifying attributes.

Moving Attributes

Command: AEATTSHOW

The **Move/Show Attribute** tool is used to move attributes of a component. To do so, choose the **Move/Show Attribute** tool from the **Modify Attributes** drop-down in the **Edit Components** panel of the **Schematic** tab, as shown in Figure 6-10; you will be prompted to select the attribute to move or pick block graphics for list. Select the attributes that you want to move and press ENTER; you will be prompted to specify the base point of the attribute. Specify the base point for the attribute and press ENTER or click on the screen; you will be prompted again to select the attributes to move. Press ENTER to exit the command. If you enter 'W' at the Command prompt, you will be able to move multiple attributes at a time by selecting the attributes using the crossing window.

***Figure 6-10** The **Modify Attributes** drop-down*

Editing Attributes

Command: AEEDITATT

The **Edit Selected Attribute** tool is used to edit the selected attribute. To do so, choose the **Edit Selected Attribute** tool from the **Edit Components** panel of the **Schematic** tab; you will be prompted to select an attribute. Select an attribute; the **Edit Attribute** dialog box will be displayed.

Enter the new name for the attribute in the edit box. Alternatively, choose the **Pick** button from the **Edit Attribute** dialog box for selecting the attribute from the current drawing. You can increase or decrease the selected attribute value by choosing the **>** and **<** buttons. Next, choose the **OK** button; the name of the selected attribute will be changed to the specified name. Press ENTER to exit the command.

Hiding Attributes

Command: AEHIDEATT

The **Hide Attribute (Single Pick)** tool is used to hide the selected attribute. To do so, choose the **Hide Attribute (Single Pick)** tool from the **Modify Attributes** drop-down in the **Edit Components** panel of the **Schematic** tab; you will be prompted to select the attribute to hide or pick the block graphics. Select the attribute to hide it. Press ENTER to exit the command. Note that if you select the graphics of a block, the **SHOW / HIDE Attributes** dialog box will be displayed. The attributes that have an asterisk (*) displayed in the **Visible** column of the **SHOW / HIDE Attributes** dialog box are visible in the drawing. To make these attributes invisible in the drawing, select these attributes in the dialog box; the asterisk will disappear from the **Visible** column indicating that the attribute is now invisible in the drawing. Similarly, if you select the attribute in the dialog box which does not have an asterisk in the **Visible** column, the asterisk will appear in this column indicating that the attribute is now visible in the drawing.

You can hide multiple attributes at a time. To do so, enter 'W' at the Command prompt; you will be prompted to select the attributes. Select the attributes by using the crossing window and then press ENTER; the attributes will hide. Press ENTER to exit the command. You can also hide multiple attributes using the **Hide Attribute (Window/Multiple)** tool.

Unhiding Attributes

Command: AESHOWATTRIB

The **Unhide Attribute (Window/Multiple)** tool is used to unhide the hidden attributes. To do so, choose the **Unhide Attribute (Window/Multiple)** tool from the **Modify Attributes** drop-down in the **Edit Components** panel of the **Schematic** tab; you will be prompted to select objects. Select graphics of the symbol block and press ENTER; the **Flip Attributes to Visible** dialog box will be displayed. Select the attributes that you want to display and choose the **OK** button in the **Flip Attributes to Visible** dialog box; the attributes will be visible on the screen. Note that multiple attributes can be selected from the **Flip Attributes to Visible** dialog box by pressing CTRL or SHIFT. If you enter 'W' at the Command prompt, you will be prompted to specify the first corner. Specify the first corner; you will be prompted to specify the opposite corner. Specify the opposite corner and then press ENTER; the **Flip Attributes to Visible** dialog box will be displayed. Select one or more attributes and choose the **OK** button; the attributes will be flipped to visible.

Adding Attributes

Command: AEATTRIBUTE

The **Add Attribute** tool is used to add a new attribute to the existing AutoCAD Electrical block. To do so, choose the **Add Attribute** tool from the **Modify Attributes** drop-down in the **Edit Components** panel of the **Schematic** tab; you will be prompted to select an object. Select the object; the **Add Attribute** dialog box will be displayed.

Specify the required options in the **Add Attribute** dialog box and choose the **OK** button in this dialog box; you will be prompted to specify the location for attribute. By default, the first location point is already selected by AutoCAD Electrical and it is 0, 0. Next, specify the second location point and click on the screen; the attribute value will be inserted into the drawing.

Squeezing an Attribute/Text

Command: AEATTSQUEEZE

The **Squeeze Attribute/Text** tool is used to reduce the attribute or text size to make it suitable for tight places. After each click on the attribute or text, the width of the attribute or text reduces by 5%. To squeeze the attribute or text size, choose the **Squeeze**

Attribute/Text tool from the **Modify Attributes** drop-down in the **Edit Components** panel of the **Schematic** tab; you will be prompted to select the attribute or text to be squeezed. Select the attribute or text; you will be prompted again to select the attribute or text. Continue the selection till it is squeezed to a required size. Next, press ENTER to exit the command.

Stretching an Attribute/Text

Command: AEATTSTRETCH

The **Stretch Attribute/Text** tool is used to expand the attribute or text size. After each click on the attribute or text, the width of the attribute or text increases by 5%. To stretch an attribute, choose the **Stretch Attribute/Text** tool from the **Modify Attributes** drop-down in the **Edit Components** panel of the **Schematic** tab; you will be prompted to select the attribute or text. Select the attribute or text; the attribute or text will be stretched. The command will continue till you press ENTER.

Changing the Attribute Size

Command: AEATTSIZE

The **Change Attribute Size** tool is used to change the height and width of the attribute text that has already been inserted into the drawing. To change the text size of the attribute, choose the **Change Attribute Size** tool from the **Modify Attributes** drop-down in the **Edit Components** panel of the **Schematic** tab; the **Change Attribute Size** dialog box will be displayed. The options in this dialog box are discussed next.

Specify the size for the attribute in the **Size** edit box. Specify the width for the attribute in the **Width** edit box. Alternatively, choose the **Pick >>** button; you will be prompted to select the attribute. Select the required attribute; the **Change Attribute Size** dialog box will be displayed again and the values of the selected attribute size and width will be displayed in the **Size** and **Width** edit boxes. By default, the **Apply** check boxes are selected. As a result, the new size and width is applied to the attribute that you select. If you clear the **Apply** check boxes in the **Change Attribute Size** dialog box, the **Single**, **By Name**, and **Type It** buttons will not be activated.

The **Single** button is used to select one attribute at a time. The **By Name** button is used to change the size and width of the same type of attributes. The **Type It** button is used to specify the name of the attribute to be matched to the selected attributes. If you do not enter any attribute name in the edit box of the **Enter Attribute Name** dialog box and choose **OK**, then it will again return to the **Change Attribute Size** dialog box. Choose the **Cancel** button to exit from this dialog box.

Rotating an Attribute

Command: AEATTROTATE

The **Rotate Attribute** tool is used to rotate the selected attribute text by 90 degrees. To do so, choose the **Rotate Attribute** tool from the **Modify Attributes** drop-down in the **Edit Components** panel of the **Schematic** tab; you will be prompted to select the attribute text to rotate. Select the attribute text; the text will be rotated by 90 degrees in counterclockwise direction. Press ENTER to exit the command. Now, if you want to move the attribute text, enter '**M**' at the Command prompt and press ENTER; you will be prompted to specify the base point. Specify the base point and press ENTER; you will be prompted to specify the destination point. Specify the destination point and press ENTER; the attribute text will be moved to the specified location.

Changing the Justification of an Attribute

Command: AEATTJUSTIFY

The **Change Attribute Justification** tool is used to change the justification of any attribute such as wire number text, component description text, and so on. To change the justification of an attribute, choose the **Change Attribute Justification** tool from the **Modify Attributes** drop-down in the **Edit Components** panel of the **Schematic** tab; the **Change Attribute/Text Justification** dialog box will be displayed,

In this dialog box, select the required justification for the attribute text from the **Select Justification** area in this dialog box; the **OK** button will be activated. Choose **OK**; you will be prompted to select the text or attribute to change the justification. Select the text or attribute; the justification of the attribute or text will be changed.

Alternatively, choose the **Pick Master** button; you will be prompted to select the master attribute or text that will be used for justification of other attributes or text. Next, select the attribute or text; the **Change Attribute/Text Justification** dialog box will be displayed on the screen again. Choose **OK**; you will be prompted to select the attribute or text for the change of justification. Select attributes or text and press ENTER; the justification of attributes or text will be changed. Alternatively, enter '**W**' at the Command prompt to select the attributes or text by using a crossing window. Press ENTER to exit the command.

Changing an Attribute Layer

Command: AEATTLAYER

The **Change Attribute Layer** tool is used to change the layer of the selected attribute. To do so, choose the **Change Attribute Layer** tool from the **Modify Attributes** drop-down in the **Edit Components** panel of the **Schematic** tab; the **Force Attribute/Text to a Different Layer** dialog box will be displayed.

After specifying the required options in the **Force Attribute/Text to a Different Layer** dialog box, choose the **OK** button in this dialog box; you will be prompted to select the attribute or text to move to layer. Select the attribute or text; the attribute or text layer will get changed to the layer that you have entered in the **Change to Layer** edit box. Press ENTER to exit the command.

TUTORIALS

Tutorial 1

In this tutorial, you will insert a ladder and its components into a drawing. Next, you will use the **Copy Component**, **Scoot**, **Move Component**, and **Delete Component** tools for copying, scooting, moving, and deleting the inserted components. refer to Figure 6-11. Also, you will hide the attributes of components. **(Expected time: 30 min)**

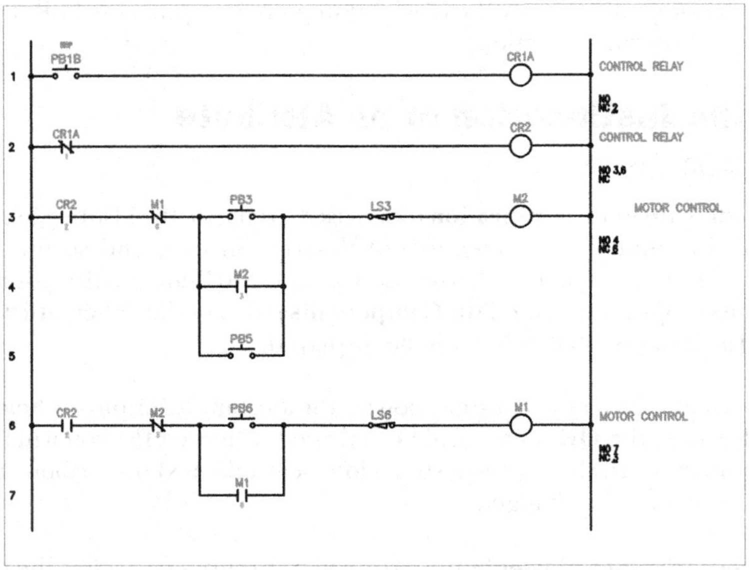

Figure 6-11 Ladder diagram for Tutorial 1

The following steps are required to complete this tutorial:

a. Create a new drawing.
b. Insert ladder in the drawing.
c. Insert components and add descriptions.
d. Copy components.
e. Hide attributes.
f. Delete components.
g. Move components.
h. Add wires.
i. Scoot components.
j. Trim wires.
k. Align components.
l. Copy the catalog data.
m. Save the drawing file.

Creating a New Drawing

1. Activate the **CADCIM** project if not already active. Next, choose the **New Drawing** button from the **PROJECT MANAGER**; the **Create New Drawing** dialog box is displayed. Enter **C06_tut01** in the **Name** edit box of the **Drawing File** area. Select the template as **ACAD_ELECTRICAL.dwt** and enter **Schematic Components** in the **Description 1** edit box.

2. Choose the **OK** button; the *C06_tut01.dwg* drawing is created in the **CADCIM** project and is displayed at the bottom of the drawing list in the **CADCIM** project. Next, move the *C06_tut01.dwg* to the *TUTORIALS* subfolder of the **CADCIM** project.

Inserting Ladder into the Drawing

1. Choose the **Insert Ladder** tool from **Schematic > Insert Wires/Wire Numbers > Insert Ladder** drop-down; the **Insert Ladder** dialog box is displayed.

2. Set the following parameters in the **Insert Ladder** dialog box:

   Width: **8.000**
   1st Reference: **1**
   1 Phase: Select this radio button
   Spacing: **1.000**
   Rungs: **7**
   Yes: Select this radio button

3. After setting these parameters, click in the **Length** edit box; the length of the ladder is automatically calculated and displayed in this edit box. Also, make sure that 0 is displayed in the **Skip** edit box of the **Draw Rungs** area.

4. Choose the **OK** button; you are prompted to specify the start position of the first rung. Enter **11,18** at the Command prompt and press ENTER; the ladder is inserted into the drawing, as shown in Figure 6-12. Next, choose **View > Zoom > In** from the menu bar to zoom in the drawing.

***Figure 6-12** Ladder with 7 rungs*

Inserting Components and Adding Descriptions

1. Choose the **Icon Menu** tool from **Schematic > Insert Components > Icon Menu** drop-down; the **Insert Component** dialog box is displayed.

2. Select the **Push Buttons** icon from the **NFPA: Schematic Symbols** area of the **Insert Component** dialog box; the **NFPA: Schematic Symbols** area is modified to the **NFPA: Push Buttons** area.

3. Select the **Push Button NO** icon from the **NFPA: Push Buttons** area; the cursor along with the component is displayed and you are prompted to specify the insertion point for the push button. Enter **11.5,18** at the Command prompt and press ENTER; the push button is placed at rung 1 and the **Insert / Edit Component** dialog box is displayed.

By default, PB1B is displayed in the edit box of the **Component Tag** area.

4. Enter **STOP** in the **Line 1** edit box in the **Description** area. Next, choose the **OK** button from the **Insert / Edit Component** dialog box; the PB1B component is inserted into the drawing.

5. Repeat step 1. Select the **Relays/Contacts** icon displayed in the **NFPA: Schematic Symbols** area of the **Insert Component** dialog box; the **NFPA: Relays and Contacts** area is displayed.

6. Select the **Relay Coil** icon in the **NFPA: Relays and Contacts** area of the **Insert Component** dialog box; you are prompted to specify the insertion point. Enter **13,18** at the Command prompt and press ENTER; the **Insert / Edit Component** dialog box is displayed. By default, CR1A is displayed in the edit box of the **Component Tag** area.

7. Enter **CONTROL RELAY** in the **Line 1** edit box in the **Description** area. Next, choose the **OK** button from the **Insert / Edit Component** dialog box; the component CR1A gets added to rung 1 of the ladder.

8. Repeat step 1. Select the **Relays/Contacts** icon displayed in the **Insert Component** dialog box; the **NFPA: Relays and Contacts** area is displayed.

9. Select the **Relay Coil** icon in the **NFPA: Relays and Contacts** area of the **Insert Component** dialog box; you are prompted to specify the insertion point. Enter **13.5,17** at the Command prompt and press ENTER. By default, CR2 is displayed in the edit box of the **Component Tag** area of the **Insert / Edit Component** dialog box.

10. Enter **CONTROL RELAY** in the **Line 1** edit box in the **Description** area. Next, choose the **OK** button from the **Insert / Edit Component** dialog box; the component CR2 gets added to rung 2 of the ladder.

11. Repeat step 1. Select the **Relays/Contacts** icon and then select **Relay NC Contact** in the **NFPA: Relays and Contacts** area of the **Insert Component** dialog box; you are prompted to specify the insertion point. Enter **11.5,17** at the Command prompt and press ENTER; the **Insert / Edit Child Component** dialog box is displayed.

12. Choose the **Drawing** button from the **Component Tag** area of the **Insert / Edit Child Component** dialog box; the **Active Drawing list for FAMILY = "CR"** dialog box is displayed. Select **CR1A** from this dialog box and choose the **OK** button; the information is displayed in the **Insert / Edit Child Component** dialog box.

    By default, **CONTROL RELAY** is displayed in the **Line 1** edit box of the **Description** area.

13. Enter **NORMALLY CLOSED** in the **Line 2** edit box of the **Description** area. Next, choose the **OK** button in the **Insert / Edit Child Component** dialog box; the component CR1A is added to the ladder rung 2.

14. Repeat step 1. Select the **Relays/Contacts** icon and then select **Relay NO Contact** in the **NFPA: Relays and Contacts** area; you are prompted to specify the insertion point. Enter

11.5,16 at the Command prompt and press ENTER; the **Insert / Edit Child Component** dialog box is displayed.

15. Choose the **Drawing** button from the **Component Tag** area of the **Insert / Edit Child Component** dialog box; the **Active Drawing list for FAMILY = "CR"** dialog box is displayed. Select **CR2** from this dialog box and choose the **OK** button; the information is displayed in the **Insert / Edit Child Component** dialog box.

16. Enter **NORMALLY OPEN** in the **Line 2** edit box of the **Description** area in the **Insert / Edit Child Component** dialog box and then choose the **OK** button from this dialog box; the CR2 control relay is inserted into the drawing in the rung 3.

17. Repeat step 1. Select the **Motor Control** icon from the **NFPA: Schematic Symbols** area in the **Insert Component** dialog box. Next, select **Motor Starter Coil** from the **NFPA: Motor Control** area; you are prompted to specify the insertion point for the component.

18. Enter **17.5,16** at the Command prompt and press ENTER; the **Insert / Edit Component** dialog box is displayed. Enter **M2** in the edit box of the **Component Tag** area. Also, enter its description as **MOTOR CONTROL** in the **Line 1** edit box of the **Description area** of the **Insert / Edit Component** dialog box. Choose the **OK** button from this dialog box; the **Motor Starter Coil** is inserted into rung 3 of the ladder.

Note

*You may need to move attributes of the inserted components by using the **Move/Show Attribute** tool.*

Next, you need to insert **2nd + Starter Contact NO** in the rung 4.

19. Choose the **Motor Control** icon from the **Insert Component** dialog box and then select **2nd+Starter Contact NO** from the **NFPA: Motor Control** area; you are prompted to specify the insertion point.

20. Enter **11.5,15** at the Command prompt and press ENTER; the **Insert / Edit Child Component** dialog box is displayed. Choose the **Drawing** button from the **Component Tag** area of the **Insert / Edit Child Component** dialog box; the **Active Drawing list for FAMILY = "M"** dialog box is displayed. Select **M2** from this dialog box and choose the **OK** button; the information is displayed in the **Insert / Edit Child Component** dialog box.

21. Enter **NORMALLY OPEN** in the **Line 2** edit box of the **Description** area in the **Insert / Edit Child Component** dialog box and then choose the **OK** button from this dialog box; the M2 starter contact is inserted into the drawing in the rung 4.

22. Repeat step 1. Select the **Motor Control** icon from the **NFPA: Schematic Symbols** area in the **Insert Component** dialog box. Next, select **Motor Starter Coil** from the **NFPA: Motor Control** area; you are prompted to specify the insertion point for the component.

23. Enter **18,13** at the Command prompt. Enter **M1** in the edit box of the **Component Tag** area. Also, enter its description as **MOTOR CONTROL** in the **Line 1** edit box of the **Description** area of the **Insert / Edit Component** dialog box. Choose the **OK** button from

the **Insert / Edit Component** dialog box; the **Motor Starter Coil (M1)** is inserted into rung 6 of the ladder.

24. Again repeat step 1. Select the **Motor Control** icon and then select **2nd+Starter Contact NC** from the **NFPA: Motor Control** area of the **Insert Component** dialog box; you are prompted to specify the insertion point for the component.

25. Enter **12.8,16** at the Command prompt and press ENTER; the **Insert / Edit Child Component** dialog box is displayed. Choose the **Drawing** button from the **Component Tag** area of the **Insert / Edit Child Component** dialog box; the **Active Drawing list for FAMILY = "M"** dialog box is displayed. Select **M1** from this dialog box and choose the **OK** button; the information is displayed in the **Insert / Edit Child Component** dialog box.

26. Enter **NORMALLY CLOSED** in the **Line 2** edit box of the **Description** area in the **Insert / Edit Child Component** dialog box and then choose the **OK** button from this dialog box; the Motor starter contact (M1) is inserted into the drawing in the rung 3.

    Next, you need to insert **Limit Switch, NC** in the rung 3.

27. Choose the **Limit Switches** icon from the **Insert Component** dialog box; the **NFPA: Limit Switches** area is displayed. Select **Limit Switch, NC** from this area; you are prompted to specify the insertion point.

28. Enter **16,16** at the Command prompt and press ENTER; the **Insert / Edit Component** dialog box is displayed. By default, **LS3** is displayed in the **Component Tag** edit box of the **Insert / Edit Component** dialog box. Also, enter **LIMIT SWITCH** and **NORMALLY CLOSED** in the **Line 1** and **Line 2** edit boxes, respectively in the **Description** area of the **Insert / Edit Component** dialog box. Choose the **OK** button; LS3 gets inserted into rung 3.

    Similarly, you need to insert **Limit Switch, NO** in the rung 1.

29. Choose the **Limit Switch, NO** from the **NFPA: Limit Switches** icon of the **Insert Component** dialog box; you are prompted to specify the insertion point.

30. Enter **18,18** at the Command prompt and press ENTER; the **Insert / Edit Component** dialog box is displayed. By default, LS1A is displayed in the edit box of the **Component Tag** area in this dialog box. Next, choose the **OK** button; LS1A is inserted into rung 1 of drawing. Figure 6-13 shows the ladder with components inserted in it.

31. Choose the **Move/Show Attribute** tool from **Schematic > Edit Components > Modify Attributes** drop-down; you are prompted to select the object. Select the **Motor Control** attribute of the **Motor Starter Coil** and press ENTER; you are prompted to specify the base point. Specify the base point at the left corner of the attribute and move the attribute, as shown in Figure 6-13.

Copying Components

1. Choose the **Copy Component** tool from the **Edit Components** panel of the **Schematic** tab; you are prompted to select a component to copy.

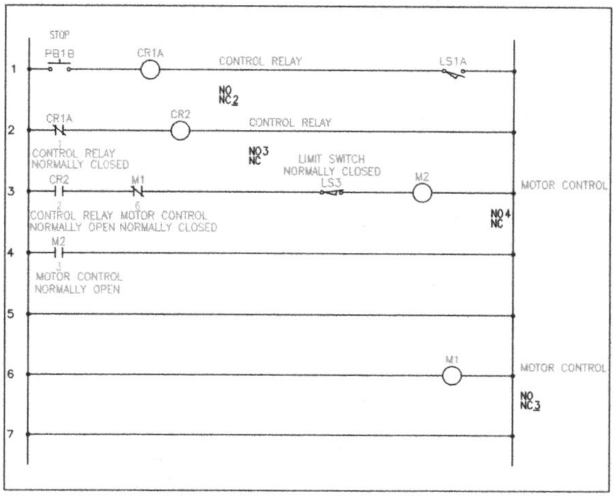

Figure 6-13 *Components inserted in the ladder*

2. Select **PB1B** (push button) placed on rung 1; you are prompted to specify the insertion point for the copied component. Enter **14,16** at the Command prompt and press ENTER; the **Insert / Edit Component** dialog box is displayed.

3. By default, PB3 is displayed in the edit box of the **Component Tag** area. Enter **DOWN** in the **Line 1** edit box of the **Description** area of the **Insert / Edit Component** dialog box. Next, choose the **OK** button in this dialog box; the component PB3 gets inserted into rung 3.

4. Similarly, right-click on CR2 placed on rung 3 and choose the **Copy Component** option from the marking menu displayed; you are prompted to specify the insertion point. Enter **12,13** at the Command prompt and press ENTER; the **Insert / Edit Child Component** dialog box is displayed.

5. Choose the **Drawing** button from the **Component Tag** area of the **Insert / Edit Child Component** dialog box; the **Active Drawing list for FAMILY = "CR"** dialog box is displayed. Select CR2 (third entry) from this dialog box and choose the **OK** button; the information is displayed in the **Insert / Edit Child Component** dialog box.

   By default, **CONTROL RELAY** is displayed in the **Line 1** edit box of the **Description** area and **NORMALLY OPEN** in the **Line 2** edit box of the **Description** area.

6. Choose the **OK** button in the **Insert / Edit Child Component** dialog box; the component CR2 is inserted into rung 6 of the drawing.

7. Right-click on PB1B that you have placed on rung 1 and choose the **Copy Component** option from the marking menu displayed; you are prompted to specify the insertion point. Enter **13,14** at the Command prompt and press ENTER; the **Insert / Edit Component** dialog box is displayed.

8. Remove description from the **Line 1** of the **Description** area in the **Insert / Edit Component** dialog box. Also, enter **PB5** in the edit box of the **Component Tag** area if not displayed. Next, choose the **OK** button in this dialog box; the PB5 is inserted on rung 5. Right-click on **PB1B** and choose the **Copy Component** option from the marking menu displayed; you are prompted to specify the insertion point.

9. Enter **14,13** at the Command prompt and press ENTER; the **Insert / Edit Component** dialog box is displayed. In this dialog box, **PB6** is displayed by default in the edit box of the **Component Tag** area. Remove the description from the **Line 1** edit box of the **Description** area and choose the **OK** button; PB6 is inserted into rung 6 in the ladder.

10. Right-click on Motor control M2 that is on rung 4 and choose the **Copy Component** option from the marking menu displayed; you are prompted to specify the insertion point. Enter **12,12** at the Command prompt and press ENTER; the **Insert / Edit Child Component** dialog box is displayed.

11. Choose the **Drawing** button from the **Component Tag** area of the **Insert / Edit Child Component** dialog box; the **Active Drawing list for FAMILY = "M"** dialog box is displayed. Select **M1** (first entry) from this dialog box and choose the **OK** button; the information is displayed in the **Insert / Edit Child Component** dialog box.

12. Enter **NORMALLY OPEN** in the **Line 2** edit box of the **Description** area in the **Insert / Edit Child Component** dialog box and then choose the **OK** button from this dialog box; the Motor starter contact (M1) is inserted into the drawing in rung 7.

13. Similarly, select M1 from rung 3 and copy it to rung 6 at **15,13**. Choose the **Drawing** button from the **Component Tag** area of the **Insert / Edit Child Component** dialog box; the **Active Drawing list for FAMILY = "M"** dialog box is displayed. Select **M2** (fourth entry) from this dialog box and choose the **OK** button; the information is displayed in the **Insert / Edit Child Component** dialog box.

14. Enter **NORMALLY CLOSED** in the **Line 2** edit box of the **Description** area in the **Insert / Edit Child Component** dialog box and then choose the **OK** button from this dialog box; the M2 starter contact is inserted into the drawing in rung 6.

15. Next, select limit switch LS3 from 3rd rung and copy it at **15,15**. Again, select LS3 and copy it at **16,13**. You will notice that the limit switch is inserted in rung 4 and rung 6. Keep the values in the **Description** area intact. Figure 6-14 shows the copied components.

Hiding Attributes

1. Choose the **Hide Attribute (Single Pick)** tool from **Schematic > Edit Components > Modify Attributes** drop-down; you are prompted to select the attributes to hide.

2. Select the DESC1 and DESC2 attributes of the following components:

   CR1A placed on rung 2,
   CR2, M1, PB3, and LS3 placed on rung 3,
   M2 and LS4 placed on rung 4,

CR2, M2, and LS6 placed on rung 6, and
M1 placed on rung 7

On doing so, the description of the components is hidden.

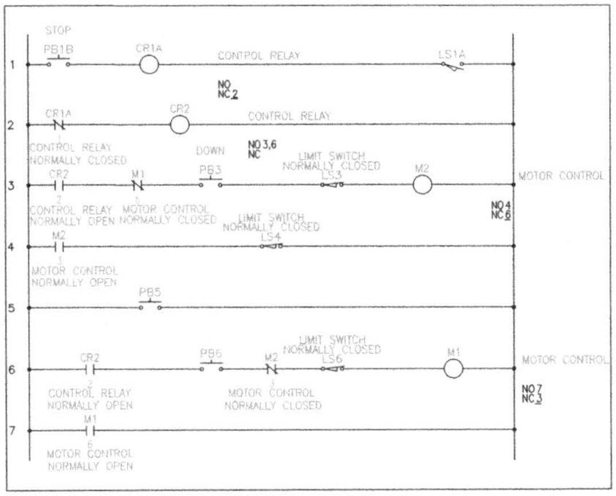

Figure 6-14 *The copied components*

3. Press ENTER to exit the command. Figure 6-15 shows the ladder and component with hidden attributes.

Deleting Components

1. Choose the **Delete Component** tool from the **Edit Components** panel of the **Schematic** tab; you are prompted to select objects.

2. Select LS1A from rung 1 and LS4 from rung 4, and then press ENTER; the **Search for / Surf to Children?** dialog box is displayed. Choose the **No** button from this dialog box; components are deleted from the drawing, as shown in Figure 6-16.

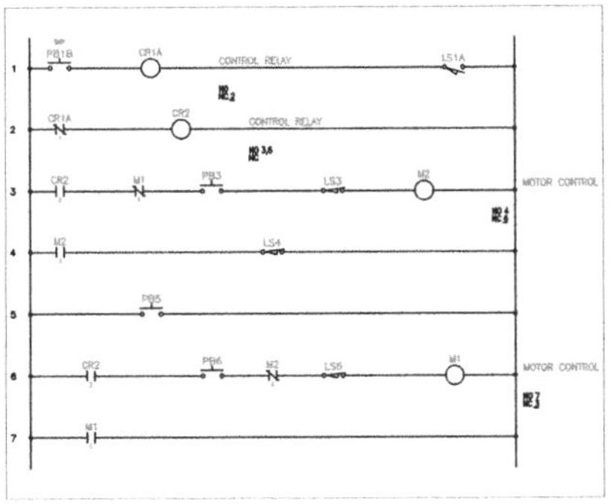

Figure 6-15 *The ladder after hiding attributes*

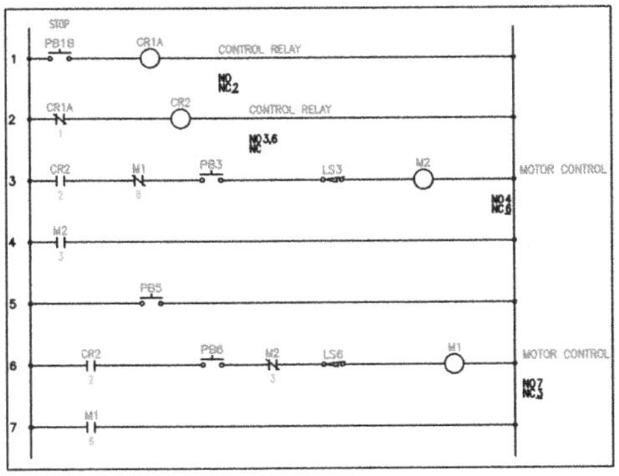

Figure 6-16 *The ladder after deleting components*

Moving Components

1. Choose the **Move Component** tool from **Schematic > Edit Components > Modify Components** drop-down; you are prompted to select the component to move.

2. Select M2 from rung 6; you are prompted to specify the insertion point for the component.

3. Enter **13,13** at the Command prompt and press ENTER; M2 is moved to a new location in rung 6, as shown in Figure 6-17.

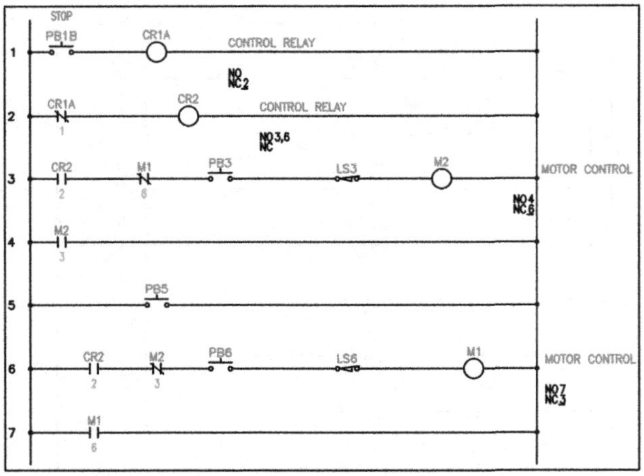

Figure 6-17 *M2 moved to a different location*

Adding Wires

1. Choose the **Wire** tool from **Schematic > Insert Wires/Wire Numbers > Wire** drop-down; you are prompted to specify the starting point of the wire at the Command prompt.

2. Enter **13.4,16** at the Command prompt and press ENTER; you are prompted to specify the wire endpoint. Next, enter **13.4,15** and press ENTER; the wire is inserted between rung 3 and 4 on the left side of the PB3.

3. Enter **14.6,16** at the Command prompt and press ENTER; you are prompted to specify the wire endpoint. Next, enter **14.6,15** and press ENTER; the wire is inserted between rung 3 and 4 on the right side of the PB3.

4. Enter **13.4,15** at the Command prompt and press ENTER; you are prompted to specify the wire endpoint. Next, enter **13.4,14** and press ENTER; the wire is inserted between rung 4 and 5.

5. Enter **14.6,15** at the Command prompt and press ENTER; you are prompted to specify the wire endpoint. Next, enter **14.6,14** and press ENTER; the wire is inserted between rung 4 and 5.

6. Enter **13.4,13** at the Command prompt and press ENTER; you are prompted to specify the wire endpoint. Next, enter **13.4,12** and press ENTER; the wire is inserted between rung 6 and 7 on the left side of the PB6.

7. Enter **14.6,13** at the Command prompt and press ENTER; you are prompted to specify the wire endpoint. Next, enter **14.6,12** and press ENTER; the wire is inserted between rung 6

and 7 on the right side of the PB6. Now, press ENTER to exit the command. Figure 6-18 shows the wires inserted in the drawing.

Scooting Components

1. Choose the **Scoot** tool from **Schematic > Edit Components > Modify Components** drop-down; you are prompted to select the component.

2. Select M2 from rung 4 and enter **14,15** at the Command prompt and then press ENTER; M2 is moved to a new location.

3. Select PB5 from rung 5 and enter **14,14** at the Command prompt and then press ENTER; PB5 is moved to a new location.

4. Select M1 from rung 7 and enter **14,12** at the Command prompt and then press ENTER; M1 is moved to a new location.

5. Press ENTER to exit the command. Figure 6-19 shows the schematic diagram after using the **Scoot** tool.

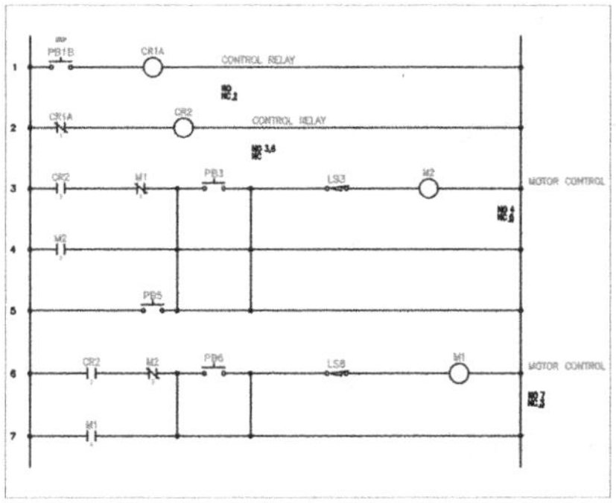

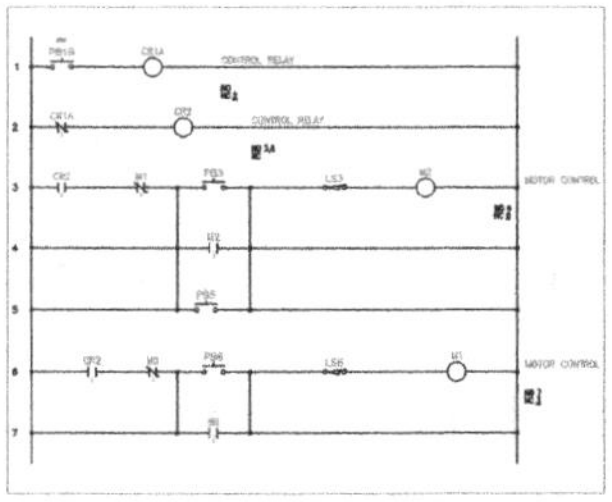

Figure 6-18 *Wires inserted in the ladder* ***Figure 6-19*** *The ladder after using the* ***Scoot*** *tool*

Trimming Wires

1. Choose the **Trim Wire** tool from the **Edit Wires/Wire Numbers** panel of the **Schematic** tab; you are prompted to select the wire to trim.

2. Select the following wires to trim: right and left portions of rung 4, rung 5, and rung 7. Next, press ENTER to exit the command.

Aligning Components

1. Choose the **Align** tool from **Schematic > Edit Components > Modify Components** drop-down; you are prompted to select the component to align horizontally or vertically.

2. Select PB1B on rung 1; an imaginary line passing through the center of PB1 is displayed and you are prompted to select objects.

3. Select CR1A placed on rung 2, CR2 placed on rung 3, and CR2 placed on rung 6 from the left of the ladder. Next, press ENTER; the components are aligned toward the left.

4. Repeat step1. Select M1 from rung 3; an imaginary line passing through the center of M1 is displayed and you are prompted to select objects. Next, select M2 placed on rung 6 and press ENTER; M1 placed on rung 3 and M2 placed on rung 6 are aligned.

5. Repeat step 1 and then select M1 placed on the right side of rung 6. Next, select M2 placed on rung 3, CR2 placed on rung 2, and CR1A placed on rung 1 and press ENTER; the selected components are aligned.

6. Repeat step 1 and then select PB3 from rung 3; you are prompted to select objects. Next, select M2 placed on rung 4, PB5, PB6, and M1 placed on rung 7 from the drawing and then press ENTER; the selected components are aligned. Figure 6-20 shows the aligned components.

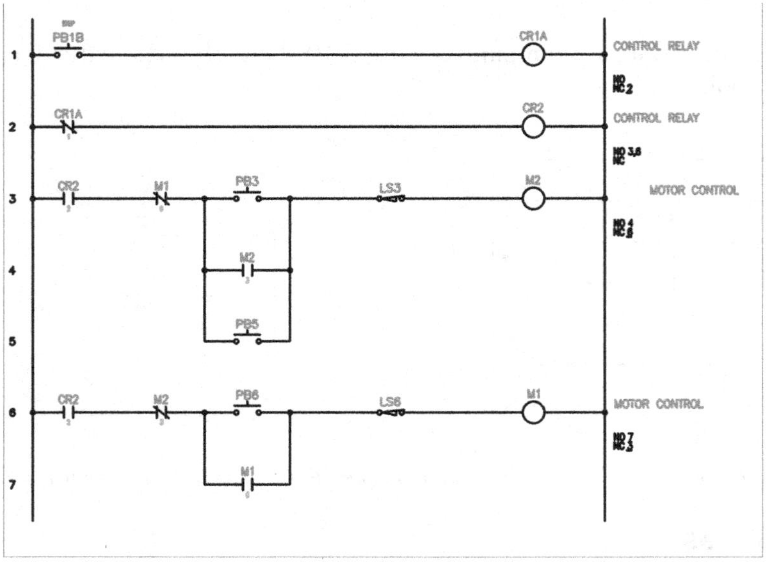

Figure 6-20 The trimmed wires and the aligned components

Copying the Catalog Data

1. Choose the **Copy Catalog Assignment** tool from **Schematic > Edit Components > Edit Components** drop-down; you are prompted to select the master component.

2. Select PB3 placed on rung 3 as the master component; the **Copy Catalog Assignment** dialog box is displayed.

3. Choose the **Catalog Lookup** button; the **Catalog Browser** dialog box is displayed.

4. Select the row which contains **800T-B6D1** in its **Catalog** column and then choose the **OK** button; the **Copy Catalog Assignment** dialog box is displayed again. In this dialog box,

AB is displayed in the **Manufacturer** edit box and **800T-B6D1** is displayed in the **Catalog** edit box.

5. Choose the **OK** button from the **Copy Catalog Assignment** dialog box; you are prompted to select the target components.

6. Select PB5 and PB6 placed on rung 5 and rung 6, respectively and press ENTER.

Note
*Choose the **OK** button from the **QSAVE** message box, if it is displayed.*

You need to check the catalog data of PB5 and PB6.

7. Right-click on PB5 and PB6 one by one; a marking menu is displayed. Next, choose the **Edit Component** option; the **Insert / Edit Component** dialog box is displayed. You will notice that the catalog data information is displayed in the **Catalog Data** area of the **Insert / Edit Component** dialog box. Choose the **OK** button in this dialog box.

Saving the Drawing File

1. Choose **Save** from the **Application Menu** to save the drawing file *C06_tut01.dwg*.

Tutorial 2

In this tutorial, you will audit the *C06_tut01.dwg* drawing of the **CADCIM** project by using the **Electrical Audit** and **Drawing Audit** tools and add catalog data to a component in this drawing. You will also surf a component in the *demo004.dwg* drawing. **(Expected time: 20 min)**

The following steps are required to complete this tutorial:

a. Open the drawing.
b. Audit the drawing using the **DWG Audit** tool.
c. Audit the drawing using the **Electrical Audit** tool.
d. Surf the component.
e. Save the drawing file.

Opening the Drawing

1. Right-click on *C06_tut01.dwg* drawing file of the **CADCIM** project in the **Projects** rollout of the **PROJECT MANAGER**; a shortcut menu is displayed. Choose the **Open** option from the shortcut menu; the drawing is opened.

Auditing the Drawing Using the DWG Audit Tool

1. Choose the **DWG Audit** tool from the **Schematic** panel of the **Reports** tab; the **Drawing Audit** dialog box is displayed.

2. Select the **Active drawing** radio button if not selected by default. Next, choose the **OK** button; the modified **Drawing Audit** dialog box is displayed.

3. Choose the **OK** button in the modified **Drawing Audit** dialog box; the **Drawing Audit** message box is displayed.

4. Choose the **OK** button in the **Drawing Audit** message box; the **Report: Audit for this drawing** dialog box is displayed.

5. Choose the **Close** button from the **Report: Audit for this drawing** dialog box.

   The visual wire indicators in the diagram are shown in red color, refer to Figure 6-21.

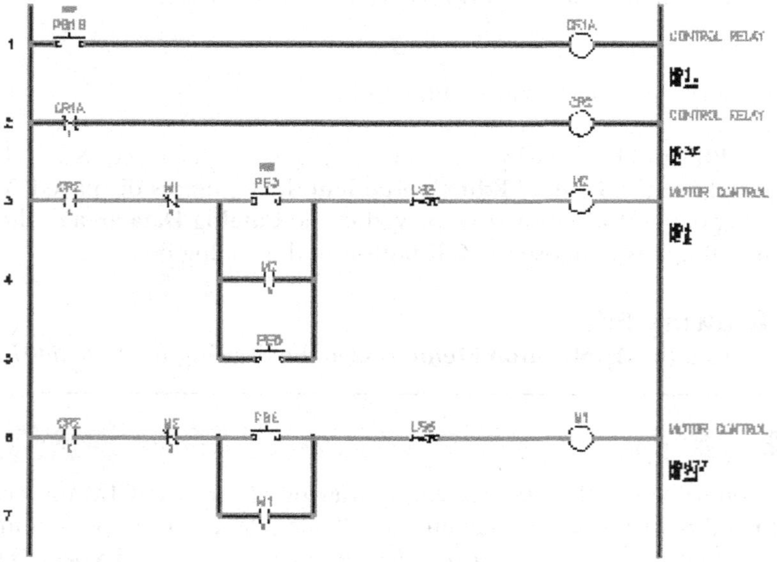

Figure 6-21 *Visual wire indicators indicating wires in the drawing*

6. Choose **View > Redraw** from the menu bar to eliminate the wire indicators.

Auditing the Drawing Using the Electrical Audit Tool

1. Choose the **Electrical Audit** tool from the **Schematic** panel of the **Reports** tab: the **Electrical Audit** dialog box is displayed.

2. Select the **Active Drawing** radio button from the **Electrical Audit** dialog box; the errors in the *C06_tut01.dwg* drawing file are displayed in the edit box located on the right of the **Active Drawing** radio button.

3. Choose the **Details** button from the **Electrical Audit** dialog box to expand it.

4. Choose the **Component - No Catalog Number** tab from the **Electrical Audit** dialog box. Next, select the PB1B push button from the **Tag Name** column of this tab. Next, choose the **Go To** button; PB1B push button is zoomed in the drawing area.

Note

*After choosing the **Go To** button, if the **QSAVE** message box is displayed, choose the **OK** button in this message box to save the changes.*

5. Right-click on the PB1B component; a marking menu is displayed. Choose the **Edit Component** option from the marking menu; the **Insert / Edit Component** dialog box is displayed.

6. Choose the **Project** button from the **Catalog Data** area of the **Insert / Edit Component** dialog box; the **Find: Catalog Assignments** dialog box is displayed. Select the **Active Project** radio button, if it is not selected.

7. Choose the **OK** button from the **Find: Catalog Assignments** dialog box; the **HPB11 / VPB11 catalog values(this project)** dialog box is displayed.

Note

*If the **QSAVE** message box is displayed, choose the **OK** button in it to save the changes.*

8. Select the **800T-A2A** catalog number from the **Catalog Number** column and choose the **OK** button from the **HPB11 / VPB11 catalog values** dialog box; AB is displayed in the **Manufacturer** edit box and 800T-A2A is displayed in the **Catalog** edit box of the **Insert / Edit Component** dialog box.

9. Next, choose the **OK** button from the **Insert / Edit Component** dialog box to save the changes and exit this dialog box. Now, you can notice 'x' on the left of the PB1B in the **Electrical Audit** dialog box.

10. Choose the **Close** button from the **Electrical Audit** dialog box.

    In this manner, you can rectify the missing catalog number error from the drawing. Now, you need to check whether an error is rectified or not.

11. Choose the **Electrical Audit** tool from the **Schematic** panel of the **Reports** tab; the **Electrical Audit** dialog box is displayed. Choose the **Active Drawing** button to display the errors in the active drawing.

12. Choose the **Details** button to expand this dialog box. Next, choose the **Component - No Catalog Number** tab from the **Electrical Audit** dialog box.

    You will notice that the selected error is corrected and is not displayed in the **Electrical Audit** dialog box. In other words, PB1B is not displayed in the list.

13. Choose the **Close** button to exit the **Electrical Audit** dialog box.

Surfing the Component

1. Add the *demo002.dwg* drawing from the **NFPADEMO** project to the **CADCIM** project as discussed in Chapter 2.

2. Make sure the **CADCIM** project is activated. Next, open the *demo004.dwg* drawing.

3. Choose the **Surfer** tool from the **Other Tools** panel of the **Project** tab; you are prompted to select tag for surfer trace. Select the **M422** tag corresponding to the ladder reference number 422; the **Surf** dialog box is displayed, as shown in Figure 6-22.

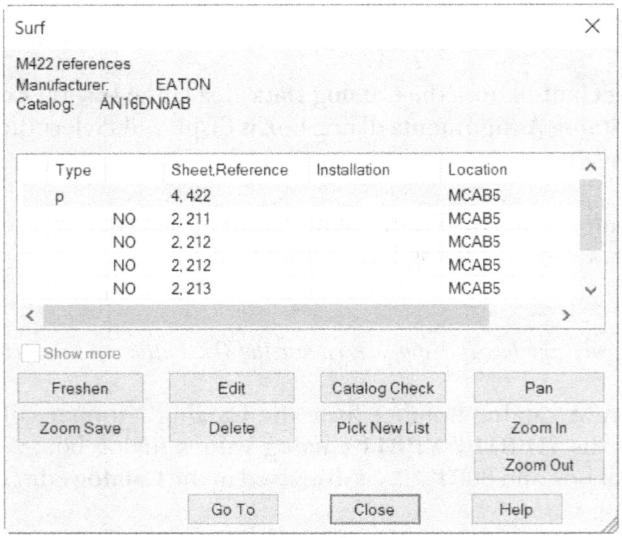

Figure 6-22 *The **Surf** dialog box*

Notice that the **Surf** dialog box displays all the five normally open contacts of M422 with manufacturer website link address.

4. Select the second entry in the **Surf** dialog box and choose **Go To**; the *demo002.dwg* drawing opens and shows the selected normally open contact of M422 zoomed in. Also, 'x' is displayed in the leftmost column of the selected entry to indicate the surfing.

Saving the Drawing File

1. Choose **Save** from the **Application Menu** to save the drawing file.

Self-Evaluation Test

Answer the following questions and then compare them to those given at the end of this chapter:

1. Which of the following commands is used to copy components?

(a) **AECOMPONENT** (b) **AECOPYCOMP**
(c) **AEWIRE** (d) **AEAUDIT**

2. The default format for component tagging is __________.

3. The __________ tool is used to change the text size and width of attributes.

4 The __________ tool is used to update or retag components.

5. The __________ tool is used to copy the selected components.

6. The __________ button of the **Electrical Auditing** dialog box is used to display the detailed information of the errors found in a project.

7. The **Electrical Audit** tool is used to correct errors in a project. (T/F)

8. You cannot print an auditing report. (T/F)

Review Questions

Answer the following questions:

1. Which of the following tabs of the **Electrical Audit** dialog box displays the missing or duplicated wire numbers in a project?

   (a) **Cable Exception** (b) **Component - No Connection**
   (c) **Wire Exception** (d) **Wire - No Connection**

2. Which of the following buttons is used to align components vertically or horizontally?

   (a) **Move** (b) **Pick Master**
   (c) **Scoot** (d) **Align**

3. By choosing the **Move Component** button, you can reposition a selected component from its current location or wire location and insert it at a new location. (T/F)

4. The auditing tools find out errors in project drawings. (T/F)

5. The **Move/Show Attribute** tool is used to move attributes. (T/F)

EXERCISES

Exercise 1

In this exercise, you will create a new drawing named *C06_exer01.dwg* and insert a ladder with Width = 5, Spacing= 1, Rungs = 6, and 1st reference = 500. Also, you will insert components in the ladder, as shown in Figure 6-23. In addition, you will use the basic editing tools such as **Copy Component** and **Align Component** to copy and align the components, respectively. You will use the **Wire** tool to insert wire between rung 500 and rung 501 and the **Trim Wire** tool to trim the wire. **(Expected time: 25 min)**

Exercise 2

In this exercise, you will audit the **NEW_PROJECT** project using the **DWG Audit** and **Electrical Audit** tools. Also, you will save the report on the desktop as *drawing_audit.txt* and *electrical_audit.txt*. **(Expected time: 15 min)**

Exercise 3

In this exercise, you will open the *C06_exer01.dwg* drawing file and save it as *C06_exer03.dwg* in the **NEW_PROJECT** project. Next, you will change ladder references, component tags, and cross-references from reference-based to sequential using the **Update/Retag** tool, refer to Figure 6-24. **(Expected time: 20 min)**

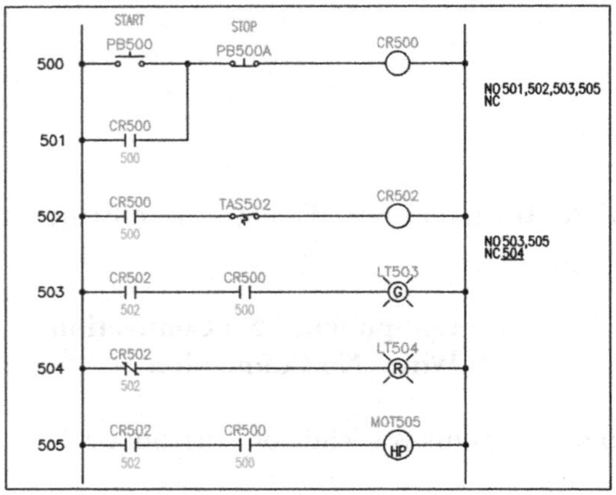

Figure 6-23 Components inserted in the ladder

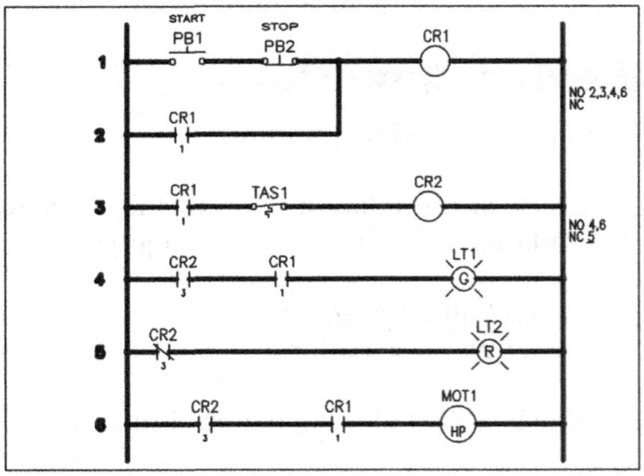

Figure 6-24 *References changed using the* ***Update/Retag*** *tool*

Hint: Enter **1** in the edit box on the right of the **Sequential** radio button in the **Change Each Drawing's Settings - Project-wide** dialog box.

Answers to Self-Evaluation Test

1. b, **2.** %F%N, **3. Change Attribute Size**, **4. Update/Retag**, **5. Copy Component**, **6. Details**, **7.** T, **8.** F

Chapter 7

Connectors, Point-to-Point Wiring Diagrams, and Circuits

Learning Objectives

After completing this chapter, you will be able to:

- *Insert connectors*
- *Edit the existing connector and connector pin numbers*
- *Insert splices*
- *Create, insert, move, and copy circuits*
- *Save circuits using WBlock*
- *Insert WBlock circuits*
- *Build a circuit using the Circuit Builder tool*

INTRODUCTION

An electrical connector is a conductive device that is used to join electrical circuits together. Typically, an electrical connector is used to connect a wire, or a group of wires at a single junction. It is designed in such a way that it separates wires easily. In AutoCAD Electrical, there are connector tools that enable you to easily place automatically built parametric connectors based on the information you specify.

In this chapter, you will learn to insert connectors in a drawing and use different connector tools to manage connector data, including pins and receptacles in the project drawings. Using these tools, you can quickly create custom connectors, add and remove pins, edit connectors for specific applications, and so on. Also, you will learn about the point-to-point wiring tools that help you in creating point-to-point wiring diagrams easily as opposed to ladder diagrams.

In the circuits section, you will learn to save repetitive circuits in a project as icons. This will help you reproduce drawings in very less time with great accuracy. These saved circuits are similar to blocks as these are also inserted as a single object.

INSERTING CONNECTORS

Command: AECONNECTOR

The **Insert Connector** tool is used to create a connector from the user-defined parameters. The connector created using this tool consists of plug, receptacle, pin numbers, and other attributes. The pin numbers are displayed on the plug side (round corners), receptacle side (square corners), or on both sides. A component tag is displayed above the connector. The connector with different components is shown in Figure 7-1.

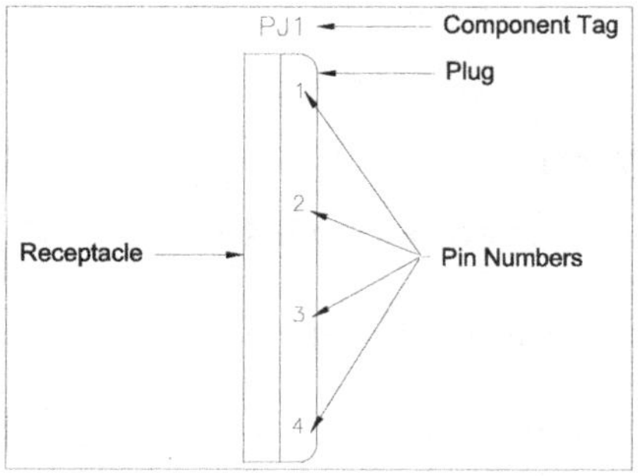

Figure 7-1 *The connector and its components*

To insert a connector in a drawing, choose the **Insert Connector** tool from the **Insert Connector** drop-down in the **Insert Components** panel of the **Schematic** tab, as shown in Figure 7-2; the **Insert Connector** dialog box will be displayed, as shown in Figure 7-3. In this dialog box, you can specify the parameters for inserting connectors.

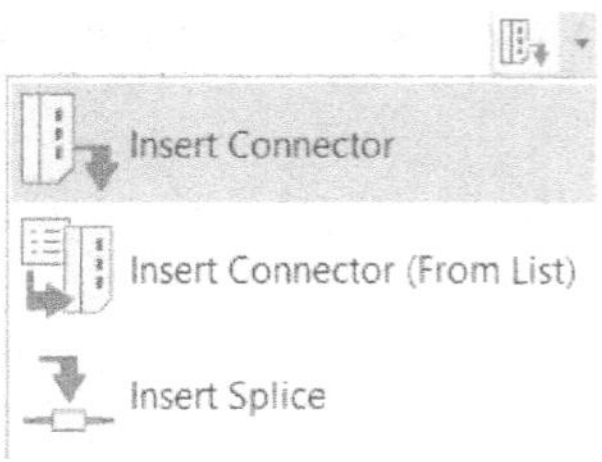

Figure 7-2 *The* ***Insert Connector*** *drop-down*

The **Layout** area is used to specify the spacing between the pins, pin count, format for the pins, and so on.

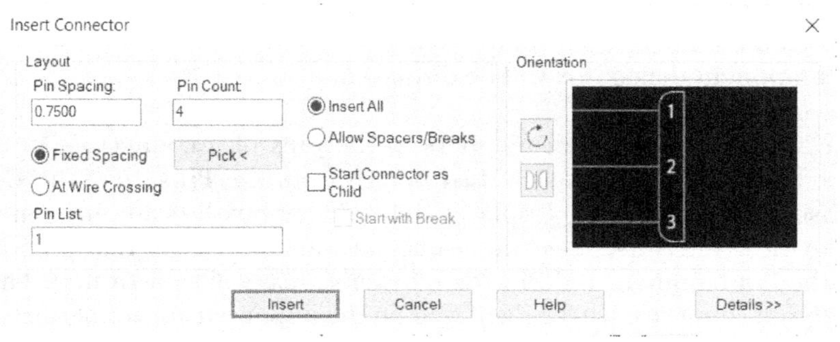

*Figure 7-3 The **Insert Connector** dialog box*

The **Allow Spacers/Breaks** radio button is used to add spacers or breaks between the pins of a connector. To add spacers or breaks, select the **Allow Spacers/Breaks** radio button and then choose the **Insert** button from the **Insert Connector** dialog box; you will be prompted to specify the insertion point for the connector. Next, specify the insertion point; the **Custom Pin Spaces / Breaks** dialog box will be displayed. In this dialog box, choose the **Insert Next Connection** tool; the next pin of the connector will be inserted without any space. Choose the **Add Spacer** button to add space between the connector pins.

The **Break Symbol Now** button is used to break a connector symbol at the current pin location. To break the connector symbol, choose the **Break Symbol Now** button; you will be prompted to specify the insertion point for the remaining connector. Specify the insertion point for the remaining part of the connector; the **Connector Layout** dialog box will be displayed. Specify the required options in the **Connector Layout** dialog box and then choose the **OK** button; the **Custom Pin Spaces / Breaks** dialog box will be displayed again. To break the connector symbol again, you need to repeat the process discussed above.

After specifying the required options, choose the **Insert** button and specify the insertion point; the connector will be inserted at the specified point and the **Insert / Edit Component** dialog box will be displayed. The options in this dialog box are discussed in the next section.

Note

*1. If you choose the **Insert** button without specifying the pin count in the **Pin Count** edit box, the **Pin count not defined** message box will be displayed informing that the connector cannot be inserted without defining the pin count.*

*2. Once the settings in the **Insert Connector** dialog box are changed, they will remain the same throughout AutoCAD Electrical sessions until modified.*

EDITING CONNECTOR

Command: AEEDITCOMPONENT

When you insert a connector in the drawing, the **Insert / Edit Component** dialog box will be displayed. You can also edit the connector at any time. To do so, choose the **Edit** tool from the **Edit Components** drop-down in the **Edit Components** panel of the **Schematic** tab; you will be prompted to select a component to edit. Select the connector to be edited; the **Insert / Edit Component** dialog box will be displayed.

The **Pins** area is used to assign pin numbers to the pins of a connector. To edit or assign pin numbers to a connector, choose the **List** button from the **Pins** area; the **Connector Pin Numbers In Use** dialog box will be displayed. This dialog box lists all pin numbers that have been assigned to the connectors. Select the connector pin number row that you want to modify; the options in the **Pin Numbers** and **Pin Descriptions** areas will be activated. The tag and pin count of the selected connector will be displayed on the upper left corner of the **Connector Pin Numbers In Use** dialog box. After specifying the required options, choose the **OK** button from this dialog box. Next, choose the **OK** button from the **Insert / Edit Component** dialog box to edit the pin numbers of the connector.

Note

If you enter a new value or modify a value in the ***Pin Numbers*** *area, the modified value will be automatically displayed in the* ***Pin List*** *area of the* ***Connector Pin Numbers In Use*** *dialog box.*

INSERTING A CONNECTOR FROM THE LIST

Command: AECONNECTORLIST

The **Insert Connector (From List)** tool is used to import connector wiring information from an external report. To do so, choose the **Insert Connector (From List)** tool from the **Insert Components** panel of the **Schematic** tab; the **Autodesk Inventor Professional Import File Selection** dialog box will be displayed. In this dialog box, select the required file and then choose the **Open** button to import the data. Note that the file should be in *.xml*, *.xls*, *.mdb*, or *.csv* format.

MODIFYING CONNECTORS

AutoCAD Electrical has various tools for modifying connectors and changing the orientation of connectors. Using these modifying tools, you can break, add, or remove the pins of a connector.

Adding Pins to a Connector

Command: AECONNECTORPIN

The **Add Connector Pins** tool is used to add additional pins to an existing connector. To do so, choose the **Add Connector Pins** tool from the **Modify Connectors** drop-down in the **Edit Components** panel of the **Schematic** tab, refer to Figure 7-4; you will be prompted to select a connector. After selecting a connector, you will be prompted to specify the insertion point for the new pin. Next, specify the insertion point; the next available pin number will be added to the connector and each pin will be added in line with the remaining connector pins. Also, you can manually specify the pin number for the connector as per your requirement. To specify the pin number for a connector manually, enter the pin number for the connector in the Command prompt and then specify the insertion point.

The **AECONNECTORPIN** command will continue until you press ESC or ENTER. Alternatively, right-click on the screen and choose the **Enter** option from the shortcut menu to exit the command.

Deleting a Connector Pin

Command: AEERASEPIN

The **Delete Connector Pins** tool is used to delete pins from a connector. To do so, choose the **Delete Connector Pins** tool from the **Modify Connectors** drop-down in the **Edit Components** panel of the **Schematic** tab, refer to Figure 7-4; you will be prompted to select a connector pin to delete. Select the connector pin; the pin number will be deleted from the connector. Next, press ENTER or ESC to exit the command or right-click on the screen to end the command.

*Figure 7-4 The **Modify Connectors** drop-down*

Moving a Connector Pin

Command: AEMOVEPIN

The **Move Connector Pins** tool is used to move pins within an existing connector. To do so, choose the **Move Connector Pins** tool from the **Modify Connectors** drop-down in the **Edit Components** panel of the **Schematic** tab; you will be prompted to select the connector pin to be moved. Select the connector pin; you will be prompted to specify a new location for the selected connector pin. Next, specify a new location for the pin; the connector pin will be moved to the specified location. Press ENTER or ESC to exit the command or right-click on the screen to terminate the command.

Swapping Connector Pins

Command: AESWAPPINS

The **Swap Connector Pins** tool is used to interchange one set of connector pin numbers with another set on an existing connector or between different connectors on the drawing. Using this tool, you can swap the pin numbers of a connector without changing locations of pins. To swap pin numbers of a connector, choose the **Swap Connector Pins** tool from the **Modify Connectors** drop-down in the **Edit Components** panel of the **Schematic** tab; you will be prompted to select the connector pin to swap. Select the connector pin; a temporary graphics will be drawn around the selected pin number, which indicates that the pins have been included in the swap list and you will be prompted to select the connector pin to swap with. Next, select the connector pin to be swapped. The **AESWAPPINS** command will continue until you press ENTER or ESC. Alternatively, right-click on the screen to end the command.

Reversing a Connector

Command: AEREVERSE

The **Reverse Connector** tool is used to reverse the direction of the connector about its horizontal or vertical axis. To reverse the direction of a connector, choose the

Reverse Connector tool from the **Modify Connectors** drop-down in the **Edit Components** panel of the **Schematic** tab; you will be prompted to select a connector to reverse. Select the connector to be reversed; the direction of the connector will be reversed. The **AEREVERSE** command will continue until you press ENTER or ESC. Alternatively, right-click to exit the command.

Note

If you reverse a receptacle connector that has no rounded corners, the appearance of graphics will remain unchanged, but the wire connection attributes will move to the other side of the connector.

Rotating a Connector

Command: AEROTATE

The **Rotate Connector** tool is used to rotate a connector by 90-degree in counterclockwise direction. To rotate a connector about its insertion point, choose the **Rotate Connector** tool from the **Modify Connectors** drop-down in the **Edit Components** panel of the **Schematic** tab; you will be prompted to select the connector to rotate. Next, select the connector; the selected connector will rotate by 90 degrees in the counterclockwise direction. Keep on selecting the connector till the appropriate orientation of the connector is achieved. You can also change the orientation of attributes of a connector by entering **H** at the Command prompt. Press ENTER or ESC to exit the command. Alternatively, right-click on the screen and choose **Enter** from the shortcut menu to exit the command.

Stretching a Connector

Command: AESTRETCH

The **Stretch Connector** tool is used to stretch a connector. Using this tool, you can increase or decrease the length of a connector. To stretch a connector, choose the **Stretch Connector** tool from the **Modify Connectors** drop-down in the **Edit Components** panel of the **Schematic** tab; you will be prompted to specify the end of the connector that you want to stretch. Next, select the end of the connector; a straight line along with the cursor will be displayed on the screen and you will be prompted to specify the second point of displacement. Specify the second point of displacement for the connector or drag the cursor downwards; the connector will be stretched.

Note

*The **Stretch Connector** tool is used only to stretch a connector. This tool cannot relocate any of its pins. To relocate its pins, use the **Move Connector Pins** button, as explained earlier in this chapter.*

Splitting a Connector

Command: AESPLIT

The **Split Connector** tool is used to split a connector into two separate parts such as parent and child or child and another child. Also, this tool is used to place a portion of a connector at a location other than the original one. To split a connector, choose the

Split Connector tool from the **Modify Connectors** drop-down in the **Edit Components** panel of the **Schematic** tab; you will be prompted to select the connector block that you want to split. Select the connector block; you will be prompted to specify the split point of the connector. Next, specify the split point; the **Split Block** dialog box will be displayed.

The **OK** button will be activated only if you specify the X and Y coordinates in the **X** and **Y** edit boxes. Specify the required options in the **Split Block** dialog box and choose the **OK** button in this dialog box; you will be prompted to specify the insertion point for the child block. Specify the insertion point; the child block will be inserted into the drawing.

Inserting Splices

Command: AESPLICE

The **Insert Splice** tool is used to connect one or more wires. To connect wires, choose the **Insert Splice** tool from the **Insert Connector** drop-down in the **Insert Components** panel of the **Schematic** tab, refer to Figure 7-2; the **Insert Component** dialog box will be displayed.

Select the **Splice** symbol from the **NFPA: Splice Symbols** area of the **Insert Component** dialog box; the splice symbol along with the cursor will be displayed and you will be prompted to specify the insertion point for the symbol. Specify the insertion point; the **Insert / Edit Component** dialog will be displayed. Next, specify the required options in this dialog box and choose the **OK** button; the splice will get inserted into the drawing, as shown in Figure 7-5.

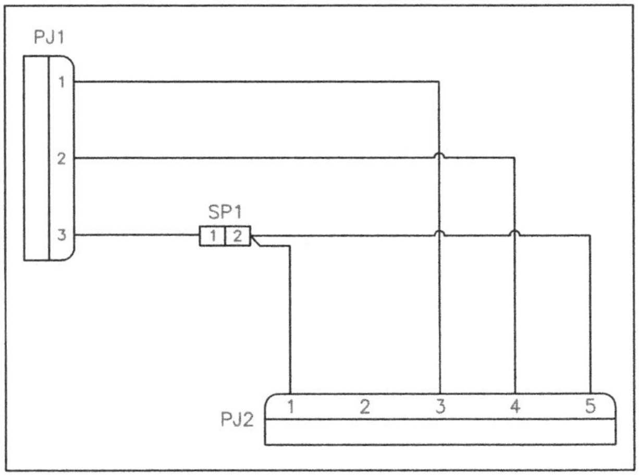

Figure 7-5 *The drawing after inserting connectors and splice*

Note

If you insert splice on an existing wire, the wire will break up. Also, the wire numbers will change.

WORKING WITH CIRCUITS

The route along which electricity flows is called electrical circuit. A circuit consists of wires, components, power supply, and so on.

In this section, you will learn to save circuits as symbols in icon menu file for repetitive use. Also, you will learn to copy the circuits, move the circuits, and so on.

Saving Circuits to Icon Menu

Command: AESAVECIRCUIT

You can save the selected portion of a circuit for future use. To save the selected portion of a circuit, choose the **Save Circuit to Icon Menu** tool from the **Circuit** drop-down in the **Edit Components** panel of the **Schematic** tab, as shown in Figure 7-6; the **Save Circuit to Icon Menu** dialog box will be displayed, as shown in Figure 7-7. The options in this dialog box are similar to those in the **Insert Component** dialog box.

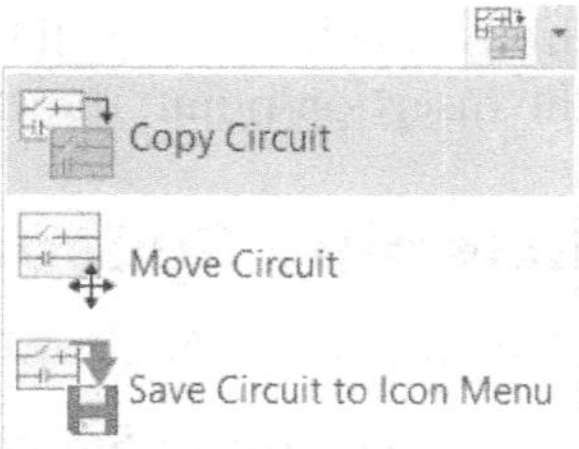

Figure 7-6 *The* ***Circuit*** *drop-down*

The options in the **Add** drop-down list are used to add a component, command, new circuit, existing circuit, and submenu icon to the menu. To add a new circuit to icon menu, click on the **Add** drop-down list displayed at the upper right corner of the **Save Circuit to Icon Menu** dialog box; different options will be displayed. Next, select the **New Circuit** option from the drop-down list; the **Create New Circuit** dialog box will be displayed.

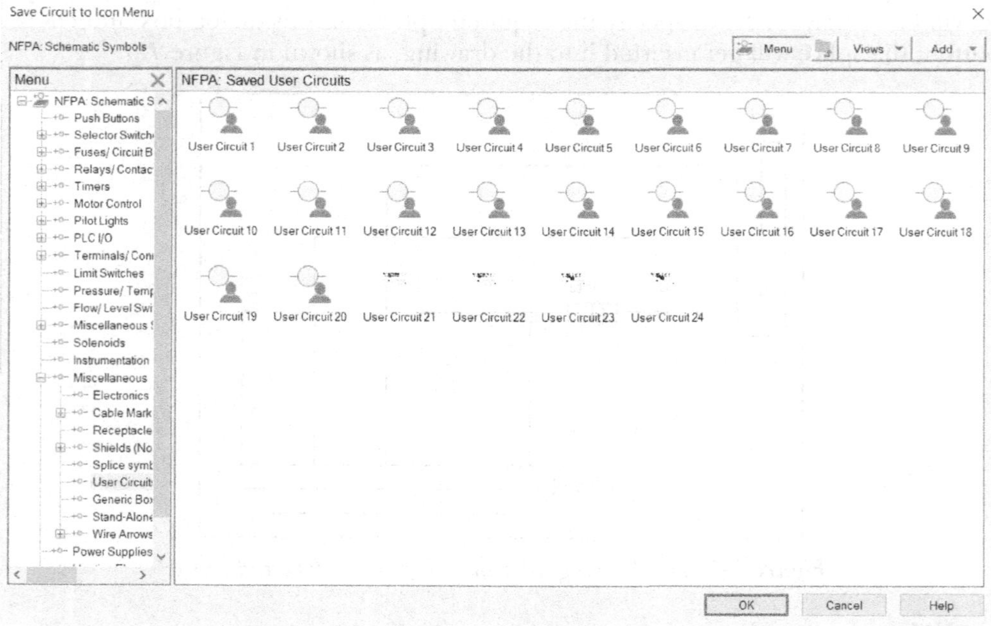

Figure 7-7 *The* ***Save Circuit to Icon Menu*** *dialog box*

After specifying the parameters in the **Icon Details** and **Circuit Drawing File** areas, the **OK** button will be activated. Choose the **OK** button; you will be prompted to specify the base point. Specify the base point in the drawing area; you will be prompted to select objects. Select individual objects or all objects at once using the crossing window and then press ENTER; the **Save Circuit to Icon Menu** dialog box will be displayed again. You will notice that the circuit is saved as an icon in the **NFPA: Saved User Circuits** area and the preview of the circuit in the form of icon is also displayed in the **Save Circuit to Icon Menu** dialog box. Next, choose the **OK** button to save the changes made in the **Save Circuit to Icon Menu** dialog box and exit from this dialog box.

Note

1. *You can also save a circuit using the* ***Save Circuit to Icon Menu*** *dialog box. To do so, right-click in the* ***NFPA: Saved User Circuits*** *area of this dialog box; a shortcut menu will be displayed. Next, choose* ***Add icon > New circuit*** *from the shortcut menu; the* ***Create New Circuit*** *dialog box will be displayed. Specify the options as per your requirement to create a new circuit and then save it in the* ***Save Circuit to Icon Menu*** *dialog box.*

2. *You can also add existing circuits to the icon menu. To do so, choose the* ***Add circuit*** *option from the* ***Add*** *drop-down in the* ***Save Circuit To Icon Menu*** *dialog box; the* ***Add Existing Circuit*** *dialog box will be displayed. You can use this dialog box to add existing circuits to the* ***Save Circuit to Icon Menu*** *dialog box.*

Inserting Saved Circuits

Command: AESAVEDCIRCUIT

The **Insert Saved Circuit** tool is used to insert the circuit that you saved using the **Save Circuit to Icon Menu** tool. Similar to blocks, the saved circuits are inserted as single blocks. To insert a circuit, choose the **Insert Saved Circuit** tool from the **Insert Circuit** drop-down in the **Insert Components** panel of the **Schematic** tab, as shown in Figure 7-8; the **Insert Component** dialog box will be displayed, as shown in Figure 7-9.

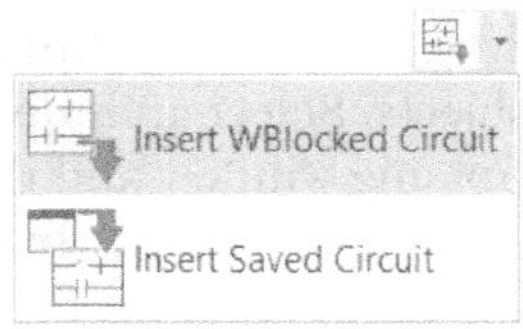

Figure 7-8 *The* ***Insert Circuit*** *drop-down*

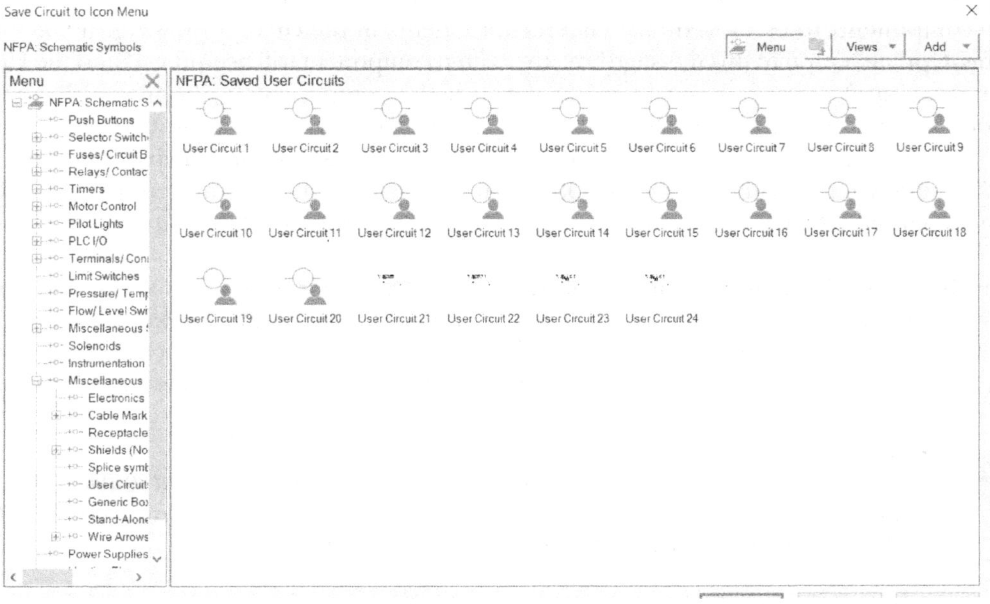

Figure 7-9 *The* ***Insert Component*** *dialog box*

Select a circuit from the **NFPA: Saved User Circuits** area of the **Insert Component** dialog box; the **Circuit Scale** dialog box will be displayed,.

After specifying the required options in the **Circuit Scale** dialog box, choose the **OK** button from this dialog box; you will be prompted to specify the insertion point for the saved circuit in your drawing. Next, specify the insertion point; the circuit will be inserted. Note that once the circuit is inserted, the components and cross-references in the circuit will also get updated.

Note

If you enter 2, 3, or any digit other than 1 in the ***Custom Scale*** *edit box and then choose the* ***OK*** *button in this dialog box, the size of the circuit will change accordingly. Also, note that when you choose the* ***Insert Saved Circuit*** *button next time, the options in the* ***Circuit Scale*** *dialog box will be slightly different. This dialog box displays the options when 2 is specified in the* ***Custom scale*** *edit box.*

Moving Circuits

Command: AEMOVECIRCUIT

The **Move Circuit** tool is used to move a selected circuit from one location to the other in a drawing. The **AEMOVECIRCUIT** command is similar to AutoCAD **MOVE** command. To move a selected circuit, choose the **Move Circuit** tool from the **Circuit** drop-down in the **Edit Components** panel of the **Schematics** tab; you will be prompted to select objects. Select an individual object or a number of objects at a time from a drawing using the crossing window and then press ENTER; you will be prompted to specify the base point or the displacement. Specify the base point or the displacement; you will be prompted to specify the second point of displacement. Next, specify the second point of displacement or the new circuit location; the circuit will be moved to a specified location. After the circuit is moved, AutoCAD Electrical will start updating components and cross-references based on their new line reference locations and the **Update Related Components?** message box will be displayed if the components to be moved has related components in the current drawing. If you choose the **Yes-Update** button in this message box, the edited component will be updated and the **Update other drawings?** message box will be displayed if the components to be moved has related components in other drawings. However, if you choose the **Skip** button, AutoCAD Electrical will skip the update process. Choose the **OK** button from the **Update other drawings?** message box; the drawings will get updated and the **QSAVE** message box will be displayed. Choose the **OK** button from the **QSAVE** message box to save the changes made while updating the drawings.

If you choose the **Task** button from the **Update other drawings?** message box, the components and cross-references of the drawings will not be updated at that time but will be saved in the project task list, which you can update later. To update the pending list, choose the **Project Task List** button from the **PROJECT MANAGER**; the **Task List** dialog box will be displayed.

Note

If the circuit being moved does not have related components in the current drawing as well as in other drawings, the circuit will move to the specified location without displaying the ***Update Related Components?*** *and* ***Update other drawings?*** *message boxes.*

The ***QSAVE*** *message box will be displayed only if you have made changes in the current drawing and have not saved it.*

Copying Circuits

Command: AECOPYCIRCUIT

The **Copy Circuit** tool is used to copy an existing circuit and then paste it to the specified location in the same drawing. Also, the copied components get automatically retagged according to the new line reference of the specified location. The **AECOPYCIRCUIT** command is similar to AutoCAD **COPY** command. To copy the circuit, choose the **Copy Circuit** tool from the **Circuit** drop-down in the **Edit Components** panel of the **Schematic** tab; you will be prompted to select the circuit to copy. Select the components and wires that you need to copy. You can select the components and wires one by one, or a number of components and wires at a time using the lasso. After selecting the objects, press ENTER; you will be prompted to specify the base point or the displacement. Next, specify the base point or displacement; you will be prompted to specify the second point of displacement. Next, specify the second point of displacement or the new circuit location point; the copied circuit will be inserted into the drawing and you will be prompted to specify another location point if you want to insert copied circuit to another point. Press ENTER to exit the command. Note that if copied circuit consists of terminals, or consists of fixed wire numbers or component tags, or the **Copy Circuit Options** dialog box will be displayed. This dialog box also consists of options to update or retain the terminal numbers of the terminal strip. Note that the options in the dialog box will change depending on the circuit you select to copy.

After specifying the required options, choose the **OK** button from the **Copy Circuit Options** dialog box; the circuit will be copied at the specified location. Note that if the circuit has source signal arrows then after choosing the **OK** button from the **Copy Circuit Options** dialog box, the **Copied Source Signal Arrow** dialog box will be displayed. Choose the **Keep Arrows** button to keep source arrows. If you do not want to keep the source arrows, choose the **Erase** button. After you choose the button from the **Copied Source Signal Arrow** dialog box, the AutoCAD Electrical will start updating components and cross-references based on the drawing property settings.

Note

1. If you want to make multiple copies of a selected circuit, enter 'M' at the Command prompt after specifying the base point or the displacement.

*2. If the circuit being copied does not have fixed wire numbers and component tags, the circuit will be copied at the specified location without displaying the **Copy Circuit Options** dialog box.*

Saving Circuits by Using WBlock

Command: WBLOCK

You can save circuits as blocks using AutoCAD **WBLOCK** command. These saved blocks can be inserted later into any drawing file. Using this command, you can save unlimited number of circuits without using the **Save Circuit to Icon Menu** tool. When you enter the **WBLOCK** command at the Command prompt, the **Write Block** dialog box will be displayed. Choose the **Pick point** button from the **Base point** area; the dialog box will disappear temporarily. Next, specify the insertion base point that will be taken as the origin point of the block's coordinate system.

After specifying the insertion base point, you need to select objects that will constitute a block. To do so, choose the **Select objects** button from the **Objects** area; the dialog box will disappear temporarily and a selection box will be displayed. You can select objects on the screen using any selection method. After completing the selection process, right-click or press the ENTER key to return to the dialog box. The number of objects selected is displayed at the bottom of the **Objects** area of the **Write Block** dialog box.

Next, enter the name and location in the **File name and path** edit box to save the circuit block. Alternatively, choose the **[...]** button on the right of the **File name and path** edit box; the **Browse for Drawing File** dialog box will be displayed. Next, specify the location for the drawing file in the **Browse for Drawing File** dialog box and enter the name of the file in the **File name** edit box. Then, choose the **Save** button from the **Browse for Drawing File** dialog box to return to the **Write Block** dialog box. By default, *C:\Users\User Name\Documents\new block* is displayed in the **File name and path** edit box. Choose the **OK** button from the **Write Block** dialog box; the circuit will be saved.

Inserting the WBlocked Circuit

Command: AEWBCIRCUIT

The **Insert WBlocked Circuit** tool is used to insert a WBlocked circuit. To do so, choose the **Insert WBlocked Circuit** tool from the **Insert Circuit** drop-down in the **Insert Components** panel of the **Schematic** tab; the **Insert WBlocked Circuit** dialog box will be displayed. Select the required WBlock circuit from the list displayed to insert into the drawing. Next, choose the **Open** button; the **Circuit Scale** dialog box will be displayed. Specify the required parameters in this dialog box and choose the **OK** button from the **Circuit Scale** dialog box; you will be prompted to specify the insertion point for the circuit. Specify the insertion point; the WBlocked circuit will be inserted into your drawing and components and cross-references in the circuit will get updated based on the drawing property settings.

BUILDING A CIRCUIT

Command: AECIRCBUILDER

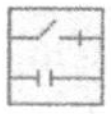

You can build motor control circuits and power feed circuits using the **Circuit Builder** tool. The circuit built using this tool includes single phase, three phase, and one-line circuits. Using this tool, you can either insert a circuit or configure a circuit by using the list of available circuit categories. The procedure to insert a circuit using this tool is discussed next.

Inserting a Circuit

Choose the **Circuit Builder** tool from the **Circuit Builder** drop-down in the **Insert Components** panel of the **Schematic** tab; the **Circuit Selection** dialog box will be displayed, as shown in Figure 7-10.

After specifying the desired options, choose the **Insert** button at the bottom of the **Circuit Selection** dialog box to insert a circuit using the settings specified in the **Circuit Selection** dialog box.

*Figure 7-10 The **Circuit Selection** dialog box*

Configuring a Circuit

If you need to configure a new circuit by modifying the circuit components in a reference circuit, select a reference circuit based on your requirement from the **Circuit Selection** dialog box, and then choose the **Configure** button from the **Circuit Selection** dialog box; you will be prompted to specify the insertion point for the circuit. Specify the insertion point for the circuit; the sketch of the circuit will be inserted with template drawing markers at the specified location and the **Circuit Configuration** dialog box will be displayed, refer to Figure 7-11. You can configure a new circuit by specifying the options in this dialog box as per your requirement.

After specifying the options in the **Circuit Configuration** dialog box, choose the **Done** button; the configured circuit will be inserted at the specified location. Note that, if you have not specified the parameters for some of the circuit elements, the **Circuit Configuration: Done** message box will be displayed warning you about the template drawing markers that were not replaced by circuit elements. Choose **Yes** to **continue** or choose **No** to return to the **Circuit Configuration** dialog box.

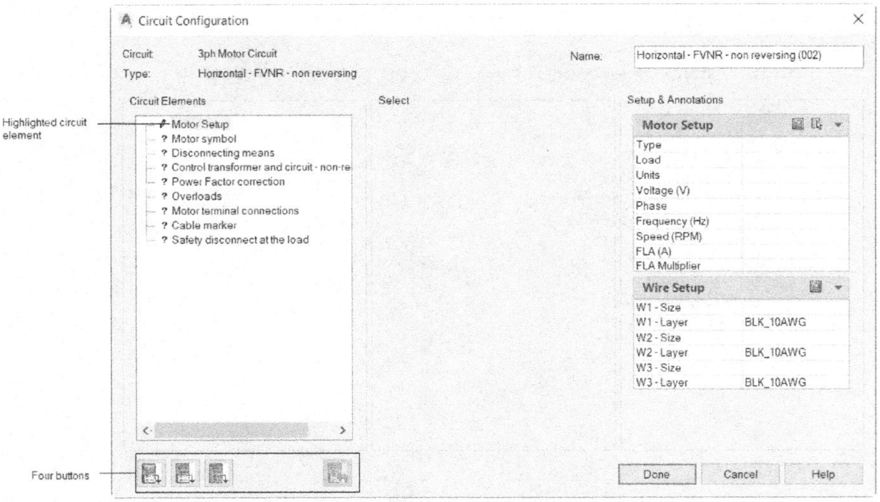

Figure 7-11 *The* ***Circuit Configuration*** *dialog box*

TUTORIALS

Tutorial 1

In this tutorial, you will create a point-to-point wiring diagram, as shown in Figure 7-12.

(Expected time: 25 min)

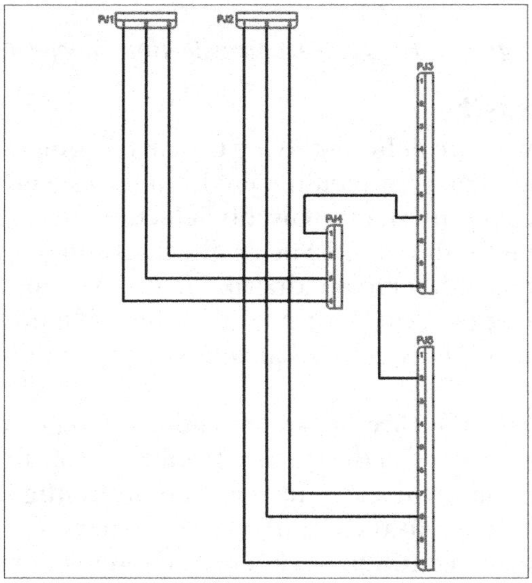

Figure 7-12 *The point-to-point wiring diagram*

The following steps are required to complete this tutorial:

a. Create a new drawing.
b. Insert connectors.
c. Copy connectors.

d. Insert the multiple wire bus.
e. Save the drawing.

Creating a New Drawing

1. Activate the **CADCIM** project as discussed in the previous chapters. Next, choose the **New Drawing** button from the **PROJECT MANAGER**; the **Create New Drawing** dialog box is displayed. Enter **C07_tut01** in the **Name** edit box of the **Drawing File** area. Select the **ACAD_ELECTRICAL.dwt** template and enter **Connector Diagram** in the **Description 1** edit box.

   Make sure the **For Reference Only** check box is cleared in the **Create New Drawing** dialog box and *C:\Users\User Name\Documents\AcadE 2022\AeData\Proj\CADCIM* is displayed in the **Location** edit box.

2. Choose the **OK** button from the **Create New Drawing** dialog box; the *C07_tut01.dwg* drawing is created in the **CADCIM** project and displayed at the bottom of the drawing list in the **CADCIM** project. Move the drawing *C07_tut01.dwg* to the *TUTORIALS* subfolder.

Inserting Connectors

1. Choose the **Insert Connector** tool from **Schematic > Insert Components > Insert Connector** drop-down; the **Insert Connector** dialog box is displayed.

2. Choose the **Details** button from the **Insert Connector** dialog box and then set the following parameters to insert the three-pin connector:

   Pin Spacing: **0.7500** **Pin Count**: **3**
   Pin List: **1**

   Select the **Plug / Receptacle Combination** radio button and the **Add Divider Line** check box from the **Type** area and then select **Plug Side** from the **Pins** drop-down list in the **Display** area. Also, make sure the **Fixed Spacing** and **Insert All** radio buttons are selected from the **Layout** area. Keep rest of the values intact.

3. Choose the **Insert** button; you are prompted to specify the insertion point for the connector. Also, preview of the connector is displayed in dashed lines along with the cursor.

4. Enter **V** and then enter **10,19** at the Command prompt. Next, press ENTER; the **Insert / Edit Component** dialog box is displayed.

5. Enter **PJ1** in the edit box of the **Component Tag** area and then choose **OK**; the three-pin connector is inserted into the drawing horizontally and the **PJ1** is displayed on the left of the connector.

6. Choose the **Insert Connector** tool from **Schematic > Insert Components > Insert Connector** drop-down; the **Insert Connector** dialog box is displayed. Next, enter **10** in the **Pin Count** edit box and keep the rest of the values intact in the **Insert Connector** dialog box.

7. Choose the **Insert** button; you are prompted to specify the insertion point. Also, the preview of the ten-pin connector is displayed along with the cursor. Enter **20,17** at the Command prompt and press ENTER; the **Insert / Edit Component** dialog box is displayed.

8. Enter **PJ3** in the edit box of the **Component Tag** area and choose **OK** from the **Insert / Edit Component** dialog box; the ten-pin connector is inserted into the drawing and **PJ3** is displayed on the top of the connector.

9. Similarly, you need to insert the four-pin connector into the drawing. To do so, enter **4** in the **Pin Count** edit box of the **Insert Connector** dialog box and keep the rest of the values intact.

10. Choose the **Insert** button; you are prompted to specify the insertion point. Enter **17,12** at the Command prompt and press ENTER; the **Insert / Edit Component** dialog box is displayed.

11. Enter **PJ4** in the **Component Tag** edit box and choose the **OK** button in the **Insert / Edit Component** dialog box; the four-pin connector is inserted into the drawing and **PJ4** is displayed on the top of the connector. Figure 7-13 shows the PJ1, PJ3, and PJ4 connectors.

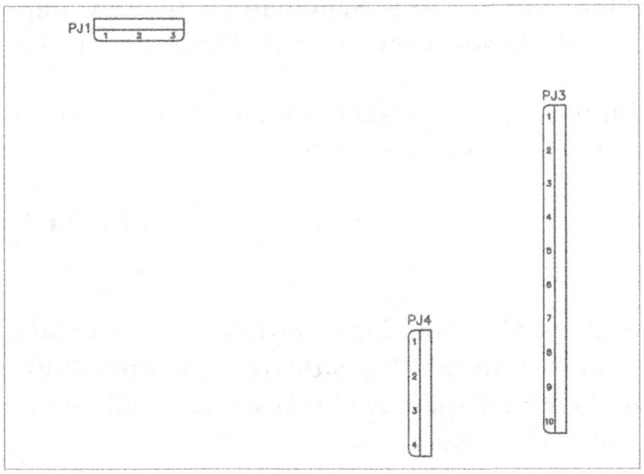

***Figure 7-13** The PJ1, PJ3, and PJ4 connectors*

Copying Connectors

1. Choose the **Copy Component** tool from the **Edit Components** panel of the **Schematic** tab; you are prompted to select the component to be copied.

2. Select the three-pin connector PJ1; you are prompted to specify the insertion point for the connector.

3. Enter **14,19** at the Command prompt and press ENTER; the **Insert / Edit Component** dialog box is displayed.

4. In this dialog box, enter **PJ2** in the edit box of the **Component Tag** area and choose the **OK** button; the PJ2 connector is inserted into the drawing.

5. Right-click on the ten-pin connector PJ3; a marking menu is displayed. Choose the **Copy Component** option from the marking menu; you are prompted to specify the insertion point for the ten-pin connector.

6. Enter **20,8** at the Command prompt and press ENTER; the **Insert / Edit Component** dialog box is displayed.

7. Enter **PJ5** in the edit box of the **Component Tag** area of the **Insert / Edit Component** dialog box and choose the **OK** button; PJ5 is inserted in the drawing, as shown in Figure 7-14.

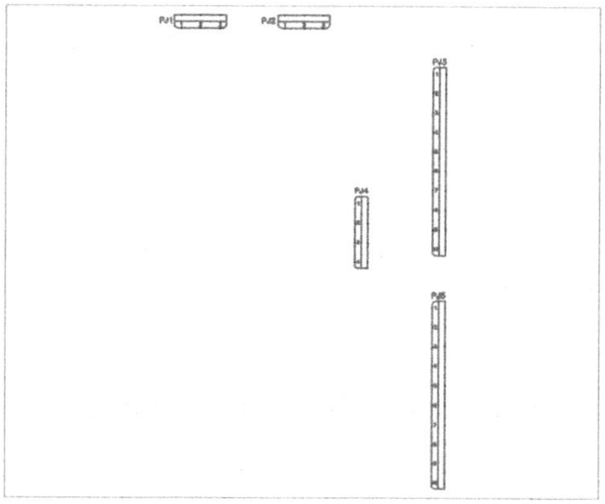

Figure 7-14 *The copied connectors inserted in the drawing*

Inserting the Multiple Wire Bus

1. Choose the **Multiple Bus** tool from the **Insert Wires/Wire Numbers** panel of the **Schematic** tab; the **Multiple Wire Bus** dialog box is displayed, as shown in Figure 7-15.

2. Set the following parameters in the **Multiple Wire Bus** dialog box:

   Spacing (Horizontal area): **0.7500**
   Spacing (Vertical area): **1.000**
   Number of Wires: **3**
   Component (Multiple Wires): Select this radio button

3. Choose the **OK** button; green crosses are displayed at the connection points of all connectors. Also, you are prompted to window-select the start wire connection points.

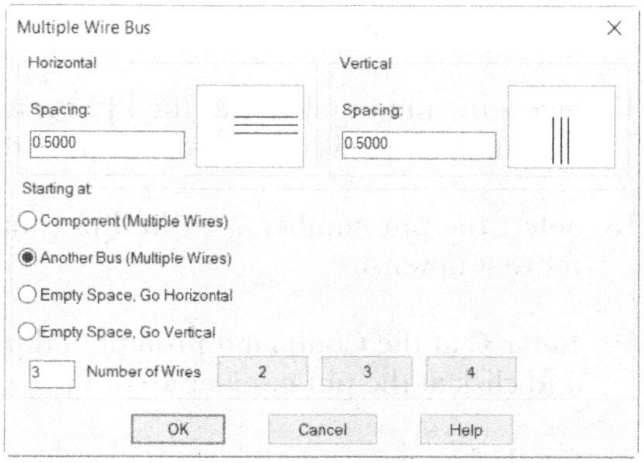

Figure 7-15 *The **Multiple Wire Bus** dialog box*

4. Select three green points of the PJ1 connector using a crossing window; red rhombus-shaped graphics are displayed at the green cross points.

5. Press ENTER and move the cursor downward and then toward right.

6. Enter **F** at the Command prompt and then press ENTER; the wires are flipped.

7. Drag the cursor toward right and join the cursor to the green connection points 2, 3, and 4 of the PJ4 connector and click on it. You will notice that wires are inserted between the PJ1 and PJ4 connectors.

8. Choose the **Multiple Bus** tool from the **Insert Wires/Wire Numbers** panel of the **Schematic** tab; the **Multiple Wire Bus** dialog box is displayed.

9. Enter **2** in the **Number of Wires** edit box. If this edit box is not available, then select any of the radio buttons in the **Starting at** area except the **Component (Multiple Wires)** radio button.

10. Select the **Component (Multiple Wires)** radio button.

11. Choose the **OK** button; you are prompted to window select the starting wire connection points. Also, green cross connection points are displayed at the pin numbers of the connectors.

12. Window select the pin numbers 2 and 3 of PJ2; red rhombus-shaped graphics are displayed at the green cross points.

13. Press ENTER and move the cursor downward and then toward the right.

14. Enter **F** at the Command prompt and press ENTER.

15. Move the cursor toward right and position it at the green connection points of pin numbers 7 and 8 of the PJ5 connector. Click at the pin number 7; a wire connecting PJ2 and PJ5 is created.

16. Choose the **Wire** tool from the **Schematic > Insert Wires/Wire Numbers > Wire** drop-down; you are prompted to specify the starting point of the wire.

17. Select the pin number 1 of the PJ2 connector and move the cursor downward. Now, move the cursor toward right to connect it to the pin number 10 of the PJ5 connector.

18. Select the pin number 2 of the PJ5 connector and move the cursor toward left and then move it upward.

19. Enter **C** at the Command prompt and press ENTER. Next, move the cursor toward right and click at the pin number 10 of PJ3.

20. Similarly, connect the wire between the pin number 1 of PJ4 and the pin number 7 of PJ3. Press ENTER to exit the command. Figure 7-16 shows point-to-point wiring diagram.

Saving the Drawing File

1. Choose **Save** from the **Application Menu** to save the drawing file.

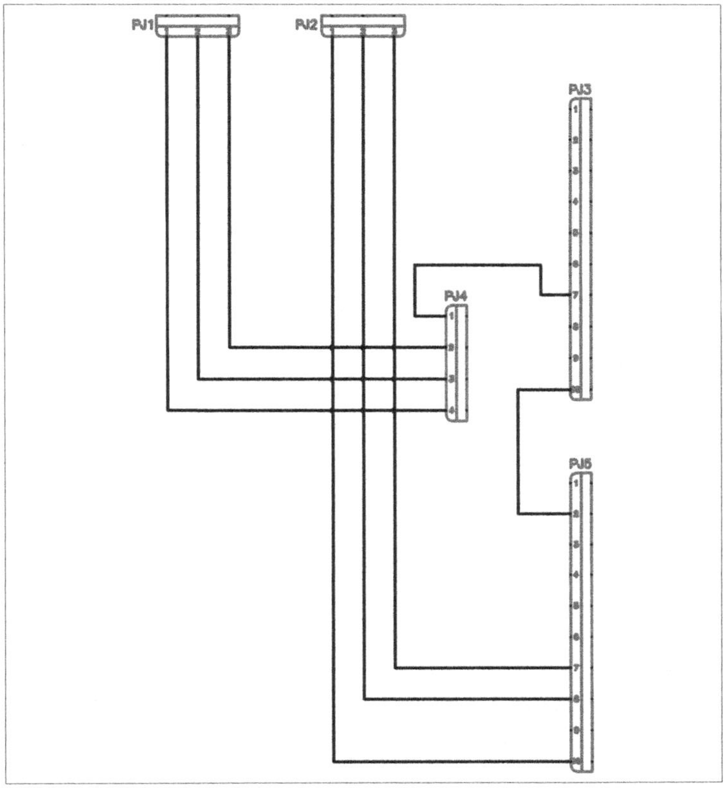

Figure 7-16 *The point-to-point wiring diagram*

Tutorial 2

In this tutorial, you will insert a user-circuit into the drawing, copy the inserted circuit, and then save the circuit, as shown in Figure 7-17. **(Expected time: 15 min)**

The following steps are required to complete this tutorial:

a. Create a new drawing.
b. Insert a three-phase ladder into the drawing.
c. Insert the saved circuit into the drawing.
d. Copy the existing circuit.
e. Save the entire circuit in the **Save Circuit to Icon Menu** dialog box.
f. Save the drawing file.

Creating a New Drawing

1. Create a drawing file with the name *C07_tut02.dwg* in the **CADCIM** project, as discussed in Tutorial 1 of this chapter.

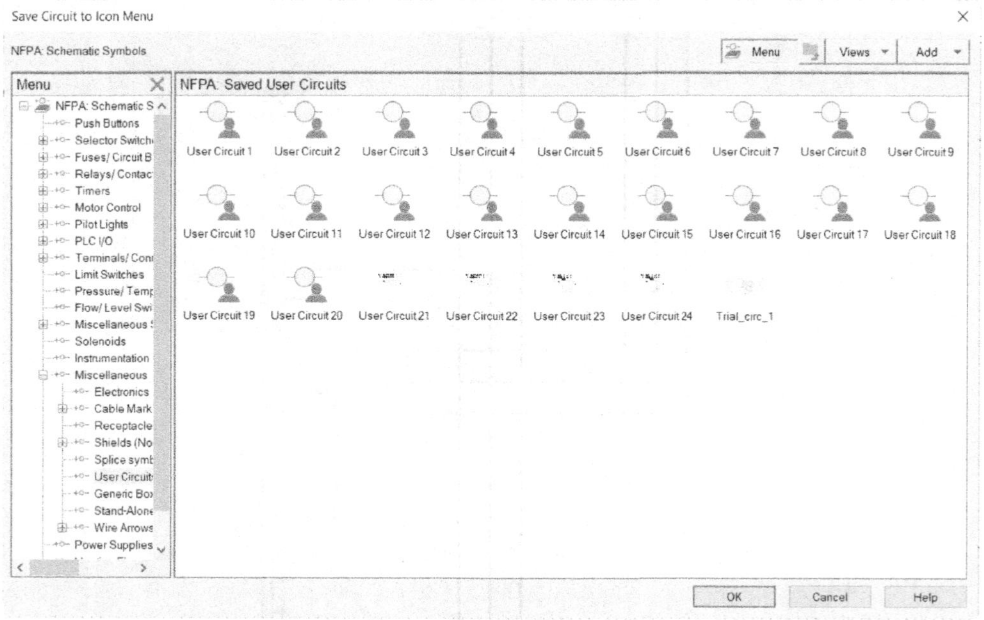

*Figure 7-17 The **Save Circuit to Icon Menu** dialog box showing the saved user-circuit*

Inserting the Three-Phase Ladder

1. Choose the **Insert Ladder** tool from **Schematic > Insert Wires/Wire Numbers > Insert Ladder** drop-down; the **Insert Ladder** dialog box is displayed.

2. Select the **3 Phase** radio button from the **Phase** area.

3. Enter **0.5** in the **Spacing** edit box of the **Phase** area.

4. Enter **0.5** in the **Spacing** edit box at the top right corner of the **Insert Ladder** dialog box.

5. Enter **1** in the **1st Reference** edit box.

   By default, 1 is displayed in the **Index** edit box.

6. Clear the **Without reference numbers** check box, if it is selected.

7. Next, enter **16** in the **Rungs** edit box and click in the **Length** edit box; you will notice that the length of the ladder is automatically calculated and is displayed in this edit box.

8. Choose the **OK** button; the dialog box is closed and you are prompted to specify the start position of the first rung. Enter **3,18** at the Command prompt and press ENTER; the three-phase ladder is inserted into the drawing, as shown in Figure 7-18.

Inserting a User-Saved Circuit

1. Choose the **Insert Saved Circuit** tool from **Schematic > Insert Components > Insert Circuit** drop-down; the **Insert Component** dialog box is displayed.

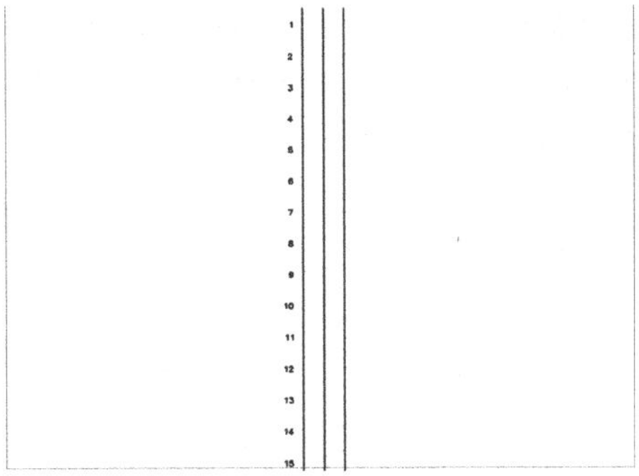

Figure 7-18 The three-phase ladder

2. Click on **User Circuit 21**; the **Insert Component** dialog box is closed and the **Circuit Scale** dialog box is displayed. In this dialog box, make sure 1.000 is displayed in the **Custom scale** edit box and the **Update circuit's text layers as required** check box is selected.

3. Choose the **OK** button to close the **Circuit Scale** dialog box; you are prompted to specify the insertion point. Place the cursor on the extreme left vertical bus of the ladder at the reference number **1** and then click on the screen; the saved circuit is inserted into the drawing, as shown in Figure 7-19.

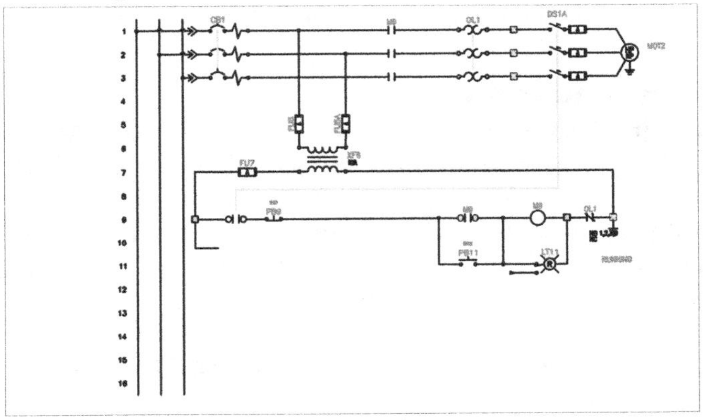

Figure 7-19 The saved circuit inserted into the drawing

Copying the Circuit

1. Choose the **Copy Circuit** tool from **Schematic > Edit Components > Circuit** drop-down; you are prompted to select objects. Select the circuit that you have inserted in the previous steps by dragging the cursor from right to left.

   You need to be careful while selecting the circuit. Do not select the wire numbering of the ladder.

2. Press ENTER; you are prompted to specify the base point. Specify the base point on the extreme left of the ladder with the reference number 1 and then move the cursor downward. Next, place the circuit on the extreme left vertical bus of the ladder with the reference number 13, as shown in Figure 7-20; the **Copy Circuit Options** dialog box is

displayed. Choose the **OK** button in this dialog box. If the **Gapped wire pointer problem** message box is displayed, choose the **OK** button in this message box.

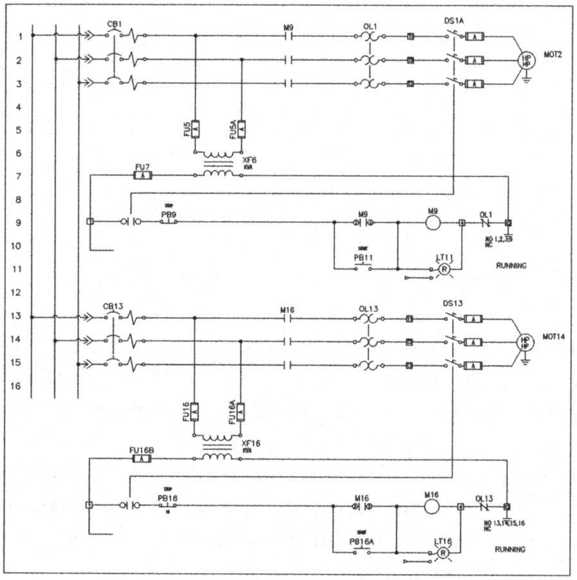

Figure 7-20 The copied circuit

Saving the Circuit to Icon Menu

1. Choose the **Save Circuit to Icon Menu** tool from **Schematic > Edit Components > Circuit** drop-down; the **Save Circuit to Icon Menu** dialog box is displayed.

2. Click on the **Add** drop-down list located at the upper right corner of the **Save Circuit to Icon Menu** dialog box; different options are displayed. Select the **New Circuit** option from it; the **Create New Circuit** dialog box is displayed.

3. Enter **Trial _circ_1** in the **Name** edit box of the **Icon Details** area.

4. Choose the **Active** button; *C07_tut02* is displayed in the **Image file** edit box.

5. Select the **Create PNG from current screen image** check box, if not selected.

6. Enter **Trial_1** in the **File name** edit box of the **Circuit Drawing File** area.

7. Choose the **OK** button; the dialog box is closed and you are prompted to specify the base point. Specify the base point on the extreme left of the ladder with the reference number 1; you are prompted to select the objects. Next, select the whole circuit and press ENTER; the **Save Circuit to Icon Menu** dialog box is displayed again.

You will notice that the circuit gets saved in the **Save Circuit to Icon Menu** dialog box, as shown in Figure 7-21.

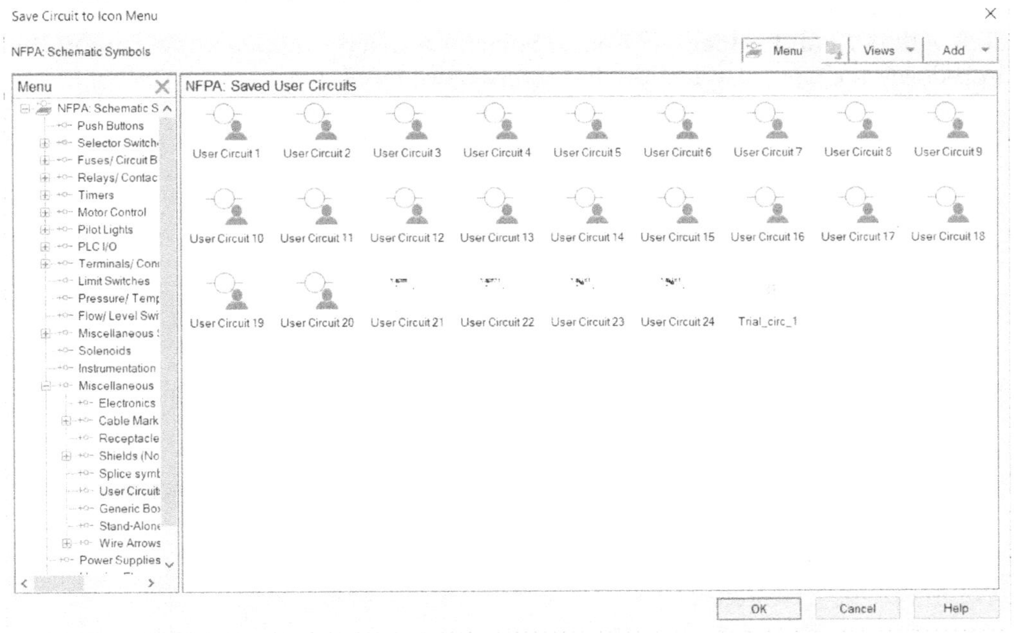

Figure 7-21 The **Save Circuit to Icon Menu** dialog box showing the saved circuit

8. Choose the **OK** button from the **Save Circuit to Icon Menu** dialog box to save the changes made and exit this dialog box.

Saving the Drawing File

1. Choose **Save** from the **Application Menu** to save the drawing file.

Tutorial 3

In this tutorial, you will configure a one-line motor circuit using the **Circuit Builder** tool, as shown in Figure 7-22. **(Expected time: 15 min)**

Figure 7-22 The one-line motor circuit

The following steps are required to complete this tutorial:

a. Create a new drawing.
b. Create a one-line motor circuit.
c. Save the drawing file.

Creating a New Drawing

1. Create a drawing file with the name *C07_tut03.dwg* in the **CADCIM** project, as discussed in Tutorial 1 of this chapter.

Creating a One-line Motor Circuit

1. Choose the **Circuit Builder** tool from **Schematic > Insert Components > Circuit Builder** drop-down; the **Circuit Selection** dialog box is displayed.

2. If the **History** area is not displayed, choose the **History >>** button at the bottom of the dialog box; the **History** area is displayed in it.

3. Click on the **+** sign at the left of the **One-line Motor Circuit** category; the circuit types related to this category are displayed below it, refer to Figure 7-23.

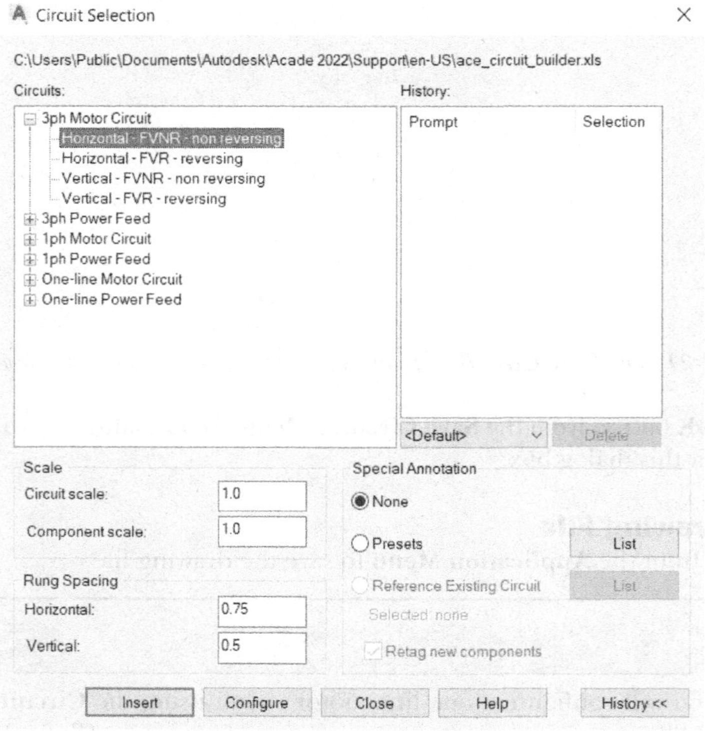

***Figure 7-23** The **Circuit Selection** dialog box*

4. Select **Horizontal - FVNR - non reversing** and choose the **Configure** button; the circuit template is displayed along with the cursor and you are prompted to specify the insertion point.

5. Place the cursor approximately at the middle of the drawing area; the template is inserted and the **Circuit Configuration** dialog box is displayed, as shown in Figure 7-24.

6. In this dialog box, make sure **Motor Setup** is selected in the **Circuit Elements** area. Next, choose the first button located at the right in the **Motor Setup** list box in the **Setup & Annotations** area; the **Select Motor** dialog box is displayed. 

7. In this dialog box, make sure **Single Phase** is selected in the **Type** drop-down list, **208** is selected in the **Voltage (V)** drop-down list, and **60** is selected in the **Frequency (Hz)**

drop-down list. Next, select the first entry from the table and choose the **OK** button to close the dialog box; the parameters for the selected motor are displayed in the **Setup & Annotations** area of the **Circuit Configuration** dialog box.

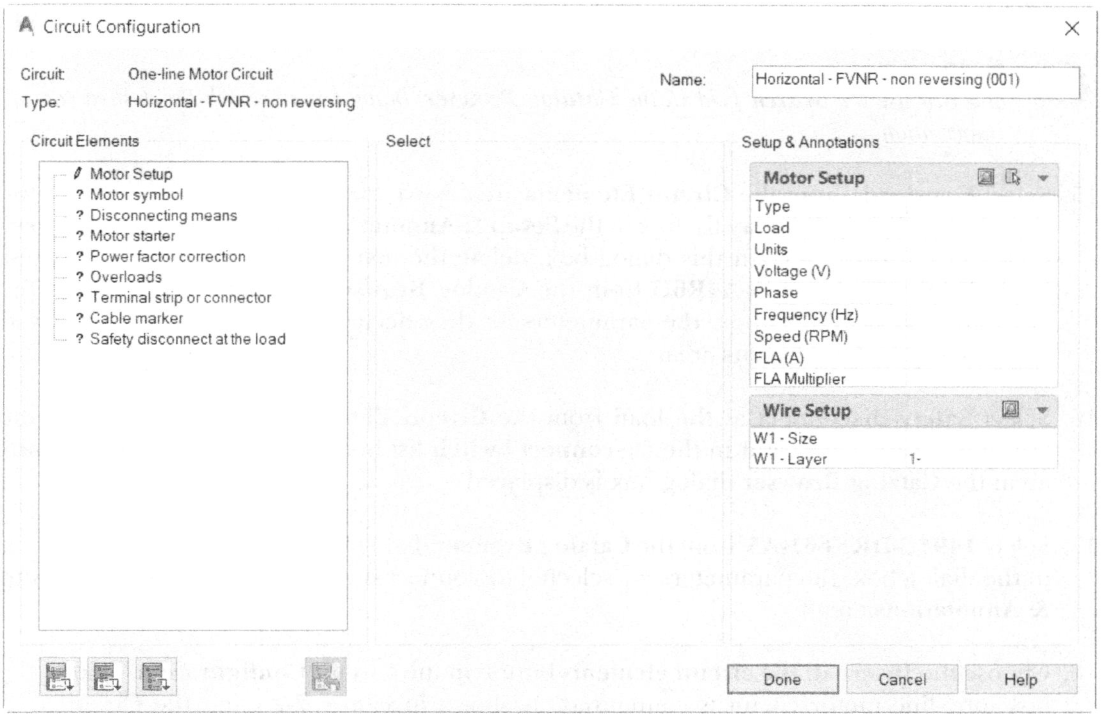

*Figure 7-24 The **Circuit Configuration** dialog box*

8. Select **Motor Symbol** from the **Circuit Elements** area and then select **None** in the **Motor** drop-down list of the **Select** area.

9. Select **Disconnecting means** from the **Circuit Elements** area. Next, select **Fuses** from the **Main Disconnect** drop-down list of the **Select** area; parameters for the fuse are displayed in the **Setup & Annotations** area.

10. Choose the button located at the right in the **Fuse** list box in the **Setup & Annotations** area; the **Catalog Browser** dialog box is displayed.

11. Delete the text in the **Search** field of the dialog box and press ENTER; all the entries for fuses are displayed in the Database grid.

12. Select **1492-FB1C30-L** from the **Catalog** column in the **Catalog Browser** dialog box and choose the **OK** button in the dialog box; the parameters for the selected fuse are displayed in the **Setup & Annotations** area.

13. Select **Motor Starter** from the **Circuit Elements** area. Next, choose the button located at the right in the **Motor Stater** list box in the **Setup & Annotations** area; the **Catalog Browser** dialog box is displayed.

14. In this dialog box, delete the text from the **Search** field and press ENTER. Next, select **193-B1R6K** from the **Catalog** column in the **Catalog Browser** dialog box and choose the **OK** button; the parameters for selected motor starter are displayed in the **Setup & Annotations** area.

Note

*You can use the **Search** field in the **Catalog Browser** dialog box to search the desired catalog information.*

15. Select **Overloads** from the **Circuit Elements** area. Next, choose the button located at the right in the **Overload Relay** list box in the **Setup & Annotations** area; the **Catalog Browser** dialog box is displayed. In this dialog box, delete the text in the **Search** field and press ENTER. Next, select **193-A4R6D** from the **Catalog Browser** dialog box and choose the **OK** button in the dialog box; the parameters for the selected overload relay are displayed in the **Setup & Annotations** area.

16. Select **Safety disconnect at the load** from the **Circuit Elements** area. Next, choose the button located at the right in the **Disconnect Switch** list box in the **Setup & Annotations** area; the **Catalog Browser** dialog box is displayed.

17. Select **1494C-DRX661-A5** from the **Catalog Browser** dialog box and choose the **OK** button in the dialog box; the parameters for selected disconnect switch are displayed in the **Setup & Annotations** area.

18. Choose the **Insert all the circuit elements** button in the **Circuit Configuration** dialog box; one-line motor circuit is configured, as shown in Figure 7-25 and the **Circuit Configuration** dialog box is displayed again. Next, choose the **Done** button in this dialog box to exit the command.

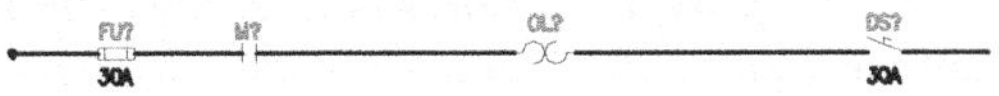

Figure 7-25 *The one-line motor circuit*

Note

You may change the component tags for the above circuit as discussed in earlier chapters.

Saving the Drawing File

1. Choose **Save** from the **Application Menu** to save the drawing file.

Tutorial 4

In this tutorial, you will insert a user-circuit into a drawing, save the circuit as a wblocked circuit, and then insert the wblocked circuit into another drawing, refer to Figure 7-26. **(Expected time: 15 min)**

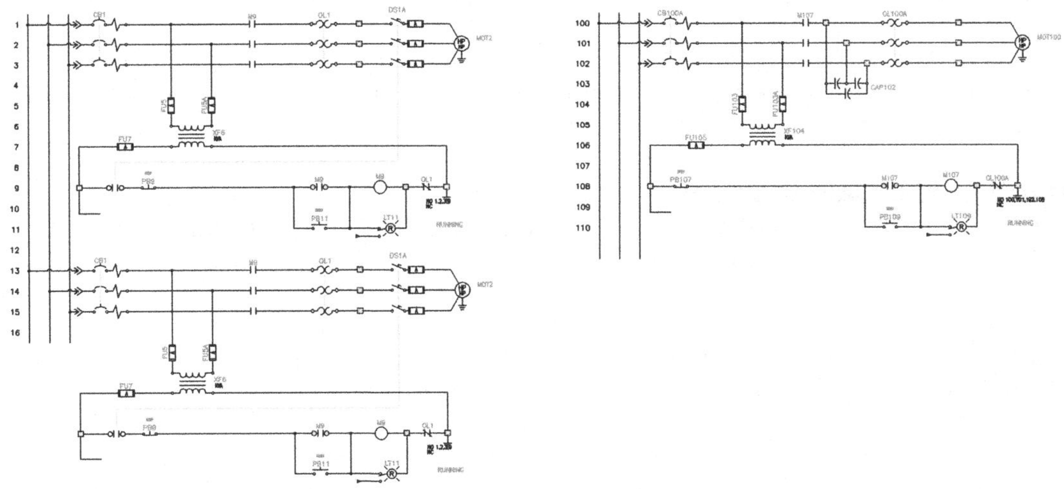

***Figure 7-26** Inserted wblocked circuit*

The following steps are required to complete this tutorial:

a. Create a new drawing.
b. Insert a three-phase ladder into the drawing.
c. Insert the saved circuit into the drawing.
d. Save the circuit as a wblocked circuit.
e. Insert the saved wblocked circuit.

Creating a New Drawing

1. Create a drawing file with the name *C07_tut04.dwg* in the **CADCIM** project, as discussed in Tutorial 1 of this chapter.

Inserting the Three-phase Ladder

1. Choose the **Insert Ladder** tool from **Schematic > Insert Wires/Wire Numbers > Insert Ladder** drop-down; the **Insert Ladder** dialog box is displayed.

2. Select the **3 Phase** radio button from the **Phase** area.

3. Enter **0.5** in the **Spacing** edit box of the **Phase** area.

4. Enter **0.5** in the **Spacing** edit box at the top right corner of the **Insert Ladder** dialog box.

5. Enter **100** in the **1st Reference** edit box.

   By default, 1 is displayed in the **Index** edit box.

6. Clear the **Without reference numbers** check box, if it is selected.

7. Next, enter **12** in the **Rungs** edit box and click in the **Length** edit box; you will notice that the length of the ladder is automatically calculated and is displayed in this edit box.

8. Choose the **OK** button; the dialog box is closed and you are prompted to specify the start position of the first rung. Enter **3,18** at the Command prompt and press ENTER; the three-phase ladder is inserted into the drawing.

Inserting the Saved Circuit

1. Choose the **Insert Saved Circuit** tool from **Schematic > Insert Components > Insert Circuit** drop-down; the **Insert Component** dialog box is displayed.

2. Click on **User Circuit 22**; the **Insert Component** dialog box is closed and the **Circuit Scale** dialog box is displayed. In this dialog box, make sure 1.000 is displayed in the **Custom scale** edit box and the **Update circuit's text layers as required** check box is selected.

3. Choose the **OK** button to close the **Circuit Scale** dialog box; you are prompted to specify the insertion point. Place the cursor on the extreme left vertical bus of the ladder at the reference number **100** and then click on the screen; the saved circuit is inserted into the drawing, as shown in Figure 7-27.

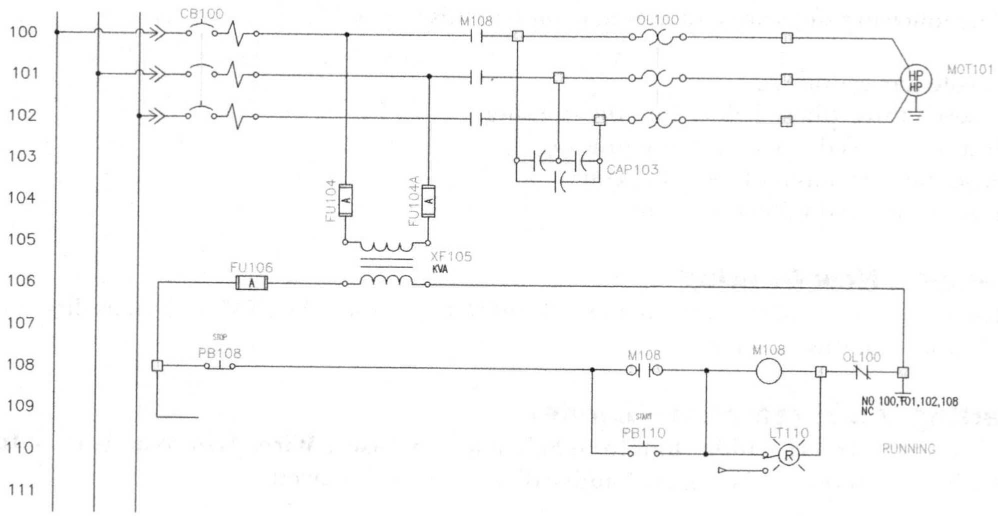

Figure 7-27 Inserted saved circuit

Saving the Circuit as a Wblocked Circuit

1. Type **Wblock** at the Command Prompt and press ENTER; the **Write Block** dialog box is displayed. In this dialog box, choose the **Pick Point** button from the **Base point** area; you are prompted to specify the insertion base point.

2. Click on the extreme left of the ladder with the reference number 100; base point is specified and the **Write Block** dialog box is displayed again. Notice that the coordinates of the base point are displayed in the **X, Y**, and **Z** edit boxes of the **Base Point** area.

3. Choose the **Select Objects** button from the **Objects** area; you are prompted to select objects. Select the inserted circuit including the ladder and press ENTER; the **Write Block** dialog box is displayed again.

   Next, you need to specify location for the wblocked circuit.

4. In the **Destination** area, enter **C:\Users\CADCIM1\Documents\customblock** in the **File name and path** edit box and then choose the **OK** button; the selected objects are converted to a wblocked circuit and the circuit is saved at the specified location.

Inserting the Saved Wblocked Circuit

1. Open the C07_tut02 drawing from the **CADCIM** project. Next, choose the **Insert Wblocked Circuit** tool from the **Insert Circuit** drop-down in the **Insert Components** panel of the **Schematic** tab; the **Insert WBlocked Circuit** dialog box is displayed, as shown in Figure 7-28.

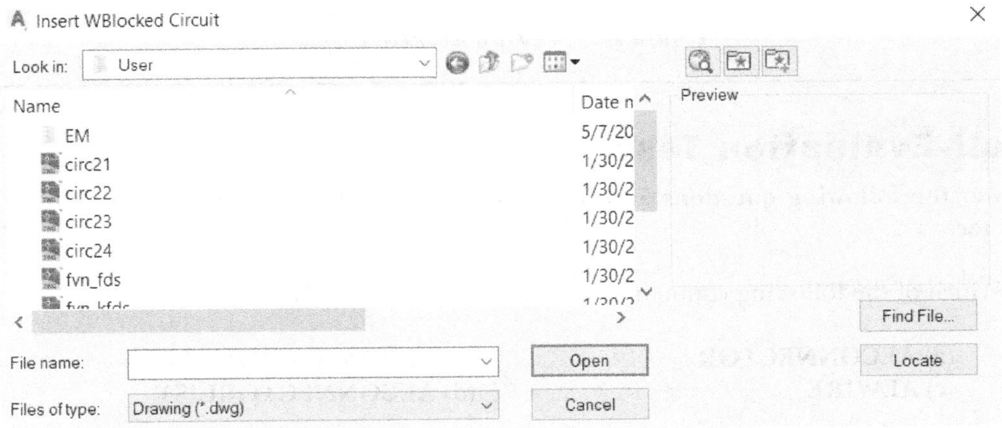

*Figure 7-28 The **Insert WBlocked Circuit** dialog box*

2. In this dialog box, navigate to the location where the wblocked circuit was saved. Next, select **customblock** from this dialog box and choose the **Open** button; the **Circuit Scale** dialog box is displayed. In this dialog box, make sure **1.000** is displayed in the **Custom scale** edit box and the **Update circuit's text layers as required** check box is selected.

3. Choose the **OK** button to close the **Circuit Scale** dialog box; you are prompted to specify the insertion point.

4. Enter **17,18** at the Command prompt and press ENTER; the saved wblocked circuit is inserted into the drawing, as shown in Figure 7-29.

Saving the Drawing File

1. Choose **Save** from the **Application Menu** to save the *C07_tut02.dwg* drawing file.

2. Also, save the *C07_tut04.dwg* drawing file.

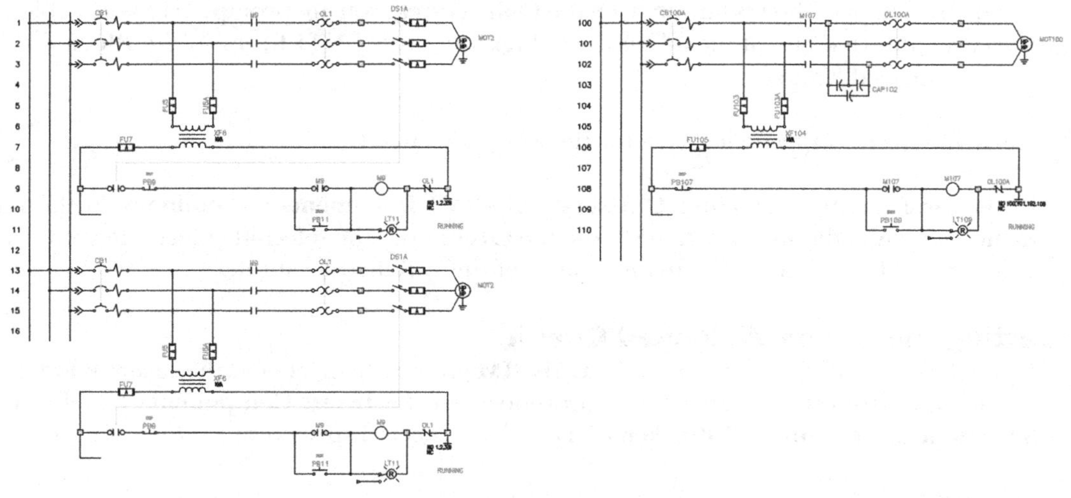

***Figure 7-29** Inserted wblocked circuit*

Self-Evaluation Test

Answer the following questions and then compare them to those given at the end of this chapter:

1. Which of the following commands is used to insert a connector?

   (a) **AECONNECTOR** (b) **AECONNECTORPIN**
   (c) **AEWIRE** (d) **AECONNECTORLIST**

2. The __________ tool is used to increase the length of a connector.

3. The __________ tool is used to insert the WBlocked circuit in a drawing.

4. Connectors are used to connect a wire or a group of wires at a single junction. (T/F)

5. When you choose the **Rotate Connector** tool, the orientation of the connector switches between the horizontal and vertical positions. (T/F)

6. The **Swap Connector Pins** tool is used to swap pin numbers and pin locations. (T/F)

Review Questions

Answer the following questions:

1. Which of the following tools is used to create a new circuit?

   (a) **Move Circuit** (b) **Copy Circuit**
   (c) **Insert Saved Circuit** (d) **Save Circuit to Icon Menu**

2. Which of the following radio buttons in the **Split Block** dialog box is used to draw straight lines at the end of the break of a connector?

   (a) **Draw it**
   (b) **Jagged Lines**
   (c) **Straight Lines**
   (d) **Allow Spacers/Break**

3. The **Scoot** tool is used to move a connector along with wires. (T/F)

4. A connector can be rotated in a clockwise direction by choosing the **Rotate Connector** tool. (T/F)

EXERCISES

Exercise 1

Create a connector diagram, as shown in Figure 7-30, and then delete the pins 7, 8, 9, 10, 11, and 14 using the **Delete Connector Pins** tool. Next, save the drawing as *C07_exer01.dwg*.

(Expected time: 20 min)

Exercise 2

Create a three-phase ladder with the following parameters: Spacing = 1.000, Spacing = 0.5000 of the **Phase** area, Rungs = 20, and Reference number = 100. Next, insert a saved circuit **User Circuit 24** from the **NFPA: Saved User Circuits** area of the **Insert Component** dialog box at rung **100**. Then, copy this circuit and place it on the rung **110**, as shown in Figure 7-31. Save the circuit in the **Save Circuit to Icon Menu** dialog box with the name **Trial_circ_2**.

(Expected time: 20 min)

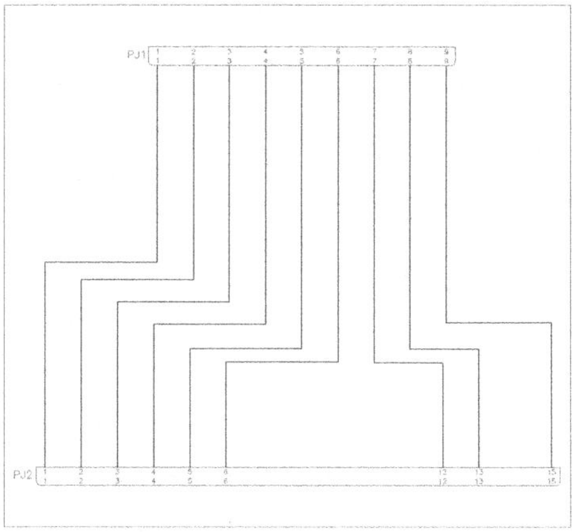

Figure 7-30 *The connector diagram for Exercise 1*

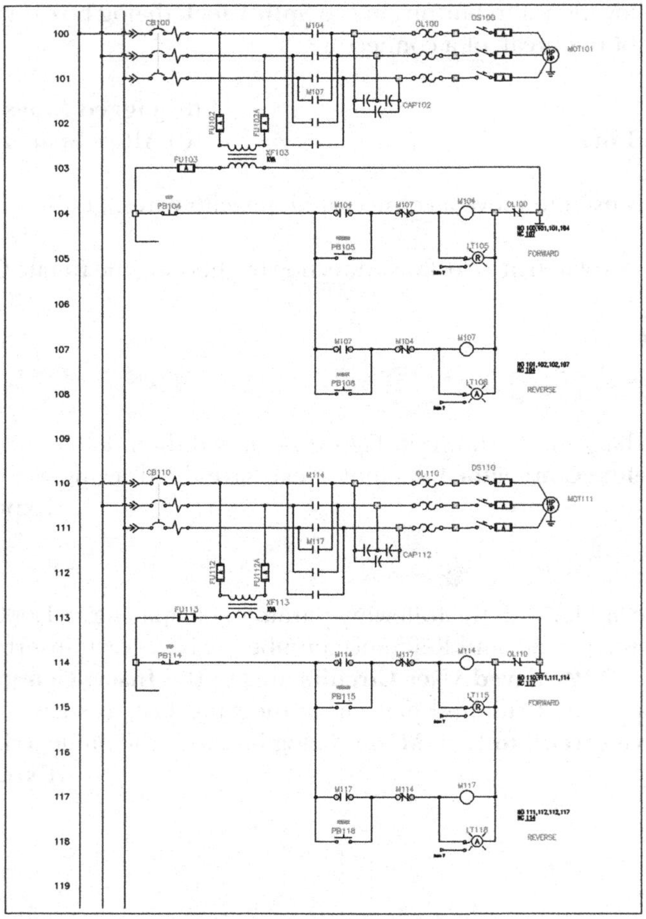

Figure 7-31 *The circuit diagram for Exercise 2*

Answers to Self-Evaluation Test

1. a, **2. Stretch Connector**, **3. Insert WBlocked Circuit**, **4.** T, **5.** T, **6.** F

Chapter 8

Panel Layouts

Learning Objectives

After completing this chapter, you will be able to:

- *Understand the WD_PNLM block file*
- *Create panel layouts from schematic list*
- *Annotate and edit a footprint*
- *Insert footprints from icon menu*
- *Insert footprints manually*
- *Insert footprints from user defined list, equipment list, and vendors menu*
- *Copy a footprint*
- *Set the panel drawing configuration and footprint layers*
- *Make an Xdata visible*
- *Insert balloons, nameplates, and DIN Rail into drawing*
- *Edit the footprint lookup database file*

INTRODUCTION

AutoCAD Electrical provides different tools to create intelligent panel layout drawings. You can design panel layouts either by using the information of the schematic drawings or without using schematic drawings. You can use footprint symbols supplied by vendors in AutoCAD format with AutoCAD Electrical.

Using the panel layout tools, you can create intelligent mechanical or panel layout drawings. The following are the key features of panel layout drawings:

1. Due to the bi-directional capabilities of panel layout drawings, the panel drawings get updated automatically whenever the schematic wiring diagrams are updated and vice-versa.

2. You can extract wire number, information of wire color or gauge, and connection sequencing data directly from the schematics and annotate it on to the panel footprint.

3. You can use AutoCAD Electrical to extract various reports from the panel layout drawings such as Bill of Material reports, panel component reports, nameplate reports, wire connection reports, and so on.

In this chapter, you will learn about WD_PNLM block file. Also, you will learn how to insert footprints using the **Schematics list**, **Icon Menu**, **Manual**, **Equipment List**, **User Defined List**, and **Manufacturer Menu** tools. This chapter also explains annotating and editing of footprints. In addition to this, you will learn about setting panel drawing configuration, copy footprint, inserting nameplates, balloons, and DIN Rail in the panel layout drawings.

THE WD_PNLM BLOCK FILE

Whenever you insert a panel component into a drawing for the first time using the **Icon Menu**, **Schematic List**, or **Manual** tools, AutoCAD Electrical first checks for the WD_PNLM block file in the drawing and reads configuration settings from the attributes that it contains. If the drawing file contains invisible WD_PNLM block, you can insert panel component into the drawing directly. But if the drawing file does not have invisible WD_PNLM block, the **Alert** dialog box will be displayed. Choose the **OK** button in the **Alert** dialog box to insert the WD_PNLM block file in the drawing. This file contains only attributes and no graphical information. A drawing is considered as a panel layout drawing if this block is present in the drawing. This block file is inserted automatically at the 0,0 location of every panel layout drawing.

Note

The WD_M and WD_PNLM blocks can be inserted in the same drawing.

CREATING PANEL LAYOUTS FROM SCHEMATIC LIST

Command: AEFOOTPRINTSCH

The **Schematic List** tool is used to insert panel footprints from the schematic component list. This list displays the names of the components that have already been inserted in the schematic drawings of an active project. Using this tool, you can link the panel footprint and schematic component. To insert a panel footprint from a schematic component list, choose the **Schematic List** tool from the **Insert Footprints** drop-down in the **Insert Component Footprints** panel of the **Panel** tab, refer to Figure 8-1; the **Schematic Components List --> Panel Layout Insert** dialog box will be displayed, as shown in Figure 8-2.

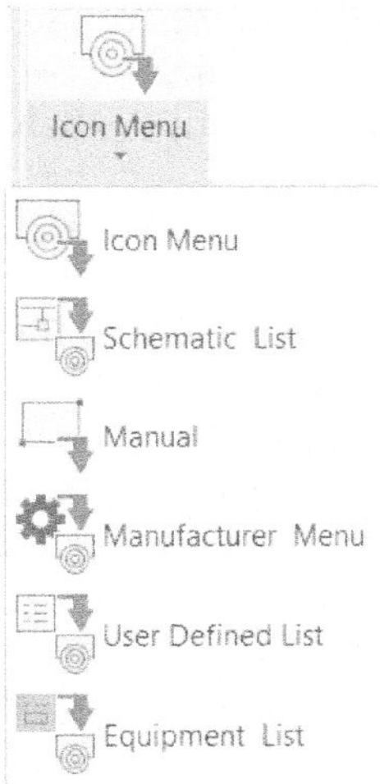

Figure 8-1 *The* ***Insert Footprints*** *drop-down*

Note

1. If you are using the ***Schematic List*** *tool for the first time in AutoCAD Electrical session, the* ***Schematic Components List --> Panel Layout Insert*** *dialog box will be displayed. Else, the* ***Schematic Components*** *dialog box will be displayed directly. The* ***Schematic Components*** *dialog box is discussed later in this chapter.*

2. The one-line components are not extracted and displayed in the schematic component list.

The **Extract component list for** area is used to extract component list data for active drawing or project.

The **Project** radio button is selected by default and is used to extract component list from the entire active project. To extract component list from the project, select the **Project** radio button and then choose the **OK** button from the **Schematic Components List --> Panel Layout Insert** dialog box; the **Select Drawings to Process** dialog box will be displayed. Next, select the drawing or drawings that you want to process and choose the **Process** button; the selected drawing(s) will be displayed in the bottom list of the **Select Drawings to Process** dialog box. Next, choose the **OK** button from the **Select Drawings to Process** dialog box; the **Schematic Components (active project)** dialog box will be displayed.

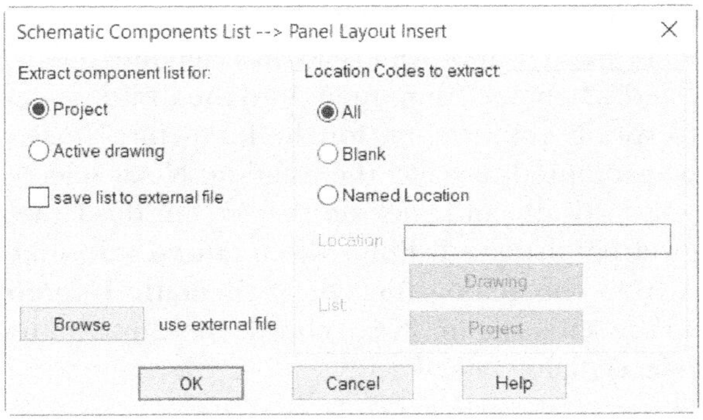

Figure 8-2 *The* ***Schematic Components List --> Panel Layout Insert*** *dialog box*

In case you extract component list from the entire project with the **save list to external file** check box selected in the **Schematic Components List --> Panel Layout Insert** dialog box and choose the **OK** button; the **Select Drawings to Process** dialog box will be displayed. Next, select the drawing or drawings that you want to process and choose the **Process** button; the selected drawing(s) will be displayed in the bottom list of the **Select Drawings to Process** dialog box. Next, choose the **OK** button from the **Select Drawings to Process** dialog box; the **Select file for Schematic – –> Panel Layout list** dialog box will be displayed. Using this dialog box, you can save component list to an external file. An external file can be saved in the comma delimited text files such as *.wd1* or *.csv*. Specify the file name in the **File name** edit box and choose the **Save** button; the **Schematic Components (active project)** dialog box will be displayed.

Select the **Active drawing** radio button from the **Schematic Components List --> Panel Layout Insert** dialog box to extract component list from the active drawing. Next, choose **OK** from the **Schematic Components List --> Panel Layout Insert** dialog box; the **Schematic Components (active drawing)** dialog box will be displayed. Note that the **Schematic Components (active drawing)** dialog box will be displayed only if the active drawing has schematic components in it. If you select the **save list to external file** check box from the **Schematic Components List --> Panel Layout Insert** dialog box and choose the **OK** button in this dialog box, the **Select file for Schematic --> Panel Layout list** dialog box will be displayed after choosing the **OK** button in the **Select Drawings to Process** dialog box, as discussed earlier. In the **Select file for Schematic --> Panel Layout list** dialog box, specify a file name in the **File name** edit box and choose the **Save** button; the **Schematic Components (active drawing)** dialog box will be displayed.

Note

1. The options in the ***Schematic Components (active project)*** *dialog box and the* ***Schematic Components (active drawing)*** *dialog box are same. If an active drawing does not have schematic data, a message will be displayed at the Command prompt indicating that no schematic data is found.*

2. One-line components are not included in the list displayed in the ***Schematic Components (active project)*** *dialog box.*

The **Insert** button in the **Schematic Components (active project)** dialog box and the **Schematic Components (active drawing)** dialog box will be activated only if the selected schematic component has catalog data and if its footprint is not inserted into the drawing already. To insert the footprint of the selected schematic component into the drawing, choose the **Insert** button; you will be prompted to specify the location for the footprint. Next, specify the location for the footprint; you will be prompted to select the rotation. Next, select the rotation by moving the cursor horizontally or vertically and click on the screen; the **Panel Layout - Component Insert/Edit** dialog box will be displayed. Enter the required information in this dialog box. Choose the **OK** button from this dialog box; the **Schematic Components (active project)** dialog box will appear on the screen again. Next, choose the **Close** button to exit the **Schematic Components (active project)** dialog box.

Note

If a footprint match is not found for the selected schematic component, then the ***Footprint*** *dialog box will be displayed after choosing the* ***Insert*** *button in the* ***Schematic Components (active project)*** *dialog box. After choosing the* ***OK*** *button from the* ***Footprint*** *dialog box, the* ***Manufacturer/ Catalog --> Footprint not found*** *dialog box will be displayed*

You can also insert multiple footprints in a single operation. To do so, press the SHIFT or CTRL key and select schematic components from the list displayed in the **Schematic Components (active project)** dialog box, and then choose the **Insert** button from this dialog box; the **Spacing for Footprint Insertion** dialog box will be displayed.

After specifying the required options in the **Spacing for Footprint Insertion** dialog box, choose the **OK** button from this dialog box; you will be prompted to specify the location for component. Specify the location; you will be prompted to select the rotation. Next, select the rotation in horizontal or vertical direction by moving the cursor horizontally or vertically and click on the screen; the **Panel Layout - Component Insert/Edit** dialog box will be displayed. Enter the required information in this dialog box and choose **OK**; a footprint will be inserted into the drawing and the above procedure will be repeated until all components present in the **Insert Order** area of the **Spacing for Footprint Insertion** dialog box are inserted. Once all components are inserted, the **Schematic Components (active project)** dialog box will be displayed again.

Note

If footprint match for the components displayed in the ***Insert Order*** *area of the* ***Spacing for Footprint Insertion*** *dialog box is not found in the footprint_lookup.mdb file or in wiring diagram tables, the* ***Footprint*** *dialog box will be displayed after choosing the* ***OK*** *button in the* ***Spacing for Footprint Insertion*** *dialog box. The options in the* ***Footprint*** *dialog box are discussed later in this chapter.*

ANNOTATING AND EDITING FOOTPRINTS

Command: AEEDITFOOTPRINT

The **Edit** tool is used to edit a footprint. Using this tool, you can make changes in the selected footprint at any time. You can edit the values such as component tag, description, installation, location codes, catalog data, and so on. You may also need to update a footprint due to changes in catalog or assembly values. To edit a footprint, choose the **Edit** tool from the **Edit Footprints** panel of the **Panel** tab; you will be prompted to select panel layout component. Select the required component; the **Panel Layout - Component Insert/Edit** dialog box will be displayed, as shown in Figure 8-3.

In the **Item Number** area, you can assign an item number to a footprint. In the **Catalog Data** area, you can assign the catalog part number to a footprint. The **Rating** area is used to specify the values for each rating attribute. This area will be activated only if the component being edited has rating attributes. This area consists of the **Rating** edit box and the **Show All Ratings** button. To specify a rating attribute, enter the rating of the attribute in the **Rating** edit box. Alternatively, choose the **Show All Ratings** button from the **Panel Layout - Component Insert/Edit** dialog box; the **View/Edit Rating Values** dialog box will be displayed. You can enter upto 12 rating attributes for a component in this dialog box. Next, choose the **OK** button from this dialog box; the rating value will be displayed in the **Rating** edit box.

In the **Component Tag** area, you can assign or edit the tag of a footprint. The **Description** area of the **Panel Layout - Component Insert/Edit** dialog box is used to enter the description of the component. You can enter upto three lines of description attribute text in the **Description** area. This area has three edit boxes: **Line 1**, **Line 2**, and **Line 3** and three buttons: **Drawing**, **Project**, and **Defaults**. The options in the **Installation / Location codes (for reports)** area are used to specify the installation, location, mount, and group codes for a component. Enter the installation, location, mount, and group codes in the **Installation**, **Location**, **Mount**, and **Group** edit boxes, respectively. Alternatively, you can specify these codes by choosing the **Drawing**, **Project**, and **Pick Like** buttons in this area.

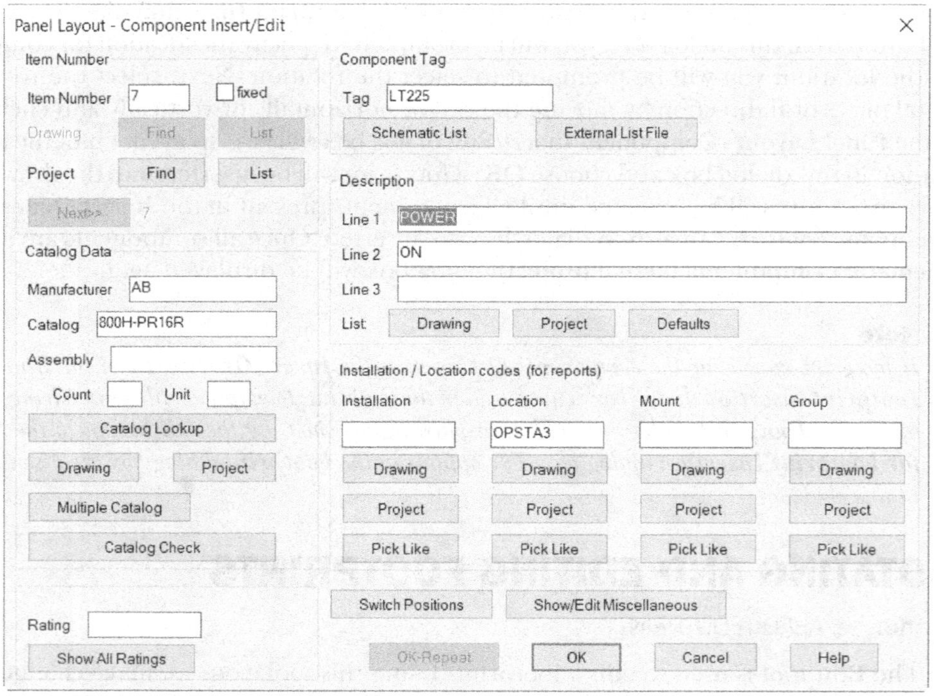

***Figure 8-3** The **Panel Layout - Component Insert/Edit** dialog box*

After specifying the required options in the **Panel Layout - Component Insert/Edit** dialog box, choose the **OK** button in this dialog box to save the changes made and to exit this dialog box.

INSERTING FOOTPRINTS FROM THE ICON MENU

Command: AEFOOTPRINT

You can insert footprints from the icon menu by using the **Icon Menu** tool. This tool is also used when you want to create panel drawings prior to schematic drawings. Also, this tool is used to insert some panel components that are not listed in schematic drawings such as nameplates, din rails, and so on. To insert a footprint from the icon menu into a panel layout drawing, choose the **Icon Menu** tool from the **Insert Component Footprints** panel of the **Panel** tab; the **Insert Footprint** dialog box will be displayed, as shown in Figure 8-4.

The options and areas in this dialog box are similar to that in the **Insert Component** dialog box. Next, select the required component from the list displayed on the left of the **Insert Footprint** dialog box or the icons shown on the right of the **Insert Footprint** dialog box; the **Panel Layout Symbols** area will get changed and various symbols of the selected icon will be displayed. For example, if you choose the **Push Buttons** from the **Menu** area or choose the **Push Buttons** icon (first row, first column) from the **Panel Layout Symbols** area, you will notice that the **Panel Layout Symbols** area is changed into the **Panel: Push Buttons** area. The dialog box displays different types of push buttons.

Next, select a symbol to insert from the **Insert Footprint** dialog box; the **Footprint** dialog box will be displayed, as shown in Figure 8-5.

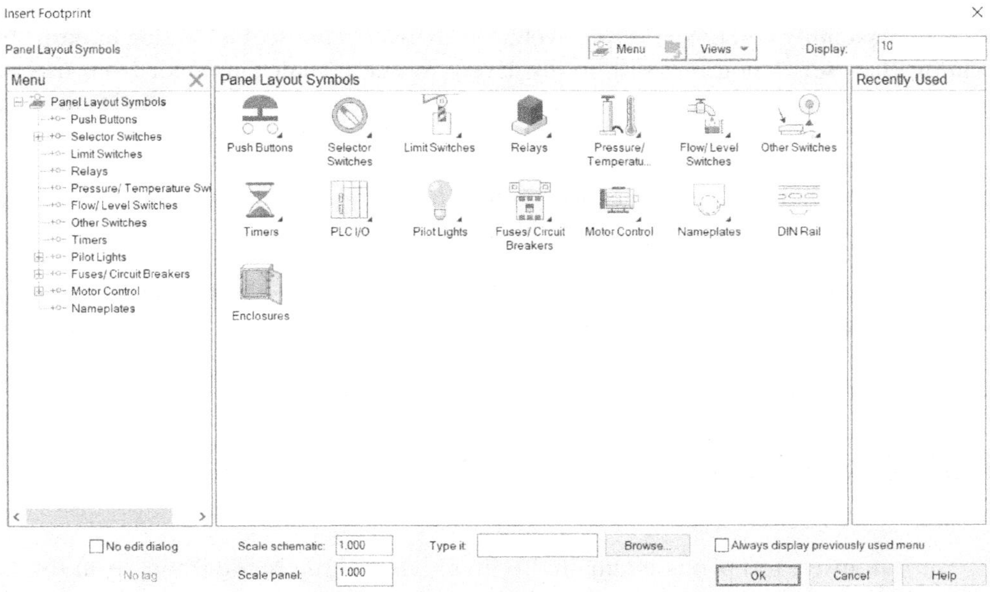

*Figure 8-4 The **Insert Footprint** dialog box*

Note

*The **Footprint** dialog box will also be displayed if a footprint match for a selected schematic component is not found or if a catalog number is not assigned to a schematic component, as discussed in the previous topic.*

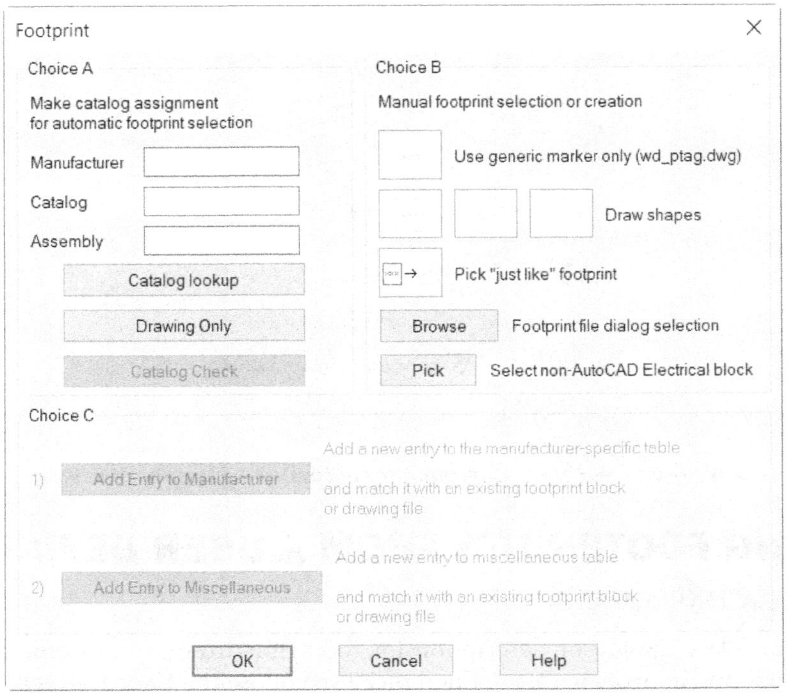

*Figure 8-5 The **Footprint** dialog box*

You can manually enter manufacturer, catalog number, and assembly values in the **Manufacturer**, **Catalog**, and **Assembly** edit boxes, respectively. Alternatively, choose the **Catalog lookup** button; the **Catalog Browser** dialog box will be displayed. Select a catalog number from this dialog box and choose the **OK** button; the catalog information will be displayed in the **Manufacturer**, **Catalog**, and **Assembly** edit boxes. You can also choose the **Drawing Only** button from this dialog box; the **catalog values (this drawing)** dialog box will be displayed. Select catalog values and choose the **OK** button from the **catalog values (this drawing)** dialog box to display values in the **Manufacturer**, **Catalog**, and **Assembly** edit boxes of the **Footprint** dialog box.

After specifying the required options in this dialog box, choose the **OK** button from this dialog box; you will be prompted to specify the location and the rotation angle for the footprint. Specify the location and the rotation angle of the footprint to be inserted as discussed earlier. Note that the footprint block will be added to the _PNLMISC lookup file.

INSERTING FOOTPRINTS MANUALLY

Command: AEFOOTPRINTMAN

You can insert footprints manually. To do so, choose the **Manual** tool from the **Insert Footprints** drop-down in the **Insert Component Footprints** panel of the **Panel** tab, refer to Figure 8-1; the **Insert Component Footprint -- Manual** dialog box will be displayed, as shown in Figure 8-6. Select the required footprint shape from this dialog box. Based on the selected footprint shape, you will be prompted to specify the position for the footprint. Specify the position; the **Panel Layout - Component Insert / Edit** dialog box will be displayed. Specify the required options in this dialog box and choose the **OK** button to insert the footprint into the drawing.

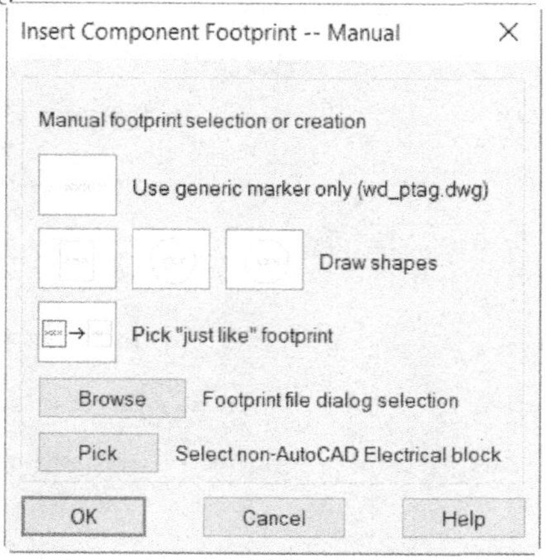

*Figure 8-6 The **Insert Component Footprint -- Manual** dialog box*

INSERTING FOOTPRINTS FROM A USER DEFINED LIST

Command: AEFOOTPRINTCAT

You can select catalog or description for footprints from a user-defined pick list. To insert a footprint from a user-defined pick list, choose the **User Defined List** tool from the **Insert Component Footprints** panel of the **Panel** tab; the **Panel footprint: Select**

and Insert by Catalog or Description Pick dialog box will be displayed, as shown in Figure 8-7. This dialog box displays the catalog data saved in the *wd_picklist.mdb* database file. The procedure to insert footprints using the **User Defined List** tool is discussed next.

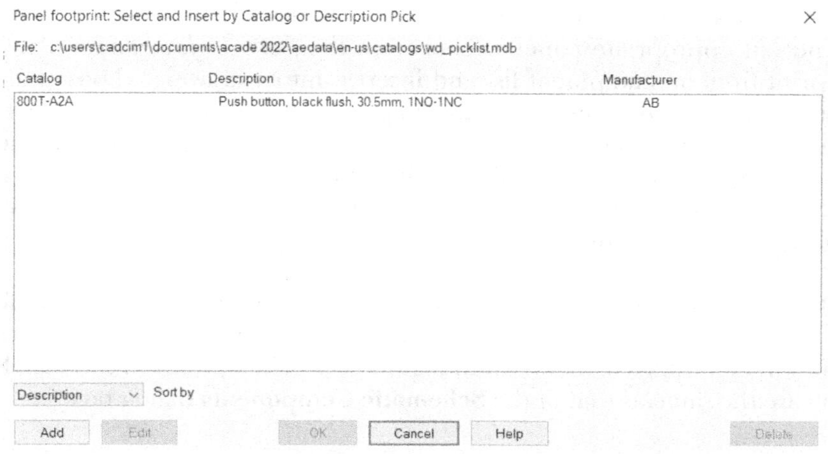

Figure 8-7 The ***Panel footprint: Select and Insert by Catalog or Description Pick*** *dialog box*

Note

The options in the ***Panel footprint: Select and Insert by Catalog or Description Pick*** *dialog box are the same as those of the* ***Schematic Component or Circuit*** *dialog box. The only difference is that a panel component is selected from the* ***Panel footprint: Select and Insert by Catalog or Description Pick*** *dialog box, whereas a schematic component is selected from the* ***Schematic Component or Circuit*** *dialog box.*

Select the required catalog number from the dialog box and choose the **OK** button; the **AutoCAD Message** message box will be displayed. Next, choose the **OK** button from this message box; the command ends and you cannot insert the footprint from the user defined list. To overcome this problem, you need to add a library path in the **Panel Footprint Libraries** library.

After adding a new path to library, choose the **User Defined List** tool again and then select the required catalog number from the **Panel Footprint: Select and Insert by Catalog or Description Pick** dialog box. Next, choose the **OK** button; you will be prompted to specify the location for footprint. Next, specify the location for footprint; you will be prompted to select the rotation. Select the rotation by moving the cursor in the horizontal or vertical direction and then press ENTER; the **Panel Layout - Component Insert/Edit** dialog box will be displayed. Enter the required information in this dialog box and choose the **OK** button; the footprint block will be inserted into the drawing.

Note

The data displayed in the ***Panel footprint: Select and Insert by Catalog or Description Pick*** *dialog box is saved in the wd_picklist.mdb access file. You can add data or edit this file by choosing the* ***Add*** *or* ***Edit*** *button or by using the* ***Microsoft Office Access****. Also, you can delete the data by choosing the* ***Delete*** *button from the* ***Panel footprint: Select and Insert by Catalog or Description Pick*** *dialog box.*

INSERTING FOOTPRINTS FROM AN EQUIPMENT LIST

Command: AEFOOTPRINTEQ

The **Equipment List** tool is used to extract data from an equipment list. It is also used to find out appropriate panel symbol from the *footprint_lookup.mdb* file. To extract a footprint from an equipment list and insert it into a drawing, choose the **Equipment List** tool from the **Insert Component Footprints** panel in the **Panel** tab; the **Select Equipment List Spreadsheet File** dialog box will be displayed. Select the required file from the dialog box and choose the **Open** button; the **Table Edit** dialog box will be displayed, as shown in Figure 8-8. Select the table or sheet in this dialog box and then choose the **OK** button; the **Settings** dialog box will be displayed, as shown in Figure 8-9.

The **OK** button in the **Settings** dialog box will be activated only if you choose the **Default settings** button. Choose the **Default settings** button and then choose the **OK** button; the **Panel equipment in** dialog box will be displayed, as shown in Figure 8-10. The options in the **Panel equipment in** dialog box are the same as that of the **Schematic Components** dialog box.

Next, select the schematic component that you want to insert in the panel layout drawing; the **Insert** button will be activated. Choose the **Insert** button; you will be prompted to specify the location for footprint. Specify the location; you will be prompted to select the rotation. Select the rotation by moving the cursor in the horizontal or vertical direction and press ENTER; the **Panel Layout - Component Insert/Edit** dialog box will be displayed. Enter the required information in this dialog box and choose the **OK** button; the footprint will be inserted into the drawing and the **Panel equipment in** dialog box will be displayed again. To insert more footprints into the drawing, you need to repeat the steps discussed above. To exit the **Panel equipment in** dialog box and to save the changes made in this dialog box, choose the **Close** button. Note that if the child components or related panel components are present in other drawings of a project then after choosing the **Close** button in the **Panel equipment in** dialog box, the **Update other drawings?** message box will be displayed. Choose the **OK** button from this dialog box to update other drawings of a project; the **QSAVE** message box will be displayed. Next, choose the **OK** button in the **QSAVE** message box to save and update the drawings accordingly.

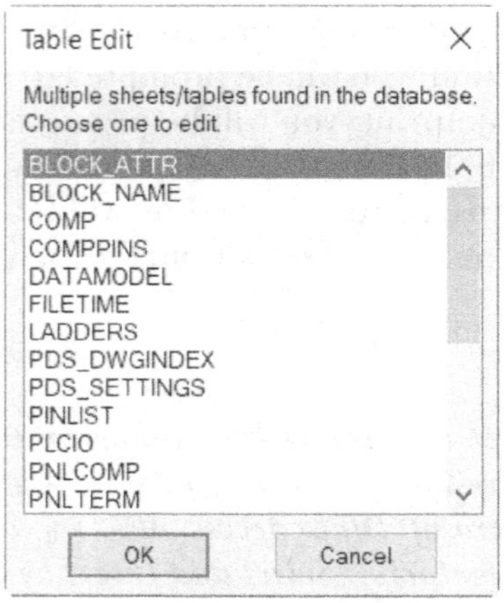

***Figure 8-8** The **Table Edit** dialog box*

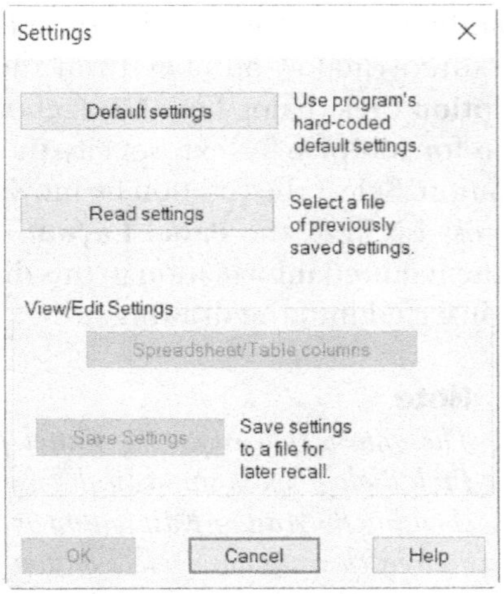

***Figure 8-9** The **Settings** dialog box*

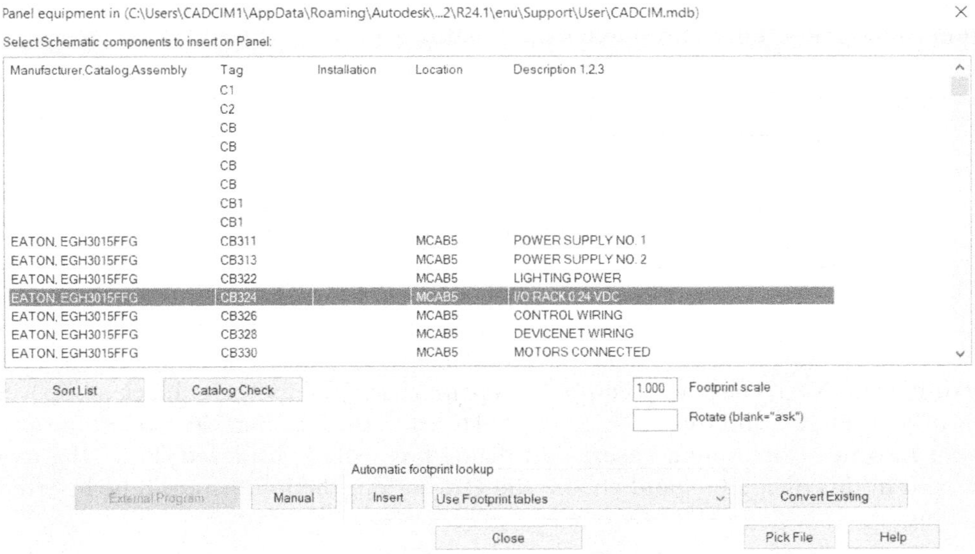

*Figure 8-10 The **Panel equipment in** dialog box*

INSERTING FOOTPRINTS FROM VENDOR MENUS

Command: AEFOOTPRINTMFG

You can insert footprints by using the vendor menus. To do so, choose the **Manufacturer Menu** tool from the **Insert Component Footprints** panel of the **Panel** tab; the **Vendor Menu Selection - Icon Menu Files (*.pnl extension)** dialog box will be displayed, as shown in Figure 8-11.

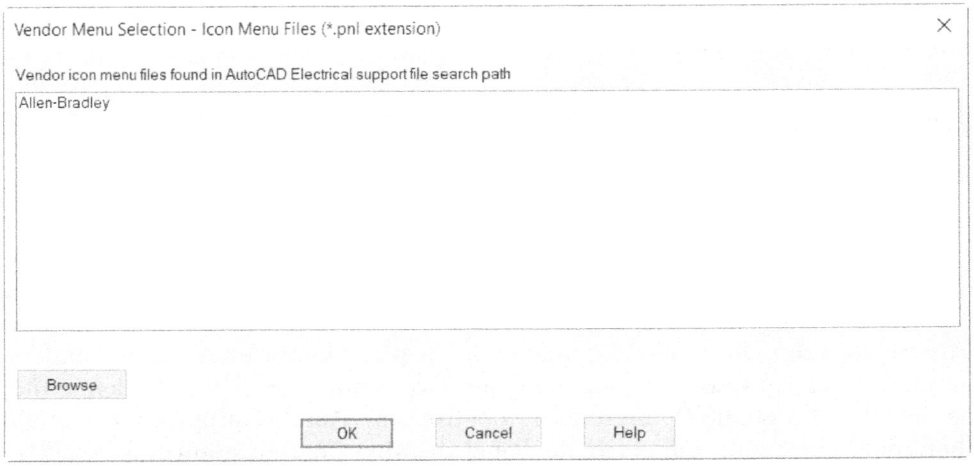

*Figure 8-11 The **Vendor Menu Selection - Icon Menu Files (*.pnl extension)** dialog box*

In this dialog box, select the required vendor icon menu file and choose the **OK** button; the **Vendor Panel Footprint** dialog box will be displayed. The options in this dialog box, except the **Vendor Menu Select...** button, are similar to the options in the **Insert Footprint** dialog box.

The **Vendor Menu Select...** button is used to select the vendor again from the **Vendor Menu Selection - Icon Menu Files (*.pnl extension)** dialog box.

Note

*If you choose the **Manufacturer Menu** tool the second time, the **Vendor Panel Footprint** dialog box will be displayed directly. This will happen only for the current session of AutoCAD Electrical. However, if you run the software again and if you select the same option, it will first display the **Vendor Menu Selection - Icon Menu Files (*.pnl extension)** dialog box and then the **Vendor Panel Footprint** dialog box.*

Next, select the required footprint icon from the **Vendor Panel Footprint** dialog box or select the name of the footprint displayed on the left of the dialog box; various types of selected icons will be displayed. Next, select the required footprint block; the footprint block along with the cursor will be displayed and you will be prompted to select the location. Next, select the location; the **Panel Layout - Component Insert/Edit** dialog box will be displayed. Enter the required information in this dialog box and choose the **OK** button; the footprint will be inserted into the drawing.

COPYING A FOOTPRINT

Command: AECOPYFOOTPRINT

The **Copy Footprint** tool is used to copy a selected footprint in an active drawing. This tool copies a balloon and a nameplate associated with the selected footprint. To copy a selected footprint, choose the **Copy Footprint** tool from the **Edit Footprints** panel of the **Panel** tab; you will be prompted to select the component. Next, select the component and specify the location and press ENTER; the **Panel Layout - Component Insert/Edit** dialog box will be displayed. Enter the required information in this dialog box and choose the **OK** button; the copied footprint will be inserted into the drawing. Note that you can also use the **COPY** command of AutoCAD for copying a footprint but the balloon and nameplate associated with the footprint will not be copied.

SETTING THE PANEL DRAWING CONFIGURATION

Command: AEPANELCONFIG

The configuration settings of a panel drawing are saved as attribute values on *WD_PNLM* invisible block. This invisible block should be present in every panel drawing. You have already learnt about WD_PNLM invisible block in this chapter. The **Configuration** tool is used to set properties of a panel drawing in addition to the settings specified in the **Drawing Properties** dialog box. This tool is also used to define panel footprint drawing defaults such as balloon setup, footprint insertion scale, footprint layer setup, and so on. To define the panel drawing defaults, choose the **Configuration** tool from the **Panel Configuration** drop-down in the **Other Tools** panel of the **Panel** tab; the **Panel Drawing Configuration and Defaults** dialog box will be displayed.

Using this dialog box, you can define the starting item number, balloon configuration, footprint layers, default settings for panel drawing functions, and so on.

MAKING THE XDATA VISIBLE

Command: AESHOWXDATA

Xdata is also called extended entity data. For some functions, AutoCAD Electrical adds invisible data to a footprint block or even to specific attributes. This invisible data is called Xdata. The **Make Xdata Visible** tool is used for converting invisible Xdata into visible attribute, which is attached to a footprint block. To do so, choose the **Make Xdata Visible** tool from the **Panel Configuration** drop-down in the **Other Tools** panel of the **Panel** tab; you will be prompted to select a footprint. Next, select the required footprint; the **Select Xdata to Change to a Block Attribute** dialog box will be displayed, as shown in Figure 8-12.

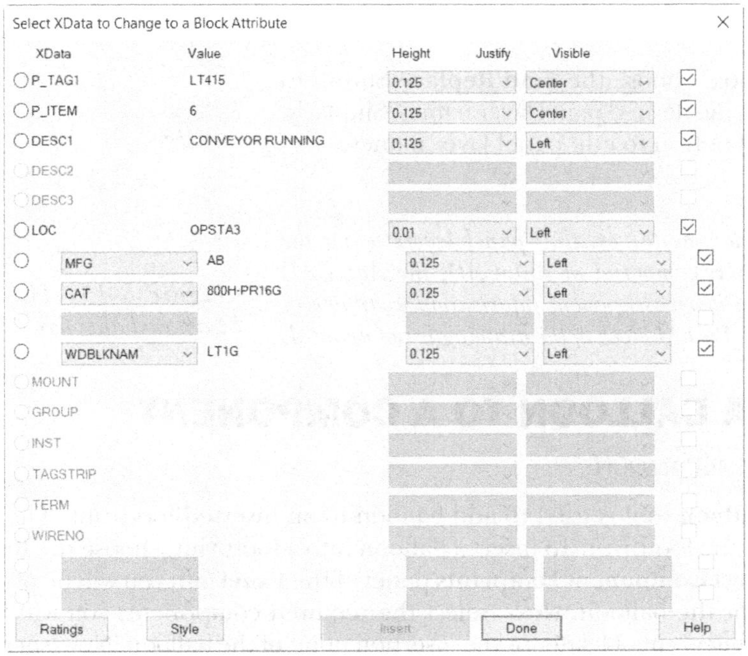

***Figure 8-12** The **Select Xdata to Change to a Block Attribute** dialog box*

Select Xdata such as P_ITEM, MFG, CAT, and so on from this dialog box and choose the **Insert** button; the **Select XData to Change to a Block Attribute** dialog box will disappear and you will be prompted to specify the location for attribute. Next, specify the location; the **Select XData to Change to a Block Attribute** dialog box will be displayed again. Repeat the above process for making the rest of XData visible on the footprint block.

The **Insert** button will be activated only if you select the XData attribute value under the **XData** column. Choose the **Insert** button; you will be prompted to specify the location for the selected XData attribute value. Next, specify the location; the XData attribute value will be inserted into the footprint block, which has been discussed earlier. Choose the **Done** button after finishing the insertion of XData attribute values into the footprint block.

RENAMING PANEL LAYERS

Command: AERENAMEPANLELLAYER

You have already learned about defining footprint layer setup for panel drawings using the **Configuration** tool in the last section. You can rename existing panel layers in panel drawings using the **Rename Layers** tool. To do so, choose the **Rename Layers** tool from the **Panel Configuration** drop-down in the **Other Tools** panel of the **Panel** tab; the **Rename Panel Layers** dialog box will be displayed, as shown in Figure 8-13.

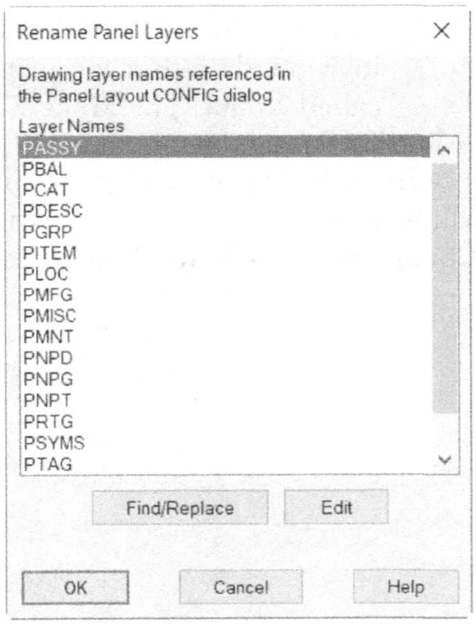

***Figure 8-13** The **Rename Panel Layers** dialog box*

In this dialog box, choose the **Find/Replace** button to find and replace the desired panel layer names. Similarly, choose the **Edit** button to edit panel layer names.

Note

*If you rename the existing panel layers using the **RENAME** command of AutoCAD, the AutoCAD electrical layer assignment information available in the WD_PNLM block of the drawing is not updated.*

ADDING A BALLOON TO A COMPONENT

Command: AEBALLOON

The **Balloon** tool is used to add balloon to an inserted footprint. The balloon is used to label the footprint. To insert a balloon into a footprint, choose the **Balloon** tool from the **Insert Component Footprints** panel of the **Panel** tab; you will be prompted to select a component for the balloon. Next, select the required component; you will be prompted to specify the starting point of leader or the insertion point of the balloon. Now, specify the insertion point of the balloon or starting point of the leader. Next, press ENTER if you want to insert a balloon without leader. Else, specify the endpoint of leader and then press ENTER; the balloon will be inserted into the footprint. Press ENTER to exit the command. Figure 8-14 shows the footprint with leader and balloon. The balloon displays the item number that you have assigned in the **Item Number** area of the **Panel Layout - Component Insert/Edit** dialog box. If you do not want to insert a leader along with the balloon, select the component and specify the start point of leader or insertion point of balloon and then press ENTER; the balloon will be inserted at the selected point. Figure 8-15 shows the footprint without leader and with balloon. Note that if you change the item number that is assigned to a footprint, balloon labels will be updated automatically.

If you add a balloon to a footprint that does not have an item number assigned to it, then on choosing the **Balloon** tool from the **Insert Component Footprints** panel of the **Panel** tab, you will be prompted to select a component. Select the component; you will be prompted to specify the starting point of a leader or insertion point of a balloon. Specify the starting point of a balloon or a leader; you will be prompted to specify the end point of the leader. Specify the endpoint of the leader and then press ENTER; the **No Item Number Match for this Catalog Part Number** dialog box will be displayed.

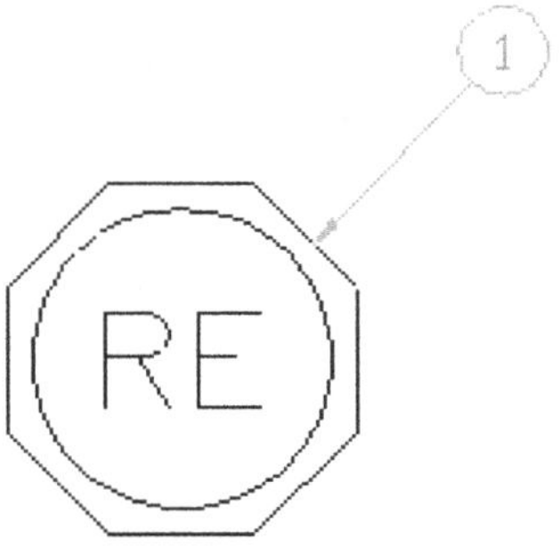

Figure 8-14 *Footprint with leader and balloon*

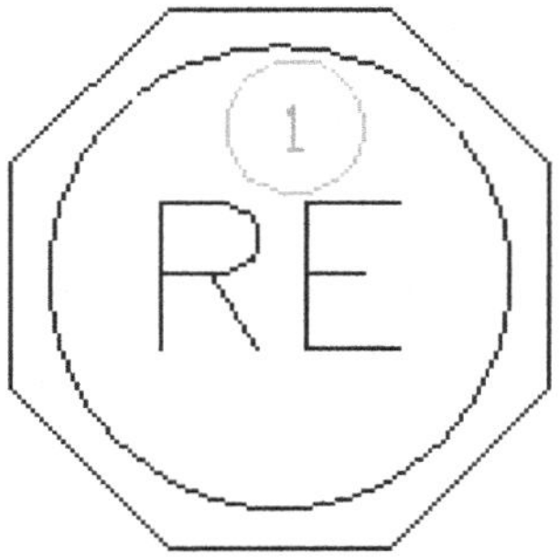

Figure 8-15 *Footprint without leader and with balloon*

In this dialog box, you can enter an item number for footprint in the **Item:** edit box. The **List** button is used to display the item numbers that have been used in the current drawing. To do so, choose the **List** button that is adjacent to **Drawing**; the **ITEM numbers in use : Drawing only** dialog box will be displayed. The list displayed in this dialog box is for reference only. You can also check for the Bill of Material by choosing the **Catalog Check** button from the **ITEM numbers in use : Drawing only** dialog box.

Now, if you choose the **List** button that is adjacent to **Project**, the **ITEM numbers in use Project-wide** dialog box will be displayed. This dialog box displays a list of item numbers assigned to the footprints in the current project. This list is for reference only. Next, choose the **Catalog Check** button from this dialog box; the Bill of Material of the selected component will be displayed in the **Bill Of Material Check** dialog box.

Note

*The **Catalog Check** button will be available only if a component has manufacturer and catalog information.*

The **Use Next >>** button is used to assign next available item number to a component.

After specifying the required options in the **No Item Number Match for this Catalog Part Number** dialog box, choose the **OK** button from this dialog box; the balloon is inserted into the footprint.

Note

*If you do not enter any item number in the **Item:** edit box and choose **OK** from the **No Item Number Match for this Catalog Part Number** dialog box, a balloon with '?' will be inserted into the footprint.*

*If you have already inserted a balloon into a footprint and chosen the **Balloon** tool from the **Insert Component Footprints** panel of the **Panel** tab, you will be prompted to select a component. Select the component; the **Existing Item Balloon** message box will be displayed. Choose the **OK** button from this message box to delete the existing balloon and to insert a new balloon into the footprint.*

ADDING MULTIPLE BALLOONS

You can also insert multiple balloons for a footprint. To do so, follow the steps given next:

(a) Right-click on the active project; a shortcut menu will be displayed. Choose the **Properties** option from the shortcut menu; the **Project Properties** dialog box will be displayed.

(b) Choose the **Components** tab and then choose the **Item Numbering** button from the **Component Options** area of this dialog box; the **Item Numbering Setup** dialog box will be displayed.

(c) In this dialog box, by default, the **Accumulate Project Wide** radio button is selected in the **Item Numbering Mode** area. Next, select the **Per-Part Number Basis (excluding ASSYCODE Combination)** radio button from the **Item Assignments** area and then choose the **OK** button from this dialog box. Next, choose **OK** in the **Project Properties** dialog box.

(d) Choose the **Edit** tool from the **Edit Footprints** panel of the **Panel** tab or choose **Panel Layout > Edit Footprint** from the menu bar; you will be prompted to select a component. Select the required component from the panel drawing; the **Panel Layout - Component Insert/Edit** dialog box will be displayed.

(e) Choose the **Multiple Catalog** button from the **Catalog Data** area; the **Multiple Bill of Material Information** dialog box will be displayed.

(f) In the **Multiple Bill of Material Information** dialog box, select a number from the **Sequential Code** drop-down list and then choose the **Catalog Lookup** button from it; the **Catalog Browser** dialog box will be displayed. Next, choose the desired part number from it; the manufacturer and catalog values will be displayed in the **Manufacturer** and **Catalog** edit boxes, respectively. Repeat this process for all part numbers assigned to a footprint. Note that you can assign 99 extra part numbers to a footprint.

(g) Enter the required item number in the **Item Number** edit box and then choose the **OK** button in the **Multiple Bill of Material Information** dialog box; the number of extra part numbers specified in the **Multiple Bill of Material Information** dialog box will be displayed on the right of the **Multiple Catalog** button in the **Panel Layout - Component Insert/Edit** dialog box. Next, choose the **OK** button from the **Panel Layout - Component Insert/Edit** dialog box.

(h) Choose the **Balloon** tool from the **Insert Component Footprints** panel of the **Panel** tab; you will be prompted to select a component to which you want to add the balloon. Next, select the component; you will be prompted to specify the starting point of a leader or insertion point of the balloon. Specify the starting point of the balloon or the starting point of the leader; you will be prompted to specify the end point of the leader. Specify the endpoint of the leader.

(i) Enter **D** at the Command prompt; you will be prompted to specify the direction of balloons. Enter a character for the direction and press ENTER; multiple balloons will be inserted on the footprint in the specified direction.

Note that if you do not specify the direction of balloons and press ENTER, the direction that was specified earlier (given in brackets in the Command prompt) will be set automatically.

RESEQUENCING ITEM NUMBERS

Command: AERESEQUENCE

You can resequence item numbers of the panel and schematic components using the **Resequence Item Numbers** tool. The order of resequencing the item numbers is: panel components, schematic components with panel representation, and then schematic components with no panel representation. Note that the item number allotted to the schematic component with panel representation is same as the related panel component.

When you resequence the item numbers, the manufacturers across the active project get sorted and consequently the item numbers of the components with the same catalog numbers are synchronized in the project.

To resequence the item numbers, choose the **Resequence Item Numbers** tool from the **Edit Footprints** panel of the **Panel** tab; the **Resequence Item Numbers** dialog box will be displayed, as shown in Figure 8-16.

After specifying the required options, choose the **OK** button in the **Resequence Item Numbers** dialog box to resequence item numbers as per the options specified in this dialog box.

Note

To revert to the item number resequencing mode used in AutoCAD Electrical 2016 or earlier version, enter the system variable AEITEMRESEQUENCEMODE in the Command Prompt and then enter 0.

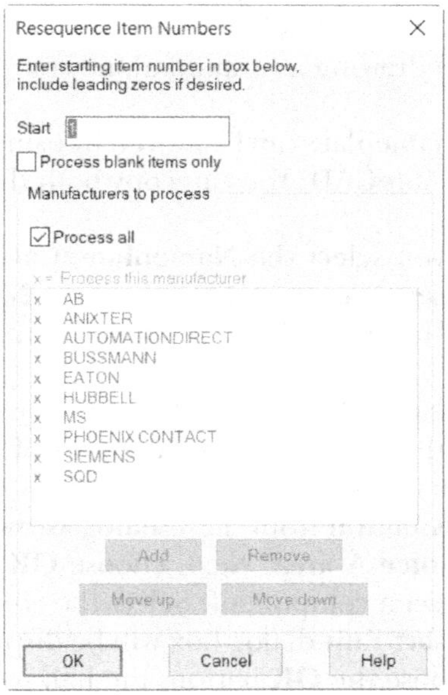

Figure 8-16 The ***Resequence Item Numbers*** *dialog box*

INSERTING NAMEPLATES

Nameplates are special type of panel footprints that are inserted into a drawing as blocks. You can link nameplates to panel footprint components or can insert them as stand-alone components. A nameplate is child of a parent footprint component. The parent-child relationship between them is established by using invisible Xdata pointers. When a nameplate is linked to a parent footprint, it will extract description, location, and other information from the associated panel component. To insert a nameplate into a drawing, choose the **Icon Menu** tool from the **Insert Component Footprints** panel of the **Panel** tab; the **Insert Footprint** dialog box will be displayed. Select the **Nameplates** icon from the **Panel Layout Symbols** area of the **Insert Footprint** dialog box; the **Panel: Nameplates** area with various types of nameplates will be displayed.

Next, select the required nameplate from the **Panel: Nameplates** area; you will be prompted to select a component for inserting the nameplate. Select the component and press ENTER; the

Panel Layout - Nameplate Insert/Edit dialog box will be displayed. The options in this dialog box are similar to that of the **Panel Layout - Component Insert/Edit** dialog box. The only difference is that in the **Panel Layout - Nameplate Insert/Edit** dialog box, you need to enter data for nameplates, whereas in the **Panel Layout - Component Insert/Edit** dialog box, you need to enter data for footprint blocks. Next, enter the required information in the **Panel Layout - Nameplate Insert/Edit** dialog box and choose the **OK** button; a nameplate will be inserted into the selected component and it will be linked to the component, as shown in Figure 8-17.

You can insert a nameplate as a stand-alone component. In this case, you need to press ENTER without selecting the component. Specify the insertion point for the nameplate; you will be prompted to select the rotation. Select the rotation by moving the cursor in the horizontal or vertical direction and press ENTER; the **Panel Layout - Nameplate Insert/Edit** dialog box will be displayed. Enter the required information in this dialog box and choose the **OK** button; the nameplate will be inserted into the drawing as a stand-alone component.

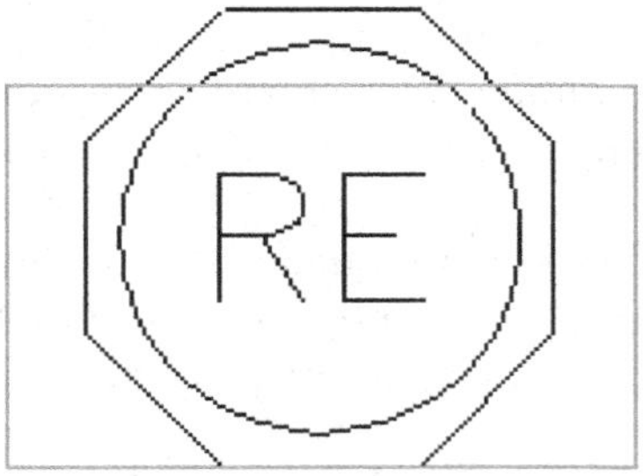

Figure 8-17 *The footprint with the nameplate inserted*

A nameplate can be moved by using the **MOVE** command of AutoCAD. You can copy both the footprint and nameplate using the **Copy Footprint** tool.

If you select the **Nameplate, Catalog Lookup** icon from the **Panel: Nameplates** area of the **Insert Footprint** dialog box; the **Nameplate** dialog box will be displayed. The options and areas in this dialog box are similar to that of the **Footprint** dialog box

You can also insert half round nameplate into a footprint. To do so, choose the **Catalog lookup** button from the **Choice A** area of the **Nameplate** dialog box; the **Catalog Browser** dialog box will be displayed. Select any Half Round part number from the **TYPE** column and choose the **OK** button from the **Catalog Browser** dialog box; part number values will be displayed in the **Choice A** area. Next, choose **OK** from the **Nameplate** dialog box; you will be prompted to select a component. Select the component and press ENTER; the **Panel Layout - Nameplate Insert/Edit** dialog box will be displayed. Enter the required information in this dialog box and choose the **OK** button; the half round nameplate will be inserted into the footprint, as shown in Figure 8-18.

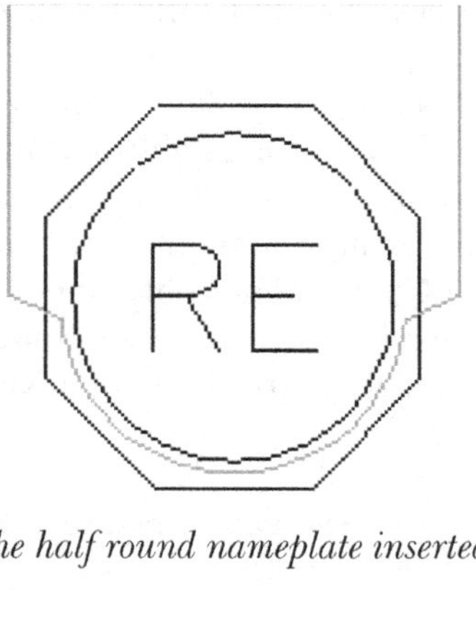

Figure 8-18 *The half round nameplate inserted into the footprint*

INSERTING DIN RAIL

The DIN Rail is a standardized 35 mm wide metal track attached to the back of an electrical panel where devices can easily be clipped or removed. It is also known as top-hat rail because it has a hat-shaped cross-section. It is widely used for mounting circuit breakers and industrial control equipment inside the equipment racks. In a DIN Rail, it is very easy to snap-on and remove hardware devices while installing, maintaining and replacing them. In addition to the popular 35 mm top-hat Rail (EN 50022, BS 5584), several other mounting rails have been standardized such as 15 mm wide top-hat Rail (EN 50045, BS 6273), 75 mm wide top-hat Rail (EN 50023, BS 5585), and G-type Rail (EN 50035, BS 5825). In this section, you will learn how the DIN Rail utility is used to create wire duct and DIN Rail objects in the panel layouts. Wire ducts and DIN Rails are frequently drawn as collection of parallel lines and equally spaced mounting holes. The DIN Rail utility simplifies the redundant tasks of creating these objects and groups them together as a single block for annotation purposes.

To insert a DIN Rail, choose the **Icon Menu** tool from the **Insert Component Footprints** panel of the **Panel** tab; the **Insert Footprint** dialog box will be displayed. Select the **DIN Rail** icon from the **Panel Layout Symbols** area of this dialog box; the **Din Rail** dialog box will be displayed, as shown in Figure 8-19.

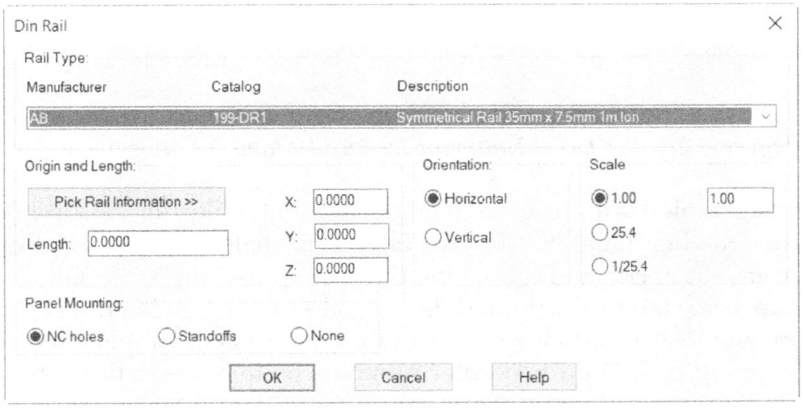

***Figure 8-19** The **Din Rail** dialog box*

After specifying the required options in the **Din Rail** dialog box, choose the **OK** button; the **Panel Layout - Component Insert/Edit** dialog box will be displayed. Enter the required information in this dialog box and choose the **OK** button; the DIN Rail will be inserted into the drawing.

Note

*1. If the **Length**, **X**, **Y**, and **Z** edit boxes are left blank, and you choose the **OK** button, the **AutoCAD Message** message box will be displayed. This message box informs you that more information is needed. Choose the **OK** button in this message box; the **Din Rail** dialog box will be displayed again.*

*2. If the length specified in the **Length** edit box for the DIN Rail is too short, and you choose the **OK** button, the **AutoCAD Message** message box will be displayed. This message box will inform you that the specified length is too short. Choose the **OK** button; the **Panel Layout - Component Insert/Edit** dialog box will be displayed. Enter the required information in this dialog box and choose the **OK** button; the DIN Rail will be inserted in the drawing. This condition is true for AB, AD, and Newark manufacturers and is not applicable to the PANDUIT manufacturer.*

EDITING THE PANEL FOOTPRINT LOOKUP DATABASE FILE

Command: AEFOOTPRINTDB

The **Footprint Database Editor** tool is used to create a new footprint lookup table and to edit an existing footprint lookup table. The default location for the footprint lookup database file is *C:\Users\User Name\Documents\Acade 2022\AeData\en-US\Catalogs*. The name of the footprint database file is *footprint_lookup.mdb*. The footprint database file contains table for each manufacturer code and the footprint lookup table name must match the manufacturer code. To create or edit a footprint database file, choose the **Footprint Database Editor** tool from the **Other Tools** panel of the **Panel** tab; the **Panel Footprint Lookup Database Editor** dialog box will be displayed, as shown in Figure 8-20.

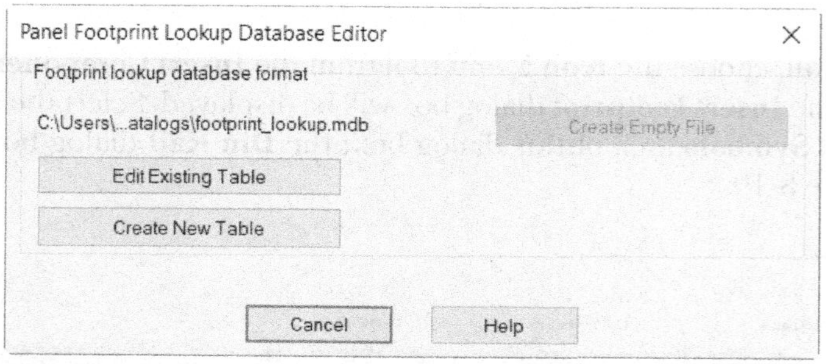

***Figure 8-20** The **Panel Footprint Lookup Database Editor** dialog box*

The **Edit Existing Table** button is used to edit an existing footprint lookup table. To do so, choose the **Edit Existing Table** button; the **Table Edit** dialog box will be displayed. Select the table that you want to edit and choose the **OK** button from the **Table Edit** dialog box; the **Footprint lookup** dialog box will be displayed. Next, select a part number from the list displayed; the **Edit Record** and **Delete** buttons will be activated. You can edit or add a record as per your requirement by choosing the **Edit Record** or **Add New** buttons, respectively. Next, choose the **OK/Save/Exit** button to save the record and to exit the **Footprint lookup** dialog box.

The **Create New Table** button is used to create a new manufacturer footprint lookup table. To do so, choose the **Create New Table** button; the **Enter New Table Name to Create** dialog box will be displayed.

In this dialog box, enter the table name in the **Table** edit box. Note that the table name should match the manufacturer code in the catalog lookup file. Next, choose the **OK** button from this dialog box; the **Footprint lookup** dialog box will be displayed. Choose the **Add New** button to add a new record; the **Add footprint record** dialog box will be displayed. The options in the **Add footprint record** dialog box have been discussed earlier. Enter the required information in the **Add footprint record** dialog box and choose the **OK** button; the details of this new footprint record will be displayed in the **Footprint lookup** dialog box. Now, you can edit this record by using different options in the **Footprint lookup** dialog box. Choose the **OK/Save/Exit** button to save the record and to exit this dialog box.

The **Create Empty File** button will be activated only if the default *Footprint_lookup.mdb* file does not exist in the assigned location. This button is used to create a blank footprint lookup file.

TUTORIALS

Tutorial 1

In this tutorial, you will extract schematic component list from the **CADCIM** project using the **Schematic List** tool. Next, you will insert components from the list in the panel drawings as footprints and then add nameplates to footprints. **(Expected time: 20 min)**

The following steps are required to complete this tutorial:

a. Create a new drawing.
b. Insert the footprint into the drawing using the **Schematic List** tool.
c. Insert nameplates.

Creating a New Drawing

1. Activate the **CADCIM** project, if it is not already activated. Click on the *TUTORIALS* subfolder in it.

2. Choose the **New Drawing** button from the **PROJECT MANAGER**; the **Create New Drawing** dialog box is displayed. Enter **C08_tut01** in the **Name** edit box of the **Drawing File** area. Select **ACAD_ELECTRICAL.dwt** as the template and enter **Panel Components** in the **Description 1** edit box.

3. Choose the **OK** button in the **Create New Drawing** dialog box; the *C08_tut01.dwg* is created in the *TUTORIALS* subfolder of the **CADCIM** project and displayed at the bottom of the drawing list in this subfolder.

Inserting Footprints

1. Choose the **Schematic List** tool from **Panel > Insert Component Footprints > Insert Footprints** drop-down; the **Alert** message box is displayed. Choose the **OK** button in this message box; a non-visible block is inserted into the drawing and the **Schematic Components List --> Panel Layout Insert** dialog box is displayed.

2. Select the **Project** radio button in the **Extract component list for:** area, if it is not already selected.

3. Select the **All** radio button in the **Location Codes to extract** area, if it is not already selected.

4. Choose the **OK** button from the **Schematic Components List --> Panel Layout Insert** dialog box; the **Select Drawings to Process** dialog box is displayed.

5. Choose the **Do All** button from the **Select Drawings to Process** dialog box; the drawings in the top list are transferred to the bottom list.

6. Choose the **OK** button from this dialog box; the **Schematic Components (active project)** dialog box is displayed.

7. Choose the **Sort List** button from the **Schematic Components (active project)** dialog box; the **Sort** dialog box is displayed. Select the **TAGNAME** option from the **Primary sort** drop-down list and choose the **OK** button; the list of schematic components is sorted out according to the tag name.

8. Scroll down the list displayed in the **Schematic Components (active project)** dialog box and search for **PB414A**, **PB422A**, and **PB425A**.

9. Press CTRL and click on rows one by one to select **PB414A**, **PB422A**, and **PB425A**.

10. Enter **0** in the **Rotate (blank = "ask")** edit box.

11. Choose the **Insert** button from the **Schematic Components (active project)** dialog box; the **Spacing for Footprint Insertion** dialog box is displayed.

12. Select the **Use uniform spacing** radio button; the **X-Distance** and **Y-Distance** edit boxes are activated.

13. Enter **0** in the **X-Distance** edit box, if it is not already displayed.

14. Enter **-3.5** in the **Y-Distance** edit box and select the **Suppress edit dialog and prompts** check box.

15. By default, **PB414A** is selected in the **Insert Order** area. Choose the **OK** button; you are prompted to specify the location for **PB414A**.

16. Enter **5,17** at the Command prompt and press ENTER; the footprints are inserted into the drawing and the **Schematic Components (active project)** dialog box appears again.

17. Choose the **Close** button from the **Schematic Components (active project)** dialog box to exit this dialog box. If the **Update other drawings** message box is displayed, choose the **Skip** button in this message box. Also, notice that the footprints of **PB414A**, **PB422A**, and **PB425A** are inserted into the drawing, as shown in Figure 8-21.

    Next, you need to move the attributes of following footprints: **PB414A**, **PB422A**, and **PB425A**.

18. Make sure the **Snap Mode** button is not chosen. Choose the **Move/Show Attribute** tool from **Schematic > Edit Components > Modify Attributes** drop-down; you are prompted to select an attribute to move. Next, select **GN** of **PB414A** and press ENTER; you are prompted to specify a base point.

19. Click on the screen near the **PB414A** component and move the cursor upward and then place the attribute **GN** on the top of **PB414A**, as shown in Figure 8-22.

Figure 8-21 *Footprints inserted into the drawing*

Figure 8-22 *Footprints after moving the attributes*

20. Similarly, move the attributes of the rest of footprints, refer to Figure 8-22. Press ENTER to exit the command.

Inserting Nameplates

1. Choose the **Icon Menu** tool from **Panel > Insert Component Footprints > Insert Footprints** drop-down; the **Insert Footprint** dialog box is displayed. 

2. Select the **Nameplates** icon from the **Panel Layout Symbols** area of the **Insert Footprint** dialog box; the **Panel: Nameplates** area is displayed.

3. Select the **Nameplate, Catalog Lookup** icon from the **Panel: Nameplates** area; the **Nameplate** dialog box is displayed.

4. Choose the **Catalog lookup** button from the **Choice A** area; the **Catalog Browser** dialog box is displayed.

5. Select the **800H-W100A** from the **CATALOG** column of the **Catalog Browser** dialog box, as shown in Figure 8-23.

6. Choose **OK** from the **Catalog Browser** dialog box; the **Nameplate** dialog box appears again. Also, the manufacturer AB is displayed in the **Manufacturer** edit box and the 800H-W100A catalog is displayed in the **Catalog** edit box.

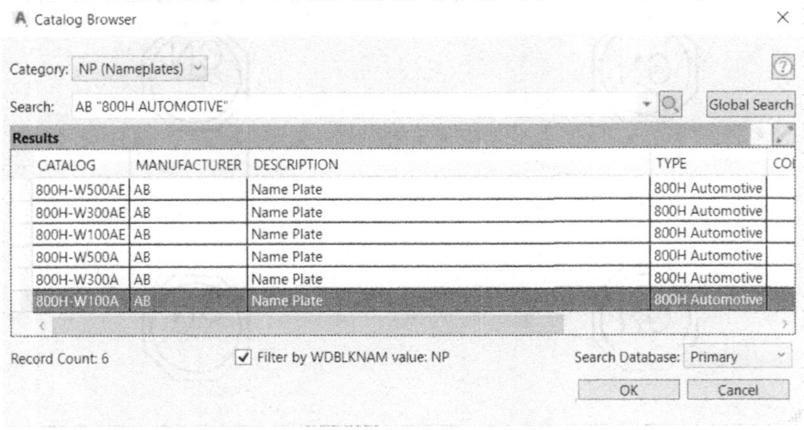

***Figure 8-23** Selecting **800H-W100A** from the **Catalog Browser** dialog box*

7. Choose **OK** from the **Nameplate** dialog box; you are prompted to select footprints. Select the footprints **PB414A**, **PB422A**, and **PB425A** from the drawing and press ENTER; the **Panel Layout - Nameplate Insert/Edit** dialog box is displayed. The description of the nameplate is displayed in the **Nameplate Description** area.

8. Choose **OK** from the **Panel Layout - Nameplate Insert/Edit** dialog box; the **Panel Layout - Nameplate Insert/Edit** dialog box is displayed again. Keep all the values intact and choose **OK**; the **Panel Layout - Nameplate Insert/Edit** dialog box is displayed again.

9. Choose **OK** from the **Panel Layout - Nameplate Insert/Edit** dialog box; nameplates are inserted into footprints, as shown in Figure 8-24.

***Figure 8-24** Nameplates inserted in footprints*

Saving the Drawing File

1. Choose **Save** from the **Application Menu** to save the drawing file.

Tutorial 2

In this tutorial, you will edit the footprints created in Tutorial 1 of this chapter and assign item numbers to them. Next, you will add balloons to footprints. **(Expected time: 15 min)**

The following steps are required to complete this tutorial:

a. Open, save, and add the drawing in the **CADCIM** project.
b. Edit the footprints and assign item numbers.
c. Insert balloons.
d. Save the drawing file.

Opening, Saving, and Adding the Drawing in the CADCIM Project

1. Open the *C08_tut01.dwg* drawing file. Save it with the name *C08_tut02.dwg*. Add it to the **CADCIM** project list, as discussed in the earlier chapters.

Editing Footprints and Assigning Item Numbers

1. Choose the **Edit** tool from the **Edit Footprints** panel of the **Panel** tab; you are prompted to select the component.

2. Select **PB414A**; the **Panel Layout - Component Insert/Edit** dialog box is displayed.

3. Enter **51** in the **Item Number** edit box of the **Item Number** area and choose the **OK** button; the **Update Related Components?** message box is displayed. Choose the **Yes-Update** button; the **Update other drawings?** message box is displayed. Choose **OK**; the **QSAVE** message box is displayed. Choose **OK**; the item number is assigned to the footprint.

Note
If the ***Mismatch Item Number Found*** *message box is displayed, choose the* ***OK*** *button.*

4. Similarly, assign 52 and **53** as item number to **PB422A** and **PB425A**, respectively.

Inserting Balloons

1. Choose the **Balloon** tool from the **Insert Component Footprints** panel of the **Panel** tab; you are prompted to select the component for balloon.

2. Select **PB414A**; you are prompted to specify the start point of the leader or the balloon insert point.

3. Select the starting point of the leader for PB414A, as shown in Figure 8-25. Note that you can select any point at the edge of a footprint.

4. Make sure the **Ortho Mode** button is not chosen. Move the cursor toward right and click at the position shown in Figure 8-26, and then press ENTER; a balloon is inserted in PB414 and you are prompted to select a component for balloon.

Figure 8-25 *Starting point of leader*

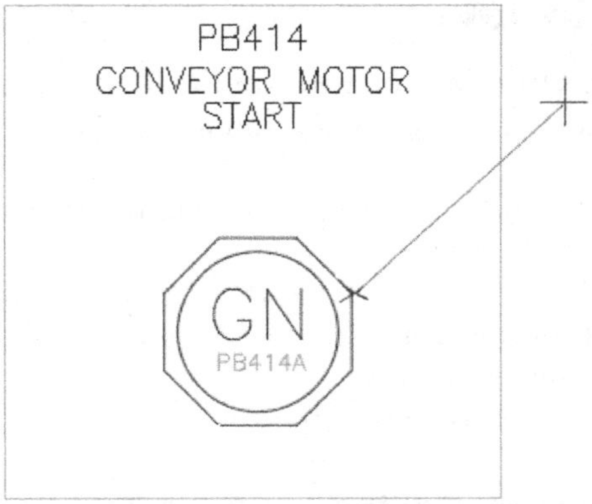

Figure 8-26 *Distance between the start and end points of the leader*

5. Select **PB422A** and repeat the above procedure for inserting balloon into it.

6. Select **PB425A** and repeat the above procedure for inserting balloon into it; the **Update Related Components?** message box is displayed. Choose the **Yes-Update** button; the **Update other drawings?** message box is displayed. Choose **OK**; the **QSAVE** message box is displayed. Choose **OK**; a balloon is inserted in PB425.

7. Press ESC or ENTER to exit the command. Figure 8-27 shows footprints with the balloons inserted.

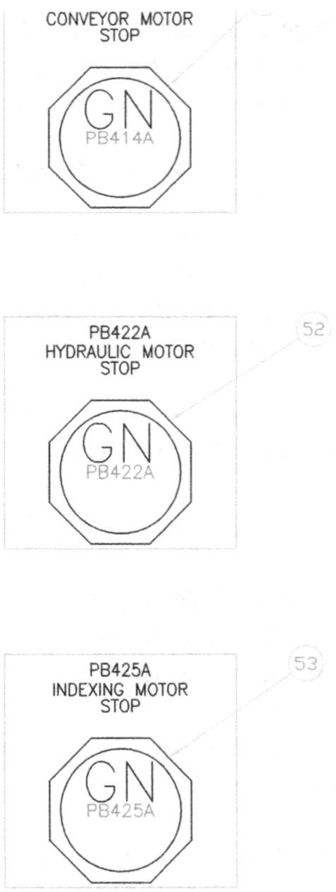

Figure 8-27 *Footprints with the balloons inserted*

Saving the Drawing File

1. Choose **Save** from the **Application Menu** to save the drawing file.

Tutorial 3

In this tutorial, you will insert a footprint manually in the drawing file, as shown in Figure 8-28, using the **Manual** tool. Also, you will make the Xdata of the footprint visible by using the **Make Xdata Visible** tool. **(Expected time: 10 min)**

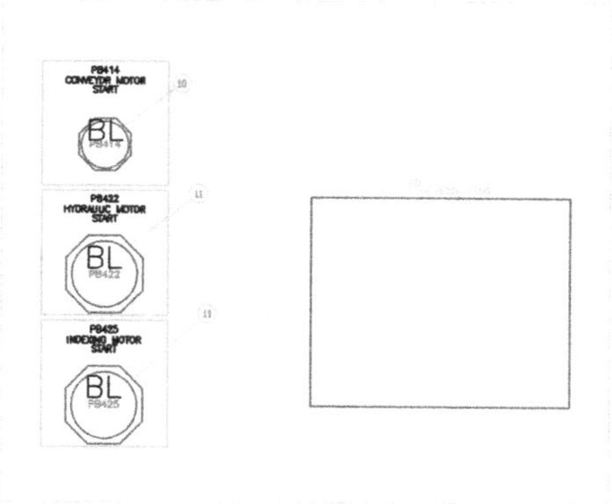

Figure 8-28 *The footprint inserted manually*

The following steps are required to complete this tutorial:

a. Open, save, and add the drawing in the **CADCIM** project.
b. Insert the footprint manually using the **Manual** tool.
c. Make the Xdata of the footprint visible.
d. Save the drawing file.

Opening, Saving, and Adding the Drawing in the CADCIM Project

1. Open the *C08_tut02.dwg* drawing file. Save it with the name *C08_tut03.dwg*. Add it to the **CADCIM** project list, as discussed in the earlier chapters.

Inserting the Footprint Manually

1. Choose the **Manual** tool from **Panel > Insert Component Footprints > Insert Footprints** drop-down; the **Insert Component Footprint -- Manual** dialog box is displayed.

2. Select the Rectangle symbol from **Draw shapes**; you are prompted to specify the first corner.

3. Next, enter **9,16** at the Command prompt and press ENTER; you are prompted to specify the opposite corner.

4. Enter **14,12** at the Command prompt and press ENTER; the **Panel Layout - Component Insert/Edit** dialog box is displayed.

5. In the **Catalog Data** area of this dialog box, enter **AB** in the **Manufacturer** edit box and **174E-E25-1756** in the **Catalog** edit box.

6. Choose **OK** from this dialog box; a footprint is inserted into the panel drawing, as shown in Figure 8-29.

Making Xdata of the Footprint Visible

1. Choose the **Make Xdata Visible** tool from **Panel > Other Tools > Panel Configuration** drop-down; you are prompted to select a footprint.

2. Select the footprint that you created manually; the **Select XData to Change to a Block Attribute** dialog box is displayed.

3. Select the **MFG** radio button and choose the **Insert** button; you are prompted to specify the location for attribute (MFG).

4. Place the MFG attribute at the top of the footprint; the **Select XData to Change to a Block Attribute** dialog box is displayed again. Now, select the **CAT** radio button and choose the **Insert** button; you are prompted to specify the location for attribute (CAT).

5. Place the CAT attribute, as shown in Figure 8-30.

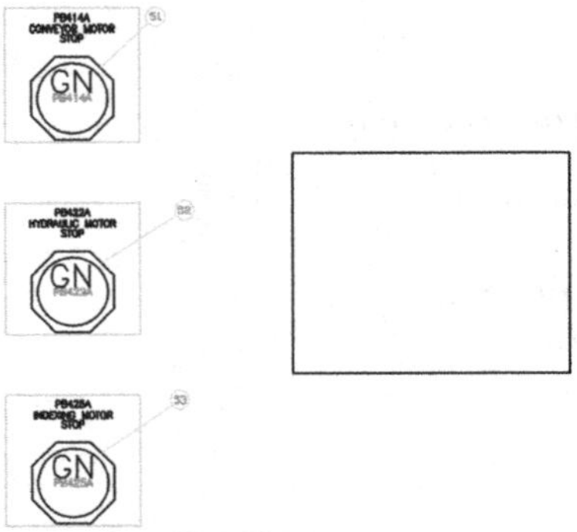

Figure 8-29 *The footprint inserted manually*

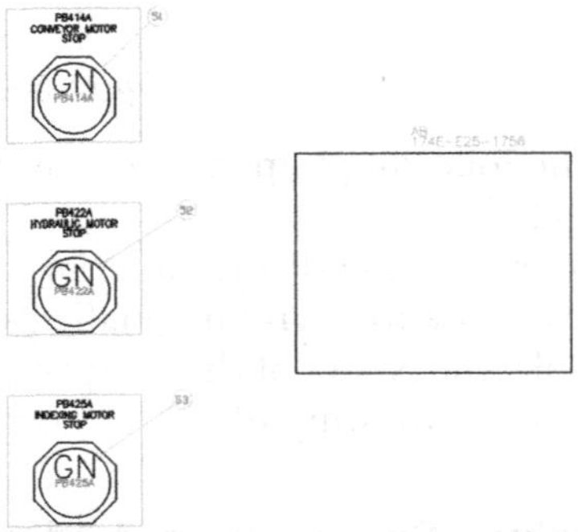

Figure 8-30 *Footprint with visible Xdata*

6. Choose the **Done** button from the **Select XData to Change to a Block Attribute** dialog box to exit the dialog box.

Saving the Drawing File

1. Choose **Save** from the **Application Menu** to save the drawing file.

Tutorial 4

In this tutorial, you will create a drawing and insert footprint in it using the **User Defined List** tool. Also, you will add a new library to the **Panel Footprint Libraries** library.

(Expected time: 15 min)

The following steps are required to complete this tutorial:

a. Create a new drawing.
b. Insert the footprint using the **User Defined List** tool.
c. Save the drawing file.

Creating a New Drawing

1. Activate the **CADCIM** project. Next, click on the *TUTORIALS* subfolder in it.

2. Choose the **New Drawing** button from the **PROJECT MANAGER**; the **Create New Drawing** dialog box is displayed. Enter **C08_tut04** in the **Name** edit box of the **Drawing File** area. Select the template as **ACAD_ELECTRICAL.dwt** and enter **Panel Components** in the **Description 1** edit box.

3. Choose the **OK** button; the *C08_tut04.dwg* file is created in the *TUTORIALS* subfolder of the **CADCIM** project and displayed at the bottom of the drawing list in the **CADCIM** project.

Inserting Footprints Using the User Defined List and Adding Library to the Panel Footprint Libraries

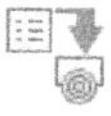

1. Choose the **User Defined List** tool from the **Panel > Insert Component Footprints > Insert Footprints** drop-down; the **Panel footprint: Select and Insert by Catalog or Description Pick** dialog box is displayed.

2. Select **800T-A2A** catalog from this dialog box; the buttons in this dialog box are activated.

3. Choose **OK** from the **Panel footprint: Select and Insert by Catalog or Description Pick** dialog box; the **AutoCAD Message** message box is displayed, as shown in Figure 8-31. Choose the **OK** button from this message box; the command ends and you cannot insert the footprint from the user defined list.

   To overcome this problem, you need to add a library path in the **Panel Footprint Libraries** library. The procedure to add a new path to libraries is discussed next.

4. Right-click on the **CADCIM** project in the **Projects** rollout of the **PROJECT MANAGER**; a shortcut menu is displayed. Next, choose the **Properties** option from the shortcut menu; the **Project Properties** dialog box is displayed.

5. In this dialog box, the **Project Settings** tab is chosen by default. Expand the **Panel Footprint Libraries** node from the **Library and Icon Menu Paths** area; the name and path of the panel footprint library is displayed.

6. Select the *C:/Users/Public/Documents/Autodesk/Acade 2022/Libs/Panel/* Library and then choose the **Add** button located on the right of the **Library and Icon Menu Paths** area in the **Project Properties** dialog box; an edit box is automatically added at the bottom of the selected library.

7. Enter **C:\Users\Public\Documents\Autodesk\Acade 2022\Libs\Panel\AB\PB-PUSHBUTTONS** in the blank edit box.

8. Choose the **OK** button in the **Project Properties** dialog box to save the changes made in this dialog box.

***Figure 8-31** The **AutoCAD Message** message box*

9. Choose the **User Defined List** tool from **Panel > Insert Component Footprints > Insert Footprints** drop-down; the **Panel footprint: Select and Insert by Catalog or Description Pick** dialog box is displayed.

10. Select the **800T-A2A** catalog from this dialog box and then choose the **OK** button; you are prompted to specify the location for AB/ABPB3.

11. Enter **15,20** at the Command prompt and press ENTER; you are prompted to select the rotation.

12. Enter **H** at the Command prompt for the horizontal orientation and then press ENTER; you are prompted to specify the rotation angle.

13. Press ENTER to accept the default value 0 for the rotation; the **Panel Layout - Component Insert/Edit** dialog box is displayed.

14. Enter **PB1** in the **Tag** edit box of the **Component Tag** area.

15. Choose the **Defaults** button from the **Description** area; the **Descriptions (general)** dialog box is displayed.

16. In the **Descriptions (general)** dialog box, select **START** that is below the **Generic descriptions for PUSH BUTTONS** heading; buttons in this dialog box are activated.

17. Choose **OK** from the **Descriptions (general)** dialog box; the description is displayed in the **Line 1** edit box of the **Description** area.

18. Choose **OK** from the **Panel Layout - Component Insert/Edit** dialog box. If the **Update other drawings** message box is displayed, choose the **Skip** button in this message box. The footprint of the **Push Button (PB1)** is inserted into the drawing, as shown in Figure 8-32.

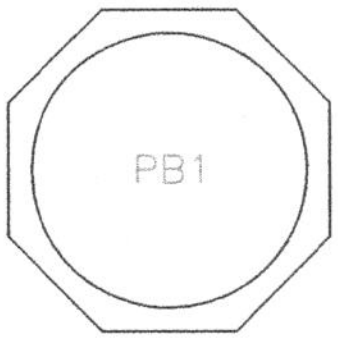

Figure 8-32 *Footprint inserted into the drawing*

Saving the Drawing File

1. Choose **Save** from the **Application Menu** to save the drawing file.

Self-Evaluation Test

Answer the following questions and then compare them to those given at the end of this chapter:

1. Which of the following commands is used to insert a footprint?

   (a) **AEFOOTPRINT** (b) **AECOMPONENT**
   (c) **AEFOOTPRINTDB** (d) **AEEDITFOOTPRINT**

2. The __________ tool is used to set the panel footprint drawing defaults.

3. The __________ tool is used to extract data from an equipment list.

4. The __________ tool is used to copy the footprint as well as the nameplate associated with it.

5. A balloon inserted into a footprint displays an __________ assigned to it.

6. The **Edit** tool in the **Edit Footprints** panel is used to edit a footprint. (T/F)

Review Questions

Answer the following questions:

1. Which of the following is a footprint database file?

   (a) wd_picklist (b) footprint_lookup.mdb
   (c) schematic_lookup (d) wd_lang1

2. Which of the following invisible blocks is used to store panel configuration settings as attributes?

   (a) wd_m.dwg (b) wd_mlrh.dwg
   (c) wd_pnlm.dwg (d) wd_mlrv.dwg

3. Which of the following tools is used to insert a footprint manually?

   (a) **Icon Menu** (b) **Schematic List**
   (c) **Manufacturer Menu** (d) **Manual**

4. The panel drawing can have a number of wd_pnlm.dwg block files. (T/F)

5. The **Mark Existing** button in the **Schematic Components (active project)** dialog box is used to identify the components that have already been inserted into the panel drawings. (T/F)

6. The **COPY** command of AutoCAD is used to copy both footprint and nameplate. (T/F)

EXERCISES

Exercise 1

In this exercise, you will insert the footprints of Pilot Lights, as shown in Figure 8-33, by using the **Schematic List** tool. Also, you will make the Xdata attached to a footprint visible by using the **Make Xdata Visible** tool. Next, you will use the **Move/Show Attribute** tool to move the attributes such as LT409, LT411, LT412, and LT413, as shown in Figure 8-33.

(Expected time: 20 min)

Exercise 2

In this exercise, you will insert the following footprints: EATON, EGH3015FFG, CB311; AB, 700-P200A1, CR101; AB, 800T-D1B, PB414A by using the **Equipment List** tool into the drawing, as shown in Figure 8-34. Also, you will move attributes such as CB311 and PB414A using the **Move/Show Attribute** tool. **(Expected time: 20 min)**

Hint: Choose the **CADCIM** file from the **Select Equipment List Spreadsheet File** dialog box and **COMP** from the **Table Edit** dialog box.

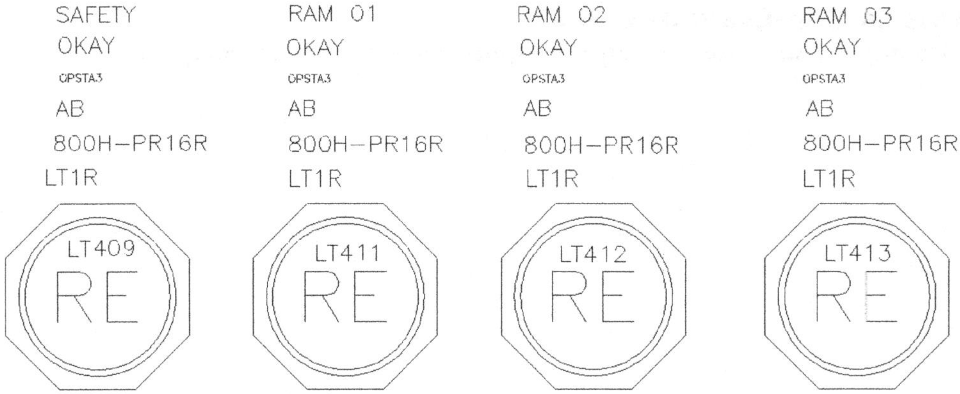

Figure 8-33 *The footprint of pilot lights inserted in the drawing using the* ***Schematic List*** *tool*

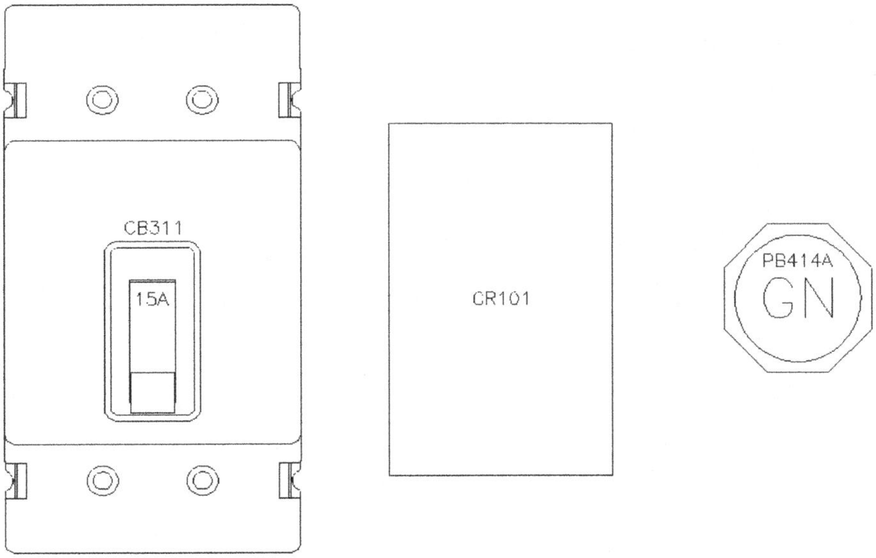

Figure 8-34 *Footprints inserted into the drawing*

Answers to Self-Evaluation Test

1. a, **2. Configuration**, **3. Insert Footprint Equipment List**, **4. Copy Footprint**, **5. item number**, **6.** T

Chapter 9

Schematic and Panel Reports

Learning Objectives

After completing this chapter, you will be able to:
- *Generate schematic and panel reports*
- *Change the report format*
- *Place reports in a drawing*
- *Save reports to external files*
- *Plot reports*
- *Generate cumulative reports*
- *Set format files for a report*

INTRODUCTION

In AutoCAD Electrical, you can manually or automatically run reports such as Bill of Materials, Wire From/To, Components, and so on. These reports consist of different fields of information and this information is as accurate as the information in your drawings. Moreover, more fields can be added to the report according to your requirement. Each report style is created with predefined categories and format. Every report can be customized to match your requirements such as adding or removing column headings, changing column order, and adding descriptive information.

AutoCAD Electrical extracts multiple fields from each report type. Using the reporting tools, you can save more time as you do not need to spend time in copying information from schematic drawings.

In this chapter, you will learn how to create reports with information extracted from the components inserted in the drawings. The formats of reports are flexible and are extracted from the information found in the drawings. Also, you will learn how to create schematic reports, panel reports, cumulative reports, edit reports, change the format of reports, save reports to external file, place report on the drawing, plot reports, and set format files for the report.

GENERATING SCHEMATIC REPORTS

Command: AESCHEMATICREPORT

In AutoCAD Electrical, you can run multiple schematic reports. To do so, choose the **Reports** tool from the **Schematic** panel of the **Reports** tab; the **Schematic Reports** dialog box will be displayed, as shown in Figure 9-1. The reports that can be extracted are displayed in the **Report Name** area of this dialog box and are discussed next.

Bill of Material Reports

By default, the **Bill of Material** option is chosen in the **Report Name** area, refer to Figure 9-1. You can extract the bill of material of a schematic drawing at any time. It reports only those components that have catalog information. This report extracts component data directly from drawing files.

Note

*The **Active drawing** radio button will be activated only if the active drawing is a part of an active project.*

After specifying the required options in the **Schematic Reports** dialog box, choose the **OK** button from this dialog box; the **Select Drawings to Process** dialog box will be displayed. In this dialog box, select the drawings to be processed and choose the **Process** button; the selected drawings will be transferred from the top list to the bottom list of the dialog box. Next, choose the **OK** button; the **Report Generator** dialog box will be displayed. This dialog box displays the result of report generation.

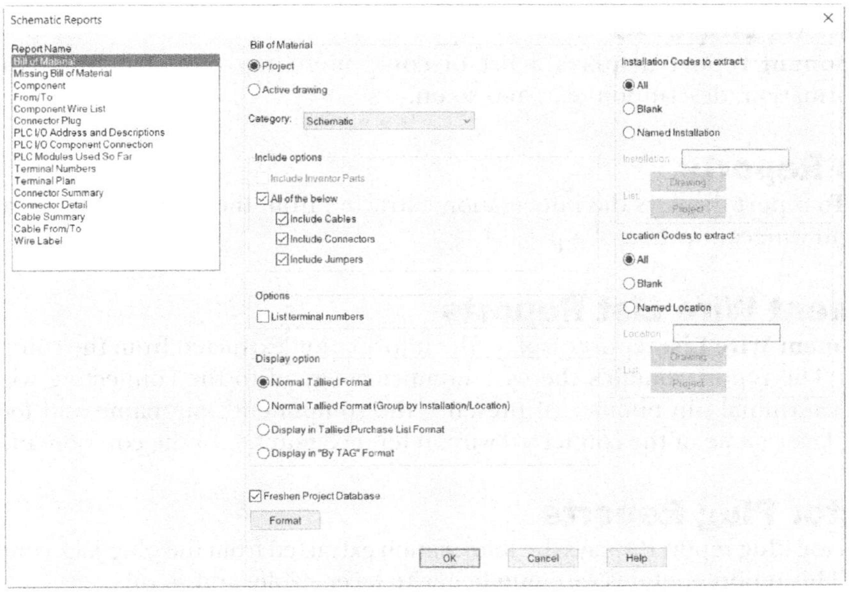

*Figure 9-1 The **Schematic Reports** dialog box*

Missing Bill of Material Reports

The **Missing Bill of Material** report displays a list of parent or stand-alone components that do not have catalog information. When you choose the **Missing Bill of Material** option from the **Report Name** area of the **Schematic Reports** dialog box, the **Schematic Reports** dialog box will be modified, as shown in Figure 9-2.

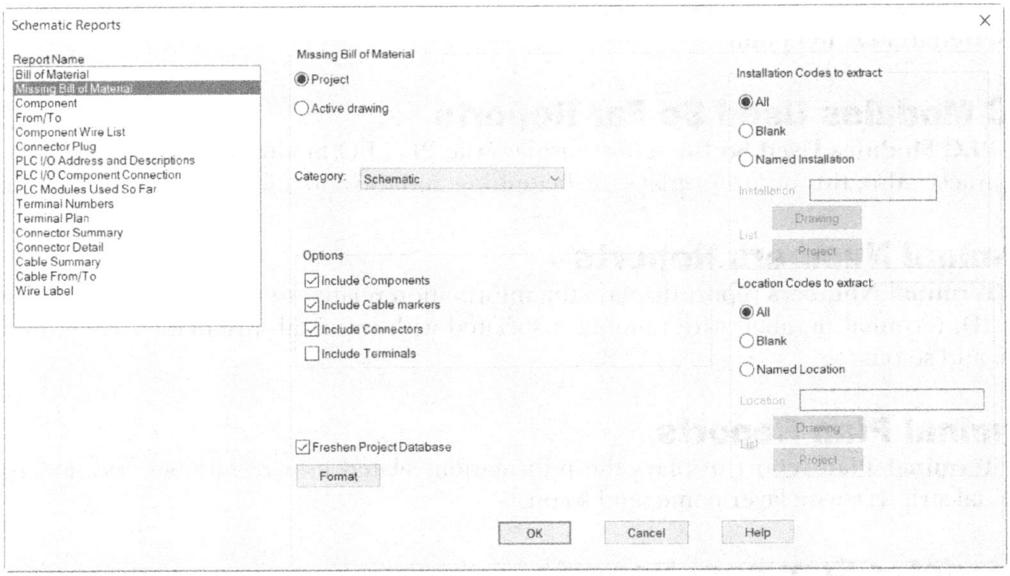

*Figure 9-2 The modified **Schematic Reports** dialog box after choosing the **Missing Bill of Material** option from the **Report Name** area*

Component Reports

The **Component** report displays a list of component tag name, manufacturer name, catalog information, description text, and so on.

From/To Reports

The **From/To** report displays the information extracted from the components, location codes, terminal, and wire connection of a project.

Component Wire List Reports

The **Component Wire List** report displays the information extracted from the component wire connection. The report contains the wire number assigned to the connecting wire, the wire connection's terminal pin number (if present), the component's tag name and location code (if present), layer name of the connected wire, reference number of the component, and so on.

Connector Plug Reports

The **Connector Plug** report displays the information extracted from the plug/jack connection and pin charts. This report contains wire number, wire layer, cable, and so on.

PLC I/O Address and Descriptions Reports

The **PLC I/O Address and Descriptions** report displays the starting and ending I/O address numbers of each PLC module. Also, the report displays five lines of description text and the connected wire number for each I/O point.

PLC I/O Component Connection Reports

The **PLC I/O Component Connection** report displays the information of components that are connected to PLC I/O points.

PLC Modules Used So Far Reports

The **PLC Modules Used So Far** report displays the PLC I/O modules that have been used in the project. Also, this report displays the beginning and the end address of the PLC module.

Terminal Numbers Reports

The **Terminal Numbers** report displays the information related to terminals such as terminal strip ID, terminal number, wire number associated with terminal, installation code, location code, and so on.

Terminal Plan Reports

The **Terminal Plan** report displays the information related to wire number, location codes, terminal strip ID, wire layer name, and so on.

Connector Summary Reports

The **Connector Summary** report displays the connector summary such as connector tag, pins used, maximum pins allowed, a list of repeated pin numbers used, and so on. This report can be run for the entire project or for a selected connector.

Connector Detail Reports

The **Connector Detail** report displays connector details such as connector tag, terminal pin number, wire number, type of the connector, location code if present, catalog, and so on. You can run this report for the entire project or for a single connector.

Cable Summary Reports

The **Cable Summary** report displays tags of cable marker, sheet number, manufacturer code, catalog, reference number, and so on. You need to run this report for the entire project.

Cable From/To Reports

The **Cable From/To** report displays the parent cable tag of the conductor, wire number, pin number, location code, installation code, and so on. You can run this report for the entire project, active drawing, or selected cables.

Wire Label Reports

The **Wire Label** report displays the list of wire labels and cable labels.

Wire Signal and Stand-alone Reference Reports

Command: AESIGNALERRORREPORT

The **Signal Error/List** tool is used to generate two types of reports: the wire signal source/destination codes report and the stand-alone reference source/destination codes report. To generate these reports, choose the **Signal Error/List** tool from the **Schematic** panel of the **Reports** tab; the **Wire Signal or Stand-Alone References Report** dialog box will be displayed, as shown in Figure 9-3.

The **Wire Signal Source/Destination codes report** radio button is selected by default. It is used to generate a report of the wire source/destination codes or generate a report for the exceptions found for the wire source/destination codes in the active project. Choose the **OK** button from the **Wire Signal or Stand-Alone References Report** dialog box; the **Wire Signal Report/Exceptions/Surf Exceptions** dialog box will be displayed, as shown in Figure 9-4.

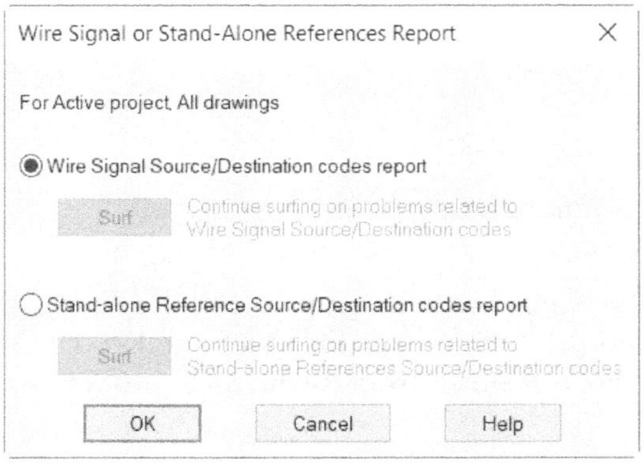

*Figure 9-3 The **Wire Signal or Stand-Alone References Report** dialog box*

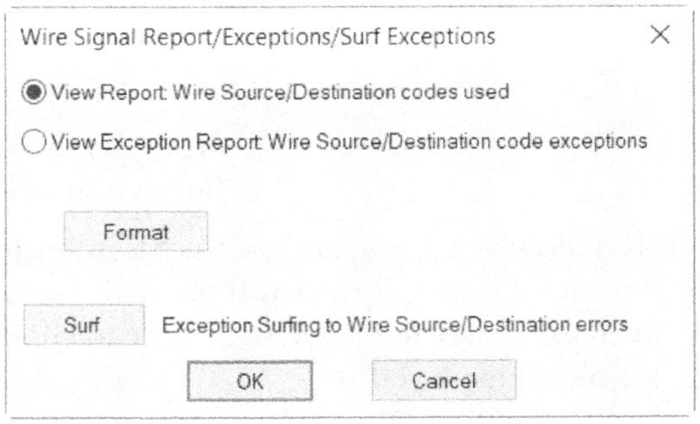

*Figure 9-4 The **Wire Signal Report/Exceptions/Surf Exceptions** dialog box*

In this dialog box, the **View Report: Wire Source/Destination codes used** radio button is selected to view the list of wire source/destination codes used in the active project in the form of a report. Choose **OK** in the **Wire Signal Report/Exceptions/Surf Exceptions** dialog box; the **Report Generator** dialog box will be displayed with the list of wire source/destination codes used in the active project.

Select the **View Exception Report: Wire Source/Destination code exceptions** radio button and then choose **OK** from the **Wire Signal Report/Exceptions/Surf Exceptions** dialog box; the **Report Generator** dialog box will be displayed with the list of signal exceptions found for the active project such as destination not found, repeated source, and so on.

If you select the **Stand-alone Reference Source/Destination codes report** radio button and choose the **OK** button in the **Wire Signal or Stand-Alone References Report** dialog box, the **Stand-alone Reference Report/Exceptions/Surf Exceptions** dialog box will be displayed. Note that if there are no stand alone references in the active project, the **AutoCAD Message** message box will be displayed, informing you that no stand alone references are found. The options in this dialog box are similar to the options in the **Wire Signal Report/Exceptions/Surf Exceptions** dialog box discussed earlier. Using these options, you can generate a report for the stand-alone reference source/destination codes in the active project or generate a report for the exceptions found for the stand-alone reference source/destination codes in the active project. The **Surf** button located below the **Wire Signal Source/Destination codes report** radio button is used for surfing the problems related to the source/destination codes in the active project.

Missing Catalog Data

Command: AEMISSINGCATREPORT

The **Missing Catalog Data** tool is used to mark the components in the active drawing that do not have catalog data. To mark the components in the active drawing, open the drawing and choose the **Missing Catalog Data** tool from the **Schematic** panel of the **Reports** tab; the **Show Missing Catalog Asssignments** dialog box will be displayed, as shown in Figure 9-5.

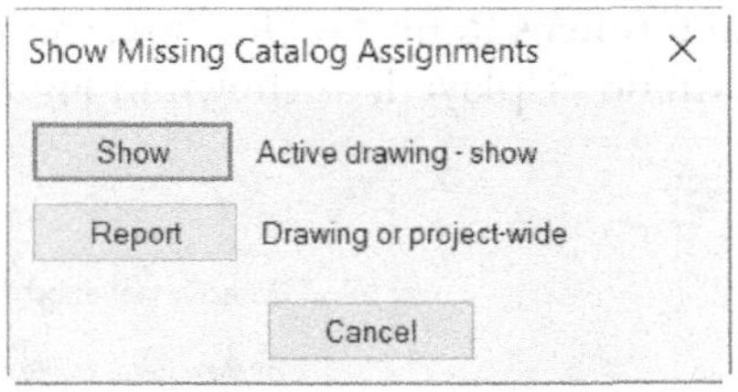

***Figure 9-5** The **Show Missing Catalog Assignments** dialog box*

In this dialog box, if you choose the **Show** button, the dialog box will diappear and the diamond shaped temporary graphics will be displayed around the components for which catalog data is not available, refer to Figure 9-6. You can choose **View > Redraw** from the menu bar to remove the graphics displayed.

If you choose the **Report** button, the **Schematic Reports** dialog box will be displayed with the **Missing Bill of Material** option selected in the **Report Name** area.

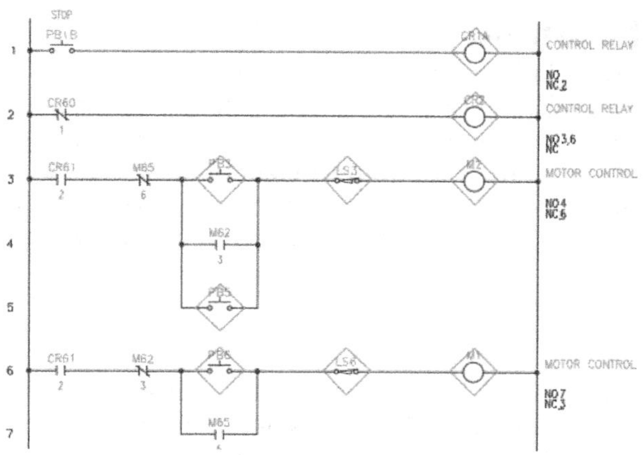

Figure 9-6 *Temporary graphics displayed*

UNDERSTANDING THE Report Generator DIALOG BOX

The **Report Generator** dialog box displays the result of the report selected from the **Report Name** area of the **Schematic Reports** dialog box. The **Report Generator** dialog box shown in Figure 9-7 displays the report when the **Bill of Material** option is chosen in the **Report Name** area of the **Schematic Reports** dialog box, refer to Figure 9-1. Note that the options, which will be available in this dialog box, depend upon the options chosen in the **Report Name** area of the **Schematic Reports** dialog box. Using the **Report Generator** dialog box, you can change the format of a report, place report in a drawing, edit report, and save report to an external file. The important buttons in this dialog box are discussed next.

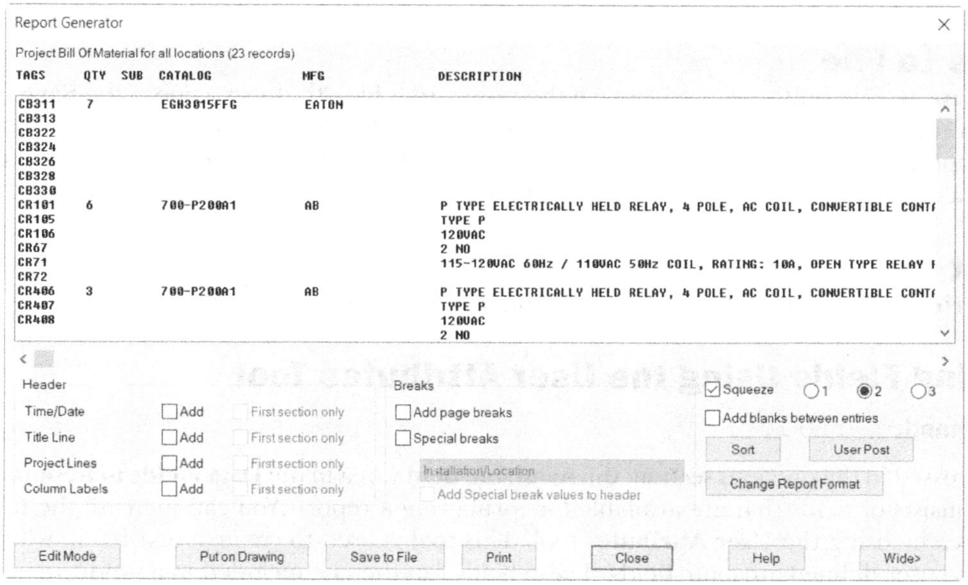

Figure 9-7 *The* ***Report Generator*** *dialog box*

Change Report Format

The **Change Report Format** button is used to change data fields and their order in the report. Also, this button is used to change the field name or justification of columns in the report to match the column settings. To change the data fields and their order in the report, choose the **Change Report Format** button from the **Report Generator** dialog box; the **### Data Fields to Report** dialog box will be displayed. Note that the prefix of the **Data Fields to Report** dialog box will change according to the option selected from the **Report Name** area of the **Schematic Reports** dialog box.

Edit Mode

The **Edit Mode** button is used to edit a report before you insert it into the drawing. To do so, choose the **Edit Mode** button from the **Report Generator** dialog box; the **Edit Report** dialog box will be displayed. In this dialog box, you can edit, delete, and move the data up and down in the report. Also, you can add new lines above or below the selected line from the catalog lookup table in the report. Note that the options in this dialog box depend on the option chosen from the **Report Name** area of the **Schematic Reports** dialog box.

Put on Drawing

The **Put on Drawing** button is used to display a report in tabular form in a drawing. To do so, choose the **Put on Drawing** button; the **Table Generation Setup** dialog box will be displayed. Specify the required options in this dialog box and choose the **OK** button to place the table in the drawing. The options in this dialog box are discussed later in this chapter. Note that the **Put on Drawing** button will not be available if you choose the **Wire Label** option from the **Report Name** area of the **Schematic Reports** dialog box.

Save to File

The **Save to File** button is used to save the report to a file. To do so, choose the **Save to File** button; the **Save Report to File** dialog box will be displayed. Using this dialog box, you can save report in multiple formats of output files such as ASCII, Excel, Access database, XML, and comma delimited. The options in this dialog box are discussed later in this chapter.

Print

The **Print** button is used to print reports from the **Report Generator** dialog box.

Adding Fields Using the User Attributes Tool

Command: AEUDA

As discussed in the previous section, the **Available fields** area in the **Data Fields to Report** dialog box consists of fields that are available for formatting a report. You can increase the fields in this area by using the **User Attributes** tool. This tool is used to create a text file in which you can add user defined attribute fields. These fields can then be included in a report to enhance it using the options in the **Data Fields to Report** dialog box.

To add fields in the **Available Field** area, choose the **User Attributes** tool from the **Miscellaneous** panel of the **Reports** tab; the **User Defined Attribute List: unnamed.wda**

dialog box will be displayed, as shown in Figure 9-8. Different columns and options in this dialog box are discussed next.

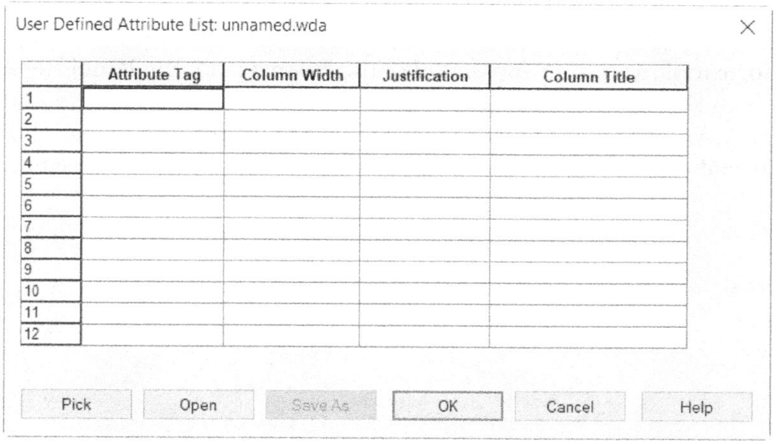

Figure 9-8 *The* ***User Defined Attribute List: unnamed.wda*** *dialog box*

The **Attribute Tag** column contains a number of cells where you can enter user-defined attributes one by one. After you enter an attribute in the **Attribute Tag** column, enter the column width value in the respective cell of the **Column Width** column. After you enter an attribute in the **Attribute Tag** column, click on the **Justification** column; a drop-down list will be displayed. Select the required option from it to justify the attribute in the report.

The **Column Title** column is used to enter a column title for an attribute. It acts as a header for the attribute field in the **Report Generator** dialog box. Choose the **Pick** button to pick an attribute from the active drawing. Choose the **Open** button to open an existing user defined attribute list file.

Choose the **Save As** button to save the user defined attribute list file with the extension .wda. The default location to save this file is *C:\Users\User name\Documents\Acade 2022\AeData\Proj\ active project name*. Note that the attributes added in this dialog box will be added in the **Available fields** area of the **Data fields to Report** dialog box only when you save this file at the default location mentioned above with the same file name as that of the active project.

Next, generate a report as discussed earlier. In the **Report Generator** dialog box, choose the **Change Report Format** button; the **Data Fields to Report** dialog box will be displayed. You will notice that the attributes added in the user-defined attribute list file are available in the **Available Fields** area of this dialog box.

GENERATING PANEL REPORTS

Command: AEPANELREPORT

Panel reports are similar to schematic reports, but they are limited to panel components, nameplates, and terminal footprint. Although schematic reports and panel reports are almost the same yet the difference lies between the two in component attributes.

Schematic components use TAG1 as the attribute name, whereas panel footprints use P_TAG1 as the attribute name.

In AutoCAD Electrical, you can run multiple panel reports at a time. To run a panel report, choose the **Reports** tool from the **Panel** panel of the **Reports** tab; the **Panel Reports** dialog box will be displayed, as shown in Figure 9-9. In this dialog box, a list of reports is available in the **Report Name** area. You can extract the reports displayed in this area as per your requirement. The **Bill of Material** option in this area is chosen by default, as shown in Figure 9-9, and it is discussed next.

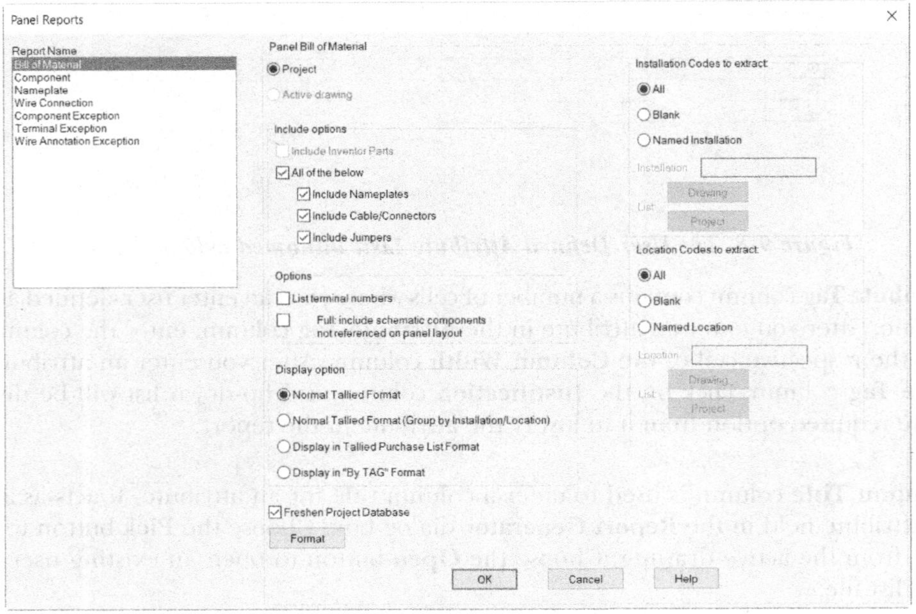

*Figure 9-9 The **Panel Reports** dialog box*

Bill of Material Report

The **Bill of Material** report extracts information from panel components, terminal footprints, nameplates, and so on.

Specify the required options in the **Panel Reports** dialog box and select the **Project** radio button, if it is not selected. Next, choose the **OK** button; the **Select Drawings to Process** dialog box will be displayed. In this dialog box, select the drawings to be processed and choose the **Process** button; the drawings will be transferred from the top list to the bottom list. Next, choose the **OK** button; the **Report Generator** dialog box will be displayed.

Note

*If you make changes in the current drawing and did not save it, then after choosing the **OK** button in the **Panel Reports** dialog box, the **QSAVE** message box will be displayed. Choose the **OK** button in this message box to save the current drawing; the **Report Generator** dialog box will be displayed.*

Now, to place the report table in the drawing, choose the **Put on Drawing** button from the **Report Generator** dialog box; the **Table Generation Setup** dialog box will be displayed. Specify the required settings in the **Table Generation Setup** dialog box and then choose the **OK** button from it; you will be prompted to specify the insertion point for the table. Next, specify the insertion point; the table will be inserted into the drawing. Rest of the options in the **Report Generator** dialog box have been discussed earlier.

GENERATING THE CUMULATIVE REPORT

Command: AEAUTOREPORT

The **Automatic Reports** tool is used to generate multiple reports at a time. Also, this tool is used to run different reports at different stages. You can use this tool to automatically place the report tables on drawings or create a number of output files. To generate a report, choose the **Automatic Reports** tool from the **Miscellaneous** panel of the **Reports** tab; the **Automatic Report Selection** dialog box will be displayed, as shown in Figure 9-10. The options in this dialog box are discussed next.

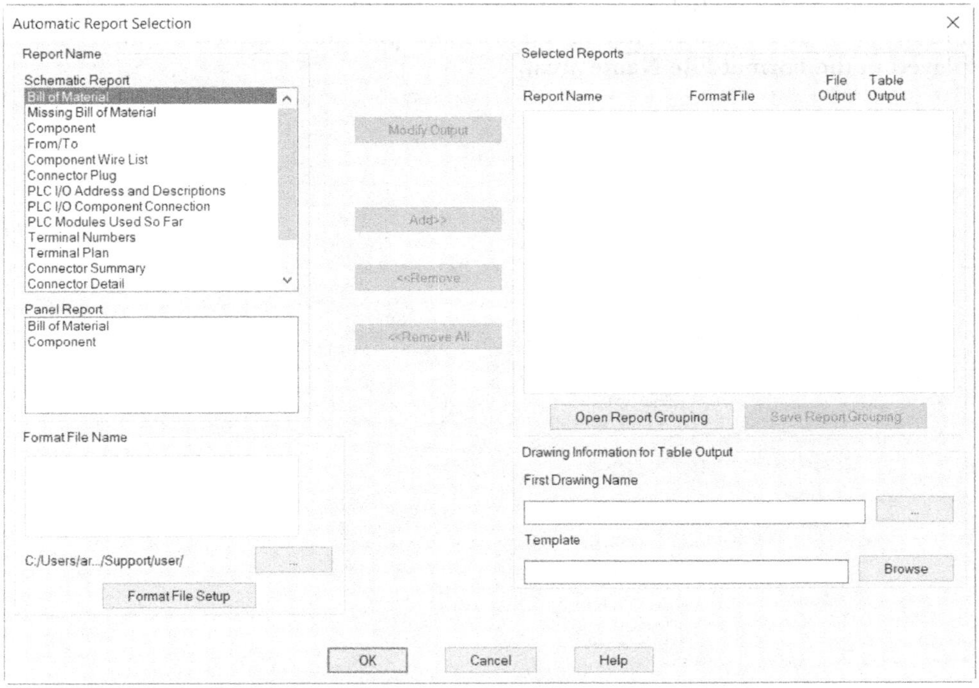

Figure 9-10 The ***Automatic Report Selection*** *dialog box*

Report Name Area

The options in the **Report Name** area are used to select the type of report to be generated. This area contains a list of schematic and panel reports that are used for automatic generation of report.

The **Schematic Report** area consists of options for generating schematic reports and the **Panel Report** area consists of options for generating panel reports.

Format File Name Area

The **Format File Name** area displays a list of format files both for schematic and panel reports. The type of format file displayed in the **Format File Name** area depends on the type of report selected from the **Schematic Report** and **Panel Report** areas.

If a report type displayed in the **Schematic Report** or **Panel Report** area does not have a format file, choose the **Format File Setup** button from the **Format File Name** area; the **Report Format File Setup** dialog box will be displayed, as shown in Figure 9-11. Using this dialog box, you can create and save the format file for the selected report type. The options in the **Report Format File Setup** dialog box are discussed in the later section. Note that the path and location of the format file is displayed below the **Format File Name** area of the **Automatic Report Selection** dialog box. To change the directory of the format file displayed in the **Format File Name** area, choose the **Browse** button located on the right of the path and location of the format file in the **Automatic Report Selection** dialog box; the **Browse For Folder** dialog box will be displayed. Specify the name of the folder in the **Browse For Folder** dialog box. Next, choose the **OK** button in the **Browse For Folder** dialog box; the new path and location of the format file will be displayed in the **Format File Name** area.

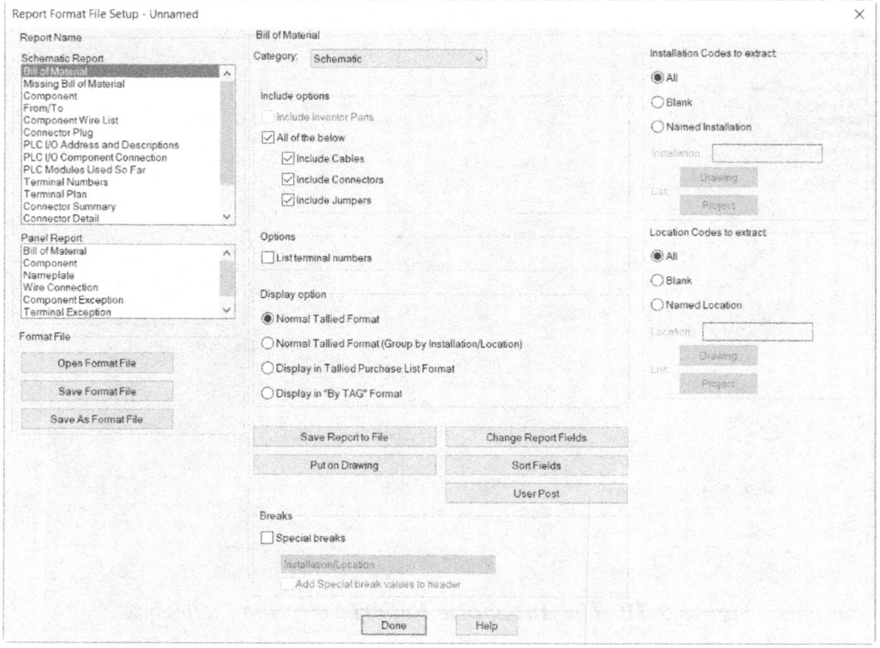

*Figure 9-11 The **Report Format File Setup** dialog box*

Add>>, <<Remove, and <<Remove All

The **Add>>** button is used to add the selected report name and format file to the **Selected Reports** area. This button will be activated only when the report name is selected from the **Report Name** area and its format file is selected from the **Format File Name** area at a time. The **<<Remove** button is used to remove the selected report from the **Selected Reports** area. The **<<Remove All** button is used to remove all reports from the **Selected Reports** area.

Modify Output

The **Modify Output** button will be activated only if you choose the report name from the **Selected Reports** area. This button is used to change the type of output of the selected report name. To do so, choose the **Modify Output** button; the **Report Output Options** dialog box will be displayed. In this dialog box, you can specify whether to set report to the file output or to the table output. After specifying the type of output in the **Report Output Options** dialog box, choose the **OK** button; the output type will be changed in the **Selected Reports** area.

Selected Reports Area

The **Selected Reports** area displays a list of reports that you want to generate. This area consists of report name, format file, file output, and table output. If 'x' is displayed under the **File Output** and **Table Output** columns, the portion will be considered for the automatic generation of report. If 'o' is displayed under the **File Output** and **Table Output** columns, the portion will not be considered for the automatic generation of report.

Save Report Grouping

The **Save Report Grouping** button will be available only when a report file is displayed in the **Selected Reports** area. This button is used to save the file that consists of report name and format file. This saved file can be retrieved and used later to run the same group of reports on the active project or on any other project. To save a group of format files, choose the **Save Report Grouping** button; the **Enter name for Automatic Report Group file** dialog box will be displayed. In this dialog box, specify a name for the file in the **File name** edit box. By default, *.*rgf* is displayed in the **Save as type** drop-down list. Next, choose the **Save** button from the **Enter name for Automatic Report Group file** dialog box to save the file and exit this dialog box.

Open Report Grouping

The **Open Report Grouping** button is used to open the saved group file of report names and format files. On choosing the **Open Report Grouping** button, the **Enter name for Automatic Report Group file** dialog box will be displayed. In this dialog box, select the required file and choose the **Open** button; the grouped report file will be displayed in the **Selected Reports** area.

Drawing Information for Table Output Area

In the **Drawing Information for Table Output** area, you can specify the name and location for the drawing file and the template to be used for creating the drawing files for reports.

After specifying the required options in the **Automatic Report Selection** dialog box, choose the **OK** button; the report of the selected report names will be created. Note that the drawing file of the report will be added at the bottom of the project drawing list.

SETTING THE FORMAT FILE FOR REPORTS

Command: AEFORMATFILE

The **Report Format File Setup** tool is used to create format files. This tool is also used to save the settings of a report to a file for later use. To do so, choose the **Report Format Setup** tool from the **Miscellaneous** panel of the **Reports** tab; the **Report Format File**

Setup dialog box will be displayed. The **Report Format File Setup** dialog box can also be invoked by choosing the **Format File Setup** button from the **Automatic Report Selection** dialog box as discussed in the previous section.

Using the **Report Format File Setup** dialog box, you can save the settings of a report to a text file with *.set* extension. You can use the report format files while generating reports using the **Reports** tool in the **Schematic** and **Panel** panel and **Automatic Reports** tool in the **Miscellaneous** panel.

Generating Component Cross-Reference Report

The **Component Cross-Reference** tool is used to display the cross-reference report on the related parent and child contacts. To display the cross-reference report on the related parent and child contacts, choose the **Component Cross Reference** tool from the **Component Cross-Reference** drop-down in the **Edit Components** panel of the **Schematic** tab, refer to Figure 9-12; the **Component Cross-Reference** dialog box will be displayed.

By default, in this dialog box, the **Active drawing (all)** radio button is selected in the **Run Cross-reference on:** area. Next, choose the **Cross-reference** button; the **Cross Reference Report** dialog box will be displayed with the cross-reference information of the components in the active drawing. If you want to check the cross-reference information of individual components, select the **Active Drawing (pick)** radio button. If you want to check the cross-reference information of the components in the project, select the **Project** radio button. If you choose the **Exception** button, the **Error Exception Report** will be displayed with the list of parent components that do not have the child contacts and the child contacts that do not have the parent component.

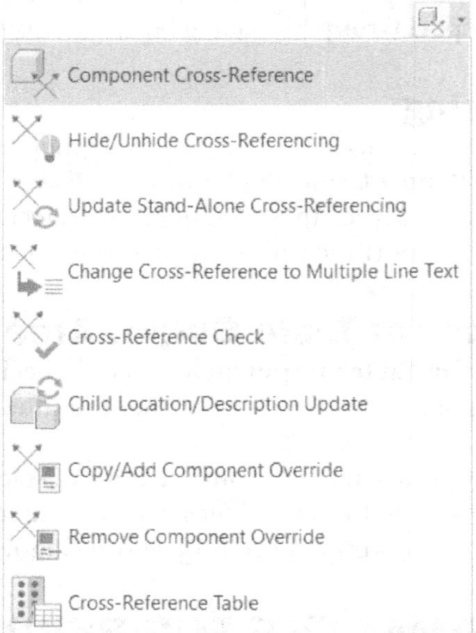

Figure 9-12 *The* ***Component Cross-Reference*** *drop-down*

TUTORIALS

Tutorial 1

In this tutorial, you will generate a schematic Bill of Material report of the **CADCIM** project and then insert a table into the drawing. Also, you will generate a schematic Component report and save the report to the **Excel Spreadsheet** format file (.xls). **(Expected time: 20 min)**

The following steps are required to complete this tutorial:

a. Create a new drawing.
b. Generate a schematic Bill of Material report and place it in the drawing.
c. Add an attribute field in the schematic Bill of Material report.
d. Generate and save the component report.
e. Close the **Excel Spreadsheet** format file (.xls).
f. Save the drawing file.

Creating a New Drawing

1. Activate the **CADCIM** project in the **PROJECT MANAGER**, if it is inactive.

2. Choose the **New Drawing** button from the **PROJECT MANAGER**; the **Create New Drawing** dialog box will be displayed. In this dialog box, enter **C09_tut01** in the **Name** edit box and **Schematic BOM** in the **Description 1** edit box. Choose the **OK** button; the *C09_tut01.dwg* file is created and displayed at the bottom of the drawing list in the **CADCIM** project. Next, move *C09_tut01.dwg* to the *TUTORIALS* subfolder of the **CADCIM** project.

Generating the Schematic Bill of Material Report and Placing it in the Drawing

1. Choose the **Reports** tool from the **Schematic** panel of the **Reports** tab; the **Schematic Reports** dialog box is displayed, as shown in Figure 9-13.

2. Choose the **Bill of Material** option from the **Report Name** area, if it is not selected. Next, specify the required options in the **Schematic Reports** dialog box, refer to Figure 9-13.

3. Choose the **OK** button from the **Schematic Reports** dialog box; the **Select Drawings to Process** dialog box is displayed. Choose the **Do All** button from this dialog box; all the drawings from the top list of the dialog box are transferred to the bottom list. Next, choose the **OK** button from this dialog box; the **Report Generator** dialog box is displayed.

Note
*If the **QSAVE** message box is displayed, choose the **OK** button from it to display the **Report Generator** dialog box.*

4. Choose the **Put on Drawing** button from the **Report Generator** dialog box; the **Table Generation Setup** dialog box is displayed. Select the **Insert New** radio button from the **Table** area and the **Include column labels** check box from the **Column Labels** area, if not already selected. Retain rest of the default values in the dialog box.

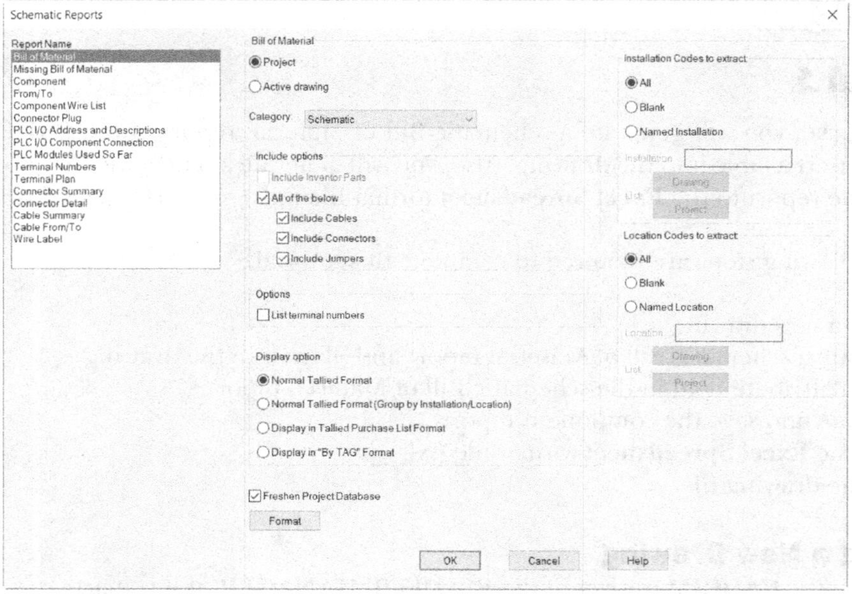

***Figure 9-13** The **Schematic Reports** dialog box*

5. Next, choose the **OK** button in the **Table Generation Setup** dialog box; you are prompted to specify the insertion point for the table. Click on the upper left corner of the title block; the Bill of Material report is inserted into the drawing and the **Report Generator** dialog box is displayed again. Choose the **Close** button in the **Report Generator** dialog box to exit.

6. Choose **View > Zoom > In** from the menu bar to zoom the table. The report is displayed in a tabular format. The zoomed view of the report in tabular format is shown in Figure 9-14.

ITEM	TAGS	QTY	SUB	CATALOG	MFG	DESCRIPTION
	CB311 CB313 CB322 CB324 CB326 CB328 CB330	7		EGH3015FFG	EATON	CIRCUIT BREAKER – E125 FRAME 3-POLE CIRCUIT BREAKER 15AMPS TYPE E125H, FIXED THERMAL & MAGNETIC TRIP 690VAC, 250VDC, 15AMPS
	CR101 CR105 CR106 CR396 CR397 CR398	6		700-P200A1	AB	P TYPE ELECTRICALLY HELD RELAY, 4 POLE, AC COIL, CONVERTIBLE CONTACTS TYPE P 120VAC 2 NO 115-120VAC 60Hz / 110VAC 50Hz COIL, RATING: 10A, OPEN TYPE RELAY RAIL MOUNT
	DS304	1		194E-A25-1753	AB	IEC LOAD SWITCH 3 POLE 194E – LOAD SWITCH 25AMPS ON-OFF BASE MOUNTED SWITCH (INCLUDES OPERATING SHAFT) 480VAC, 25AMPS
	FU309	1		FRS-R-15	BUSSMANN	DUAL ELEMENT FUSE – CLASS RK5 TIME DELAY, CURRENT LIMITING 600VAC 15AMPS
	FU307 FU307A	2		FRS-R-5	BUSSMANN	DUAL ELEMENT FUSE – CLASS RK5 TIME DELAY, CURRENT LIMITING 600VAC 5AMPS
	LS396 LS397 LS398	3		MS-50L	MS	
	LT405 LT413 LT416	3		800H-PR16G	AB	GREEN PILOT LIGHT – STANDARD, ROUND OILTIGHT 30.5mm 120VAC XFMR PLASTIC LENS CORROSION RESISTANT
	LT399 LT401 LT402 LT403	4		800H-PR16R	AB	RED PILOT LIGHT – STANDARD, ROUND OILTIGHT 30.5mm 120VAC XFMR PLASTIC LENS CORROSION RESISTANT

***Figure 9-14** The Zoomed view (Partial) of the Bill of Material report*

Adding an Attribute Field in the Schematic Bill of Material Report

1. Choose the **User Attributes** tool from the **Miscellaneous** panel of the **Reports** tab; the **User Defined Attribute List: unnamed.wda** dialog box is displayed, as shown in Figure 9-15.

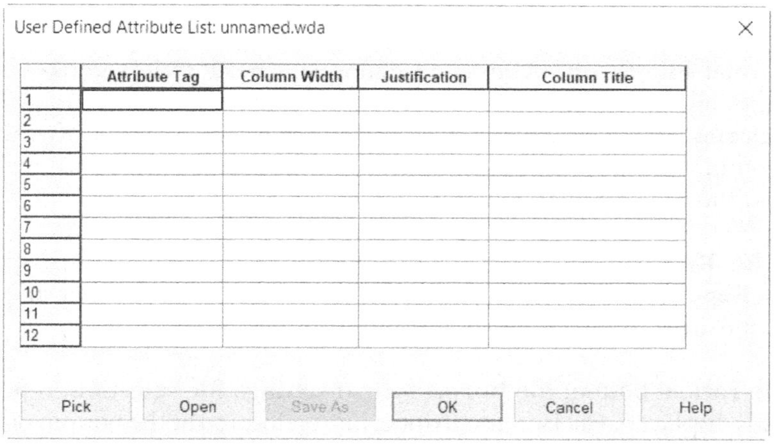

***Figure 9-15** The **User Defined Attribute List: unnamed.wda** dialog box*

2. Enter **FAMILY** in the first cell of both the **Attribute Tag** and **Column Title** columns of this dialog box.

3. Click in the first cell of the **Justification** column; a drop-down list is displayed. Select the **Top Left** option from it.

   Even if you leave this option blank, by default the **Top Left** option will be selected.

   As the **Column Width** column of the **User Defined Attribute List: unnamed.wda** dialog box is left blank, column width for the attribute is restricted to 24 characters.

4. Choose the **Save As** button; the **Create File** dialog box is displayed. Browse to *C:\Users\User name\Documents\Acade 2022\AeData\Proj\CADCIM* from the **Save in** drop-down list and then make sure **CADCIM** is displayed in the **File Name** edit box. Next, choose **Save** from the dialog box; a **CADCIM** text file is created at the specified location with the *.wda* extension.

5. Choose the **OK** button in the **User Defined Attribute List: unnamed.wda** dialog box to close it. Next, choose the **Report Format Setup** tool from the **Miscellaneous** panel of the **Reports** tab; the **Report Format File Setup - Unnamed** dialog box is displayed. Make sure that the **Bill of Material** option is selected in the **Schematic Report** list in this dialog box.

6. Next, choose the **Change Report Fields** button from the **Report Format File Setup - Unnamed** dialog box; the **Bill of Material Data Fields to Report** dialog box is displayed. You will notice that the **Family** field is added to the **Available Fields** area of this dialog box.

7. Select **Family** from the **Available fields** area; it automatically moves to the **Fields to report** area.

Note

*You can change the order of fields in the **Field to report** area by using the **Move Up** and **Move Down** buttons available below it.*

8. Choose the **OK** button to close this dialog box. Next, choose the **Done** button from the **Report Format File Setup - Unnamed** dialog box to close it; the **Save Report Format Changes** message box is displayed. Choose the **Yes** button in the message box; the **Select Bill of Material *.set format file** dialog box is displayed. Now, specify the desired name and location in this dialog box to save the report format changes in the form of file and then choose the **Save** button to close the dialog box.

9. Choose the **Reports** tool from the **Schematic** panel of the **Reports** tab; the **Schematic Reports** dialog box is displayed. Choose the **Bill of Material** option from the **Report Name** area, if it is not already chosen.

10. Choose the **Format** button; the **Report format settings file selection, report type=BOM** dialog box is displayed. Click on the **Browse** button located at the bottom of the dialog box; the **Select BOM*.set format file** dialog box is displayed. In this dialog box, browse to the location that is specified in step 8 and select the saved format file. Next, choose the **Open** button; the name and location of the format file is displayed on the right of the **Format** button.

11. Choose the **OK** button from the **Schematic Reports** dialog box; the **Select Drawings to Process** dialog box is displayed. In this dialog box, choose the **Do All** button; all the drawings from the top list of the dialog box are transferred to the bottom list. Choose the **OK** button from this dialog box; the **QSAVE** message box is displayed. Choose the **OK** button from it; the **Report Generator** dialog box is displayed.

    You will notice that the **Family** attribute field is added in the **Report Generator** dialog box.

12. Choose the **Close** button in the **Report Generator** dialog box to close it.

Generating and Saving the Component Report

1. Choose the **Reports** tool from the **Schematic** panel of the **Reports** tab; the **Schematic Reports** dialog box is displayed.

2. Choose the **Component** option from the **Report Name** area of this dialog box.

3. Accept the default values in this dialog box and choose the **OK** button in the **Schematic Reports** dialog box; the **Select Drawings to Process** dialog box is displayed.

4. Choose the **Do All** button from the **Select Drawings to Process** dialog box; all the drawings from the top list are transferred to the bottom list of this dialog box.

5. Choose the **OK** button from the **Select Drawings to Process** dialog box; the **Report Generator** dialog box is displayed.

Note

*If the **QSAVE** message box is displayed, choose the **OK** button from it to display the **Report Generator** dialog box.*

6. Choose the **Sort** button from the **Report Generator** dialog box; the **Sort** dialog box is displayed.

7. Select **TAGNAME** from the **Primary sort** drop-down list, **ITEM** from the **Secondary sort** drop-down list, and **MFG** from the **Third sort** drop-down list. Next, choose the **OK** button from the **Sort** dialog box; the report displayed in the **Report Generator** dialog box is sorted accordingly.

8. Choose the **Save to File** button from the **Report Generator** dialog box; the **Save Report to File** dialog box is displayed.

9. Select the **Excel spreadsheet format (.xls)** radio button and choose the **OK** button in this dialog box; the **Select file for report** dialog box is displayed.

10. Browse to *C:\Users\User Name\Documents\Acade 2022\AeData\Proj\CADCIM* from the **Save in** drop-down list and then enter **COMPONENT** in the **File name** edit box. By default, *. *xls* is displayed in the **Save as type** edit box. Do not change this file extension.

11. Choose the **Save** button from the **Select file for report** dialog box; the **Optional Script File** dialog box is displayed.

12. Choose the **Close - No Script** button to return to the **Report Generator** dialog box.

13. Choose the **Close** button to exit the **Report Generator** dialog box.

14. Open the saved *COMPONENT.xls* file from *C:\Users\User Name\Documents\Acade 2022\AeData\Proj\CADCIM*. The excel file is shown in Figure 9-16.

Closing the Excel Spreadsheet Format File

1. Choose **Close** from the **Office Button** or **File > Exit** from the menu bar to exit the **COMPONENT**.xls file.

Saving the Drawing File

1. Choose **Save** from the **Application Menu** to save the drawing file *C09_tut01.dwg*.

2. Choose **Close > Current Drawing** from the **Application Menu** to close the drawing file.

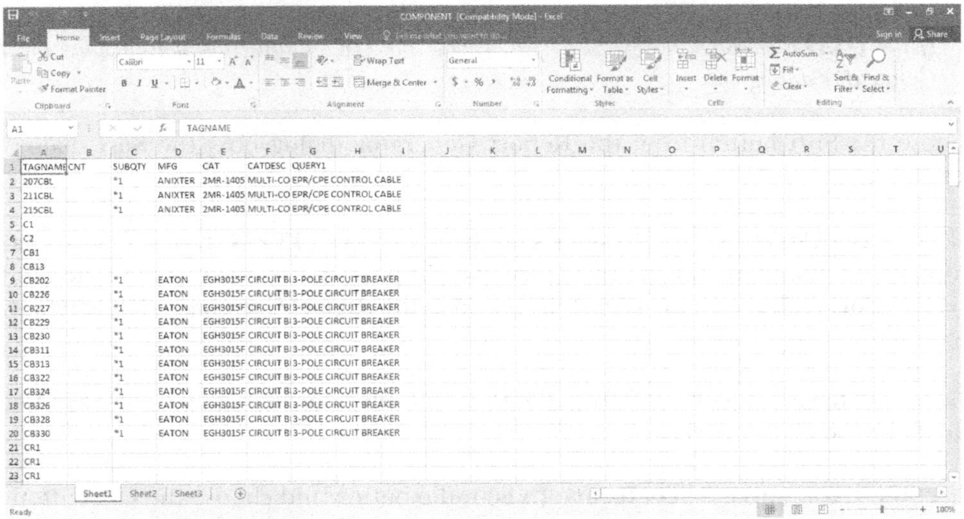

Figure 9-16 Component report saved in an Excel file

Tutorial 2

In this tutorial, you will create and save the format files for schematic components and panel components of the **CADCIM** project. Next, you will use the **Automatic Reports** tool to group the format files and generate the report. **(Expected time: 25 min)**

The following steps are required to complete this tutorial:

a. Create format file for schematic components.
b. Create format file for panel components.
c. Create group for automatic report generation.
d. Generate automatic report.

Creating Format File for Schematic Components

In this section, you will create a format file for schematic components using the **Report Format Setup** tool.

1. Make sure the **CADCIM** project is activated and the *C09_tut02.dwg* file is opened. Next, choose the **Report Format Setup** tool from the **Miscellaneous** panel of the **Reports** tab; the **Report Format File Setup - Unnamed** dialog box is displayed.

2. Choose the **Component** option from the **Schematic Report** area of this dialog box; the modified **Report Format File Setup-Unnamed** dialog box is displayed, as shown in Figure 9-17.

3. Choose the **Change Report Fields** button from the **Report Format File Setup-Unnamed** dialog box; the **Component Data Fields to Report** dialog box is displayed.

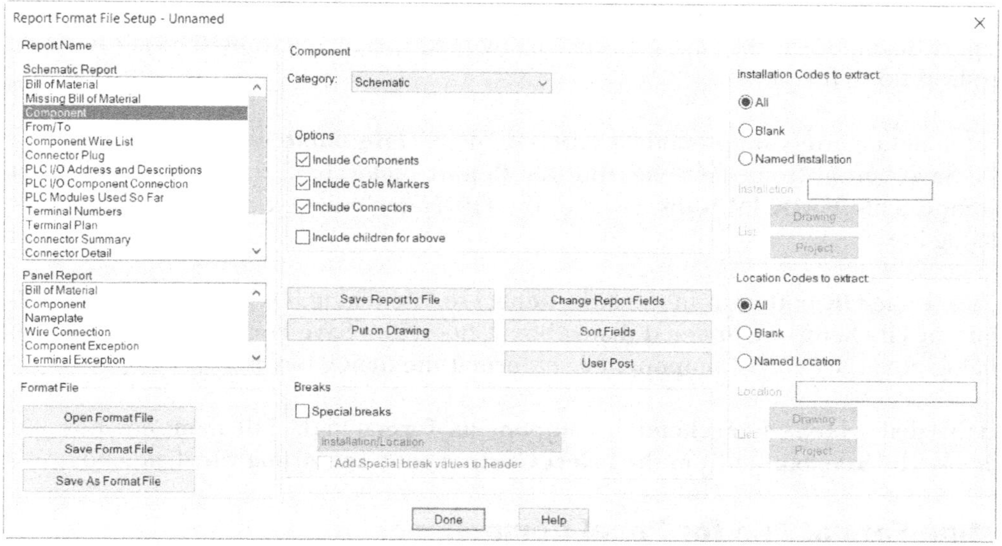

*Figure 9-17 The modified **Report Format File Setup - Unnamed** dialog box*

4. Choose the **FILENAME** option from the **Available fields** area; the **FILENAME** is shifted to the **Fields to report** area.

5. Make sure **FILENAME** is highlighted in the **Fields to report** area and then move it to the bottom using the **Move Down** button, as shown in Figure 9-18.

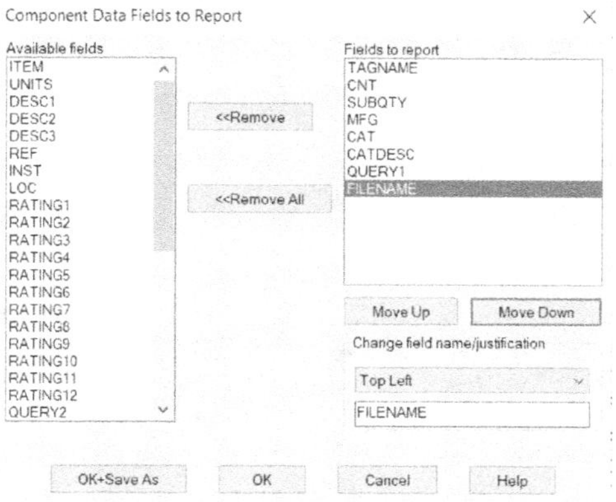

*Figure 9-18 The **FILENAME** option moved to the bottom in the **Fields to report** area*

6. Choose the **OK** button in the **Component Data Fields to Report** dialog box.

7. Choose the **Save Report to File** button from the **Report Format File Setup - Unnamed** dialog box; the **Save Report to File** dialog box is displayed.

8. Select the **Excel spreadsheet format [.xls]** check box from the **Excel Spreadsheet** area in the **Save Report to File** dialog box; the options below this check box are activated. Next, select the **Include project values** and **First section only** check boxes.

9. Choose the **Browse** button located at the right of the **File Name** edit box; the **Select file for report** dialog box is displayed. In this dialog box, specify the desired location in the **Save in** drop-down list.

10. Enter **Schematic Components_CADCIM** in the **File name** edit box and then choose the **Save** button from the **Select file for Report** dialog box; the path of the **Schematic Components_CADCIM** is displayed in the **File Name** edit box of the **Save Report to File** dialog box.

11. Choose the **OK** button from the **Save Report to File** dialog box; you return to the **Report Format File Setup - Unnamed** dialog box. Choose the **Save Format File** button from this dialog box; the **Select Component *.set format file** dialog box is displayed.

12. In this dialog box, enter **schematic components_format** in the **File name** edit box and then choose the **Save** button from the **Select Component *.set format file** dialog box.

Creating Format File for Panel Components

In this section, you will create a format file for schematic components using the **Report Format Setup** tool.

1. Make sure the **Report Format File Setup - XX** dialog box is opened. Choose the **Component** option from the **Panel Report** area of this dialog box; the modified **Report Format File Setup-Unnamed** dialog box is displayed, as shown in Figure 9-19.

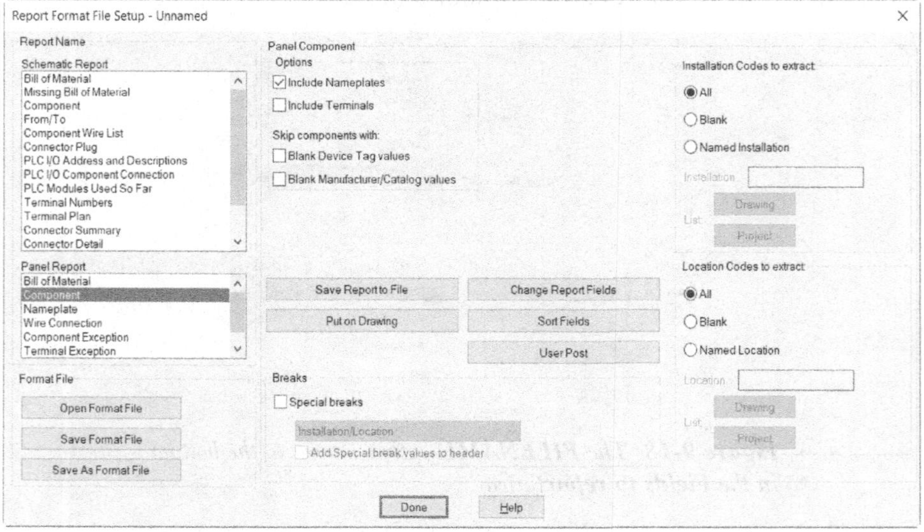

***Figure 9-19** The modified **Report Format File Setup - Unnamed** dialog box*

2. Choose the **Change Report Fields** button from the **Report Format File Setup-Unnamed** dialog box; the **Panel Component Data Fields to Report** dialog box is displayed.

3. Choose the **FILENAME** option from the **Available fields** area; it is shifted to the **Fields to report** area.

4. Make sure **FILENAME** is highlighted in the **Fields to report** area and then move it to the bottom using the **Move Down** button, as shown in Figure 9-20.

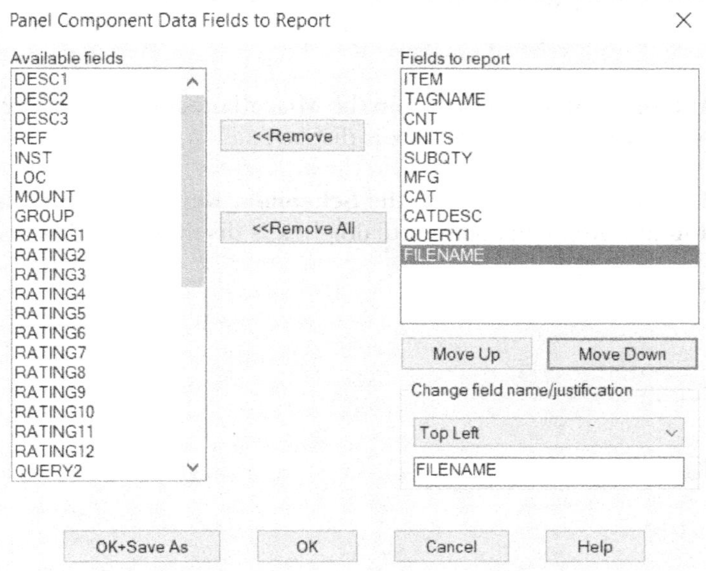

***Figure 9-20 FILENAME** moved to the bottom in the **Fields to report** area*

5. Choose the **OK** button in the **Panel Component Data Fields to Report** dialog box; you return to the **Report Format File Setup - XX** dialog box.

6. Choose the **Put on Drawing** button from the **Report Format File Setup - Unnamed** dialog box; the **Table Generation Setup** dialog box is displayed.

7. Select the **Include time/date**, **Include project information**, and **Include title line** check boxes from the **Title** area of this dialog box. Also, make sure that the **Insert New** radio button from the **Table** area, the **Calculate automatically** radio button from the **Column Width** area, and the **Rows for Each Section** check box from the **Row Definition** area are selected.

8. Choose the **Pick** button from the **First new Section Placement** area; you are prompted to specify the insertion point. Click on the upper left corner of the drawing area.

   Note that the table that will be generated in the automatic report generation process will have this point as its upper left corner.

9. Choose the **OK** button from the **Table Generation Setup** dialog box; you return to the **Report Format File Setup - XX** dialog box.

10. Choose the **Save As Format File** button from this dialog box; the **Select Panel Component *.set format file** dialog box is displayed. In this dialog box, enter **panel components_format** in the **File name** edit box and then choose the **Save** button from the **Select Panel Component *.set format file** dialog box.

11. Choose the **Done** button from the **Report Format File Setup - Unnamed** dialog box.

Creating Group for Automatic Report Generation

In this section, you will create group for generating automatic report of all the components.

1. Choose the **Automatic Reports** tool from the **Miscellaneous** panel of the **Reports** tab; the **Automatic Report Selection** dialog box is displayed.

2. Choose the **Component** option from the **Schematic Report** area of this dialog box; the modified **Automatic Report Selection** dialog box is displayed, as shown in Figure 9-21.

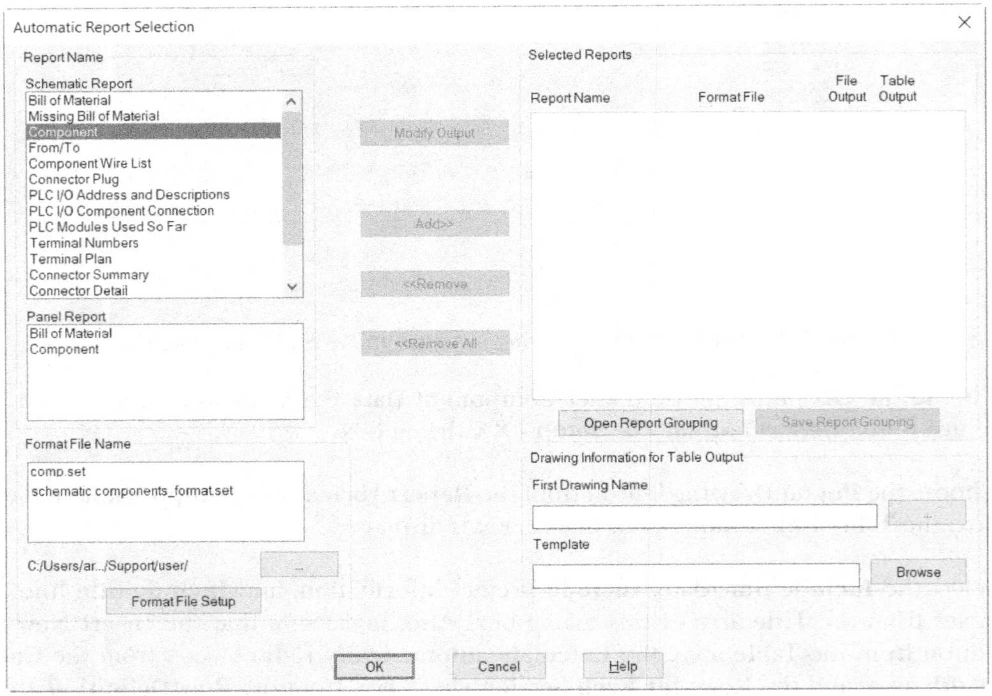

*Figure 9-21 The modified **Automatic Report Selection** dialog box*

3. Select **schematic components_format.set** from the **Format File Name** area of this dialog box; the **Add>>** button is activated. Choose the **Add>>** button; a row is added to the **Selected Reports** area of the dialog box.

   You will notice that x is displayed in the **File Output** column of the added row. This is because the file output is selected in the *schematic components_format.set* format file.

4. Choose the **Component** option from the **Panel Report** area of this dialog box; the modified **Automatic Report Selection** dialog box is displayed.

5. Select **panel components_format.set** from the **Format File Name** area of this dialog box; the **Add>>** button is activated. Choose the **Add>>** button; a row is added to the **Selected Reports** area of the dialog box, as shown in Figure 9-22.

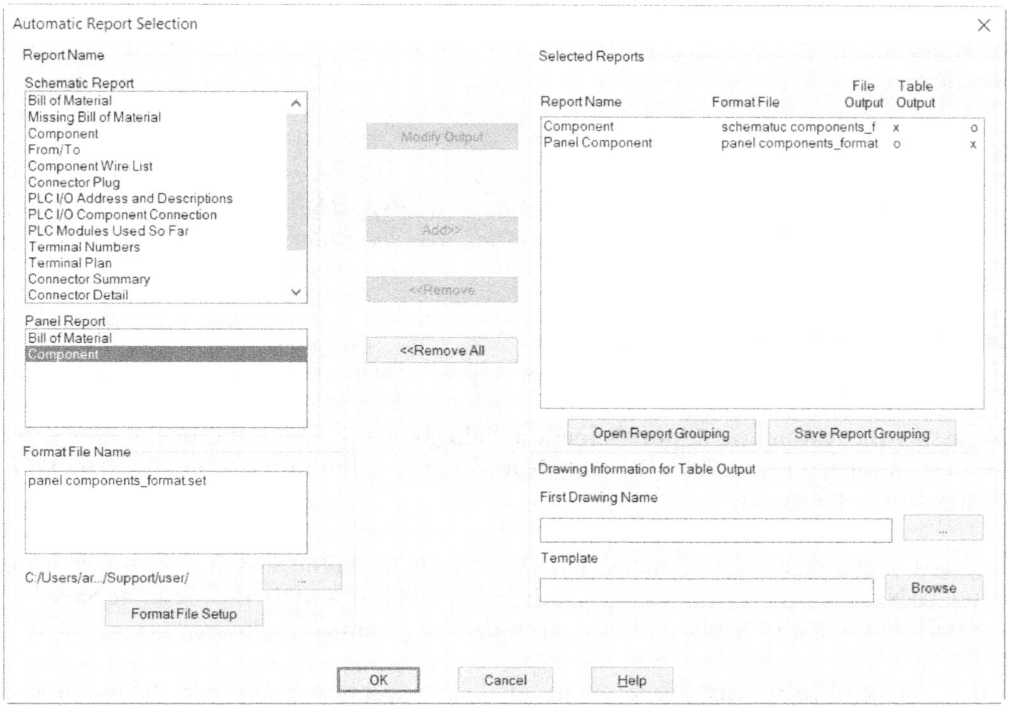

Figure 9-22 The modified ***Automatic Report Selection*** *dialog box*

You will notice that x is displayed in the **Table Output** column of the added row. This is because the table output is selected in the *panel components_format.set* format file.

6. Choose the **Save Report Grouping** button located below the **Selected Reports** area of the dialog box; the **Enter name for Automatic Report Group file** dialog box is displayed, as shown in Figure 9-23.

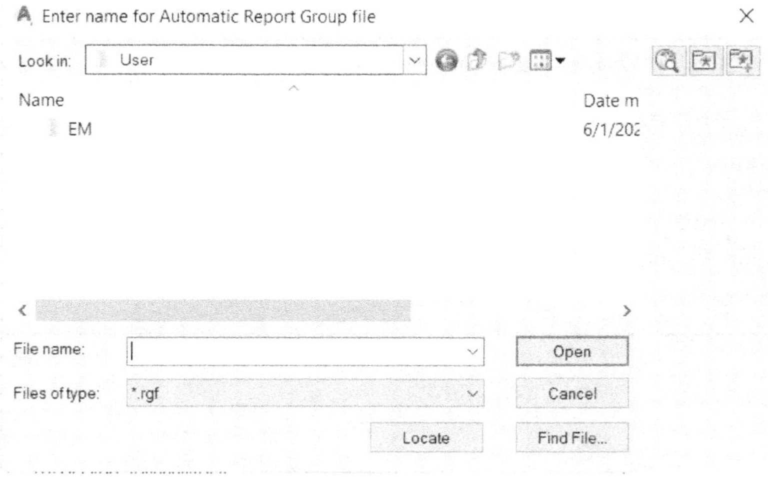

Figure 9-23 The ***Enter name for Automatic Report Group file*** *dialog box*

7. In this dialog box, enter **Schematic & Panel Components** in the **File name** edit box and then choose the **Save** button; the *Schematic & Panel Components.rgf* file is created at the default location.

Note
For generating automatic report of schematic and panel components of other projects, choose the ***Open Report Grouping*** *button from the* ***Automatic Report Selection*** *dialog box and then select the* ***Schematic & Panel Components.rgf*** *file from the* ***Enter name for Automatic Report Group file*** *dialog box.*

Generating Automatic Report

In this section, you will generate the automatic report of schematic and panel components.

1. Make sure the **Automatic Report Selection** dialog box is opened. Choose the Browse button located on the left of the **First Drawing Name** edit box; the **Enter Name for First Drawing** dialog box is displayed.

2. In this dialog box, make sure that CADCIM is displayed in the **Save in** drop-down list. Also, enter **PANEL COMPONENTS** in the **File name** edit box and then choose the **Save** button; the path of the file is displayed in the **First drawing name** edit box.

3. Now, you need to specify the template in the **Template** edit box. To do so, choose the **Browse** button; the **Select template** dialog box is displayed. In this dialog box, select **ACAD_ELECTRICAL.dwt** from the list displayed and then choose the **Open** button; the name and path of the template is displayed in the **Template** edit box.

4. Choose the **OK** button from the **Automatic Report Selection** dialog box; the automatic report for schematic as well as panel components is generated at the specified location.

Note
If the ***QSAVE*** *message box is displayed, choose the* ***OK*** *button from this message box.*

5. Open the windows explorer and browse to the location where you have saved the *Schematic Components_CADCIM.xls* file. Next, open this file.

6. Open the ***PANEL COMPONENTS*.dwg** file created in the **CADCIM** project during the automatic report generation process. You will notice that panel components are listed in the table.

7. Choose **Close > Current Drawing** from the **Application Menu** to close the the *C09_tut03.dwg* file.

Self-Evaluation Test

Answer the following questions and then compare them to those given at the end of this chapter:

1. Which of the following commands is used to create schematic reports?

   (a) **AESCHEMATICREPORT** (b) **AEAUDIT**
   (c) **AEPANELREPORT** (d) **AEAUDITDWG**

2. The __________ dialog box is used to change the report format and to export the report to the drawing or to an external file.

3. When you choose the **Put on Drawing** button from the **Report Generator** dialog box, the __________ dialog box is displayed.

4. If you choose the __________ report from the **Report Name** area of the **Schematic Reports** dialog box, then the report of the component wire connection will be extracted and displayed in the **Report Generator** dialog box.

5. The **Panel Reports** tool is used to create and specify the panel-based reports. (T/F)

Review Questions

Answer the following questions:

1. Which of the following buttons in the **Report Generator** dialog box is used to modify a report?

   (a) **Surf** (b) **User Post**
   (c) **Sort** (d) **Edit Mode**

2. Which of the following commands is used to create panel reports?

   (a) **AESCHEMATICREPORT** (b) **AEAUDIT**
   (c) **AEPANELREPORT** (d) **AEAUDITDWG**

3. The **Panel Bill of Material** report displays the report of panel components and nameplates only. (T/F)

4. You can change the order of data fields in the **Data Field to Report** dialog box. (T/F)

5. In AutoCAD Electrical, the report generation of schematic drawings is a time consuming process. (T/F)

EXERCISES

Exercise 1

In this exercise, you will create a new drawing as *C09_exer01.dwg* and then generate the schematic Missing Bill of Material report of the **NEW_PROJECT** project. Next, you will save the report to the Access database (.mdb) as *missing_bom*. Also, you will generate a Panel Bill of Material report of the **NEW_PROJECT** project and place it on *C09_exer01.dwg* drawing.

(Expected time: 20 min)

Exercise 2

In this exercise, you will use the **Report Format Setup** tool for creating format files for the schematic Bill of Material and panel Bill of Material and then save them as *bom.set* and *panel_bom.set* respectively, in the **NEW_PROJECT** project. Also, you will save the schematic Bill of Material report and panel Bill of Material report to a file in the **Excel Spreadsheet (.xls)** format as *bom.xls* and *panel_bom.xls,* respectively. Next, you will use the **Automatic Reports** tool to group the format files and for generating the reports. **(Expected time: 30 min)**

Answers to Self-Evaluation Test

1. a, **2. Report Generator**, **3. Table Generation Setup**, **4. Component Wire List**, **5. T**

Chapter 10

PLC Modules

Learning Objectives

After completing this chapter, you will be able to:

- *Insert parametric and nonparametric PLC modules*
- *Edit parametric and nonparametric PLC modules*
- *Insert individual PLC I/O points*
- *Create and modify the parametric PLC modules*
- *Create PLC I/O wiring diagrams using the spreadsheet data*
- *Map the spreadsheet information*
- *Modify and save the new setting configuration*
- *Tag components and wire numbers based on PLC I/O address*

INTRODUCTION

A programmable logic controller (PLC) or programmable controller is a digital computer that is used to automatically regulate the industrial process. For example, you can use programmable logic controller to control machinery on assembly lines of a factory. This controller is designed to meet a range of industrial activities such as multiple inputs and output arrangements, extended temperature ranges, providing resistance to vibration, and so on. The programs designed to regulate the machine operation are usually stored in a battery-backed or non-volatile memory. A PLC is an instance of a real time system because output results are needed to be produced in response to input conditions within a time-bound period, else it will result in an unintended operation. Over the years, the functionality of PLC has evolved to accommodate sequential relay control, motion control, process control, distributed control systems, and networking. In PLC, microprocessor controlled interface is inbuilt and is designed to control or monitor some other I/O functions. Being an industrial computer control system, it always monitors the state of input devices and makes decisions on the basis of custom program for controlling the state of devices connected as outputs.

In this chapter, you will learn how to use PLC I/O module functions, insert parametric and nonparametric PLC modules, and create custom PLC I/O modules.

INSERTING PARAMETRIC PLC MODULES

Command: AEPLCP

AutoCAD Electrical can generate PLC I/O modules in different graphical styles through parametric generation technique on demand. These modules are generated by a database file and a library of symbol blocks. PLC I/O modules can be inserted as independent symbols. PLC modules behave like other schematic components. These modules are AutoCAD blocks containing attributes for connection points, tagging, and so on.

To insert a PLC module, choose the **Insert PLC (Parametric)** tool from the **Insert PLC** drop-down in the **Insert Components** panel of the **Schematic** tab, as shown in Figure 10-1; the **PLC Parametric Selection** dialog box will be displayed, as shown in Figure 10-2. Using this dialog box, you can select the PLC module and its graphics that you want to insert in a drawing.

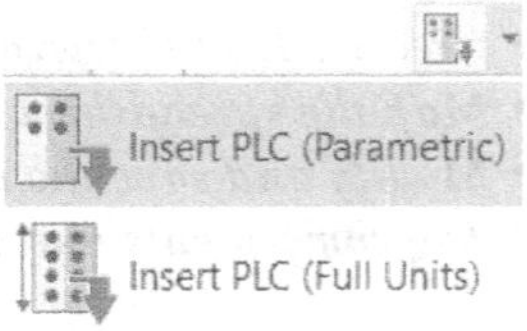

***Figure 10-1** The **Insert PLC** drop-down*

The Manufacturer Catalog tree, shown in Figure 10-3, displays a list of PLC modules. This list can be filtered by selecting the manufacturer, series, and type of PLC. The data displayed in the Manufacturer Catalog tree is stored in the *ACE_PLC.mdb* database file.

You can select a PLC module from the Manufacturer Catalog tree. To do so, click on the + sign on the required manufacturers module; PLC modules of the selected manufacturer will be displayed. Select the required module from the module list; the detailed information of the selected module will be displayed at the lower part of the **PLC Parametric Selection** dialog box, as shown in Figure 10-4.

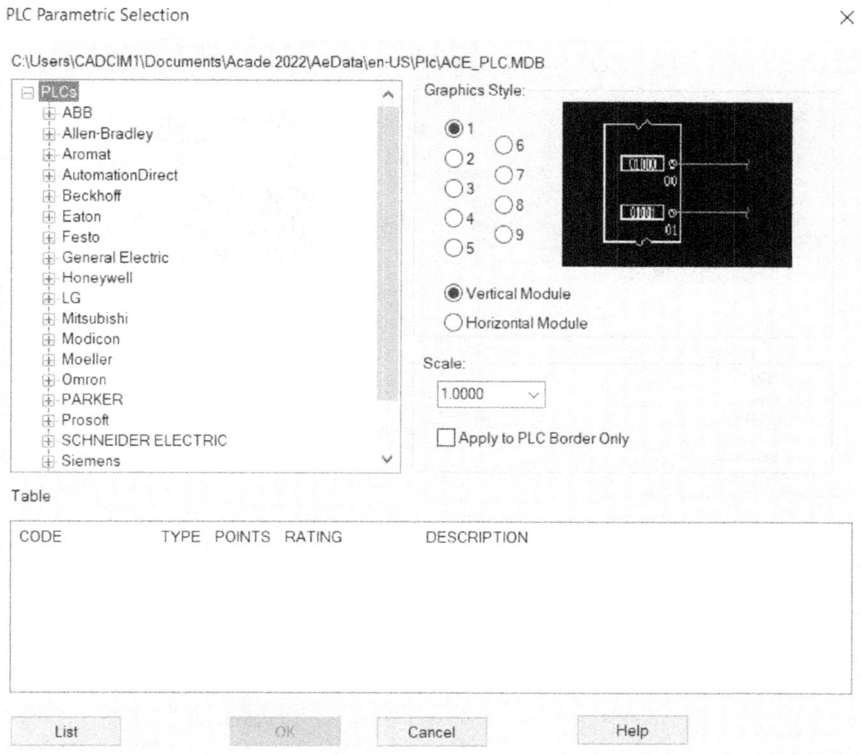

Figure 10-2 *The* ***PLC Parametric Selection*** *dialog box*

The **Graphics Style** area displays the graphical styles of a PLC module. This area consists of five (1-5) pre-defined PLC styles and four (6-9) user-defined PLC styles. You can insert a PLC module into a drawing vertically or horizontally by selecting the **Vertical Module** or **Horizontal Module** radio button from the **Graphics Style** area.

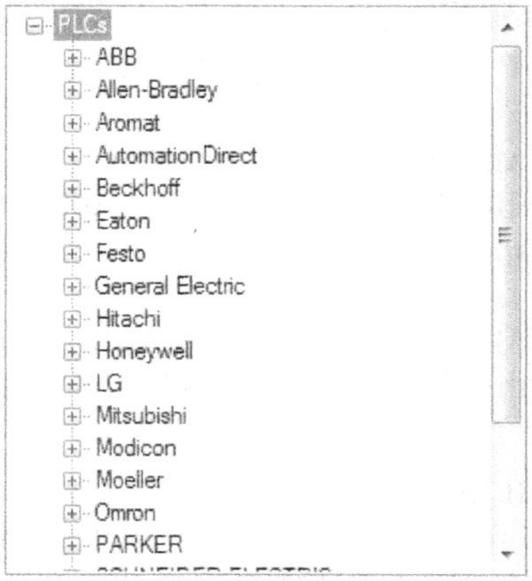

Figure 10-3 *The Manufacturer Catalog tree*

After selecting the required module, graphics style, and scale, choose the **OK** button from the **PLC Parametric Selection** dialog box; the outline of the module will be displayed with an 'X' at the wire connection point of the topmost I/O point and you will be prompted to specify the insertion point for the PLC module. Specify the insertion point; the **Module Layout** dialog box will be displayed.

After specifying the required options in the **Module Layout** dialog box, choose the **OK** button from this dialog box; the **I/O Point** dialog box will be displayed. In this dialog box, enter the required rack and slot values in the **RACK** and **SLOT** edit boxes, respectively.

***Figure 10-4** The **PLC Parametric Selection** dialog box displaying the details of the selected module*

Note

*The options in the **I/O Point** dialog box depend on the PLC module selected from the **PLC Parametric Selection** dialog box.*

After specifying the required values in the **I/O Point** dialog box, choose the **OK** button from this dialog box; the **I/O Address** dialog box will be displayed. Using this dialog box, you can specify the address for the first I/O point.

After specifying the required options in the **I/O Address** dialog box, choose the **OK** button in this dialog box; the selected PLC module will be inserted into the drawing.

If the module does not fit into a single ladder, you can break the module and continue inserting the module in other area of the same drawing or in another drawing. Also, you can add spacers for skipping I/O point of a module. To break a module or add a spacer, select the **Allow spacers/breaks** radio button from the **I/O points** area of the **Module Layout** dialog box and then choose the **OK** button; the **I/O Point** dialog box will be displayed. In this dialog box, specify the required options and then choose the **OK** button; the **Custom Breaks/Spacing** dialog box will be displayed, as shown in Figure 10-5.

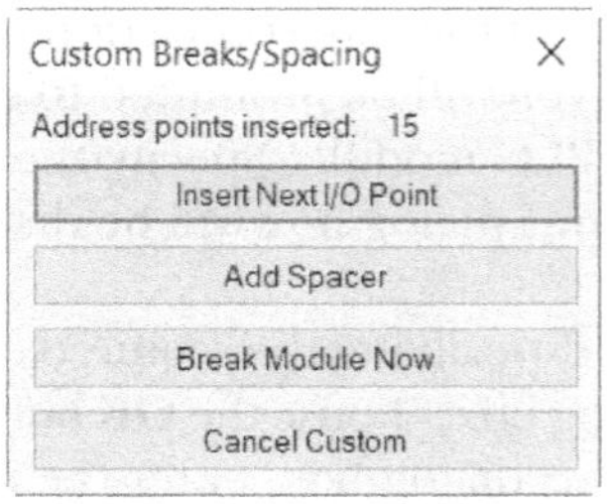

***Figure 10-5** The **Custom Breaks/Spacing** dialog box*

The **Break Module Now** button is used to break a module at the current I/O point as many times as you need. To break a module, choose the **Break Module Now** button from the **Custom Breaks/Spacing** dialog box; you will be prompted to specify the insertion point for the required PLC module. Specify the insertion point; the **Module Layout** dialog box will be displayed again. Specify the required options in this dialog box and choose the **OK** button; the **Custom Breaks/ Spacing** dialog box will be displayed again. The **Cancel Custom** button is used to insert the remaining I/O points in a PLC module without any breaks or spacers.

Now, if you want to stop the insertion process to move to a different drawing or to execute a command, then you must first break the module. To do so, select the **Allow spacers/breaks** radio button in the **Module Layout** dialog box and then choose **OK**; the **I/O Point** dialog box will be displayed. Enter the required information in this dialog box and choose **OK**; the **I/O Addresses** dialog box will be displayed. Choose **OK** from this dialog box; the **Custom Breaks/ Spacing** dialog box will be displayed. In this dialog box, choose the **Break Module Now** button; you will be prompted to specify the insertion point for the module. Next, press ESC to cancel the insertion of the module; the **Data Saved** message box will be displayed. The **Data Saved** message box informs you that the remaining part of the module has been saved. Note that AutoCAD Electrical will save this information for the current session only. Choose the **OK** button from the **Data Saved** message box. Now, you can execute any number of commands or operations before you continue inserting the PLC module.

Now, if you choose the **Insert PLC (Parametric)** tool from the **Insert Components** panel of the **Schematic** tab, the **Continue "Broken" Module** dialog box will be displayed, as shown in Figure 10-6. In this dialog box, the **Continue Module** button is used to continue the previously broken module. The **Start New Module** button is used to start a new module and discard the remainder of the broken module's data. Choose the **Cancel** button to exit the **Continue "Broken" Module** dialog box. After choosing this button, the broken module's data is saved. You can insert the broken module later.

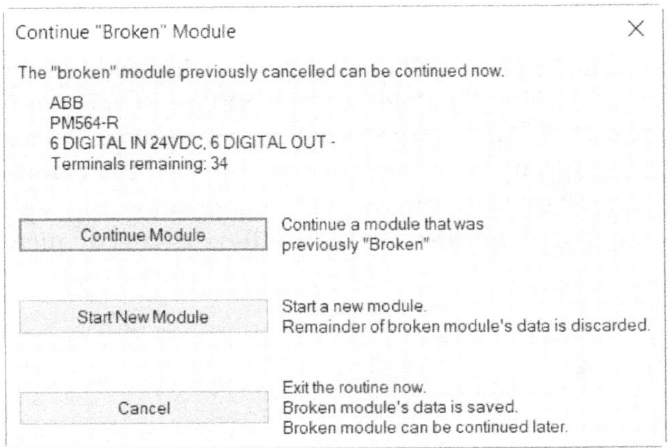

Figure 10-6 *The* ***Continue "Broken" Module*** *dialog box*

Note

If you choose the ***Start New Module*** *button or exit AutoCAD Electrical, you will lose the saved data.*

INSERTING NONPARAMETRIC PLC MODULES

Command: AEPLC

The **Insert PLC (Full Units)** tool is used to insert a PLC module into a drawing as a single unit. If you insert a PLC module into a drawing using this tool, the underlying rungs of a ladder will break and then reconnect. Using this tool, you can insert PLC I/O points as a complete PLC module or as independent symbols into the drawing. To insert a PLC module, choose the **Insert PLC (Full Units)** tool from the **Insert Components** panel of the **Schematic** tab; the **Insert Component** dialog box will be displayed, as shown in Figure 10-7.

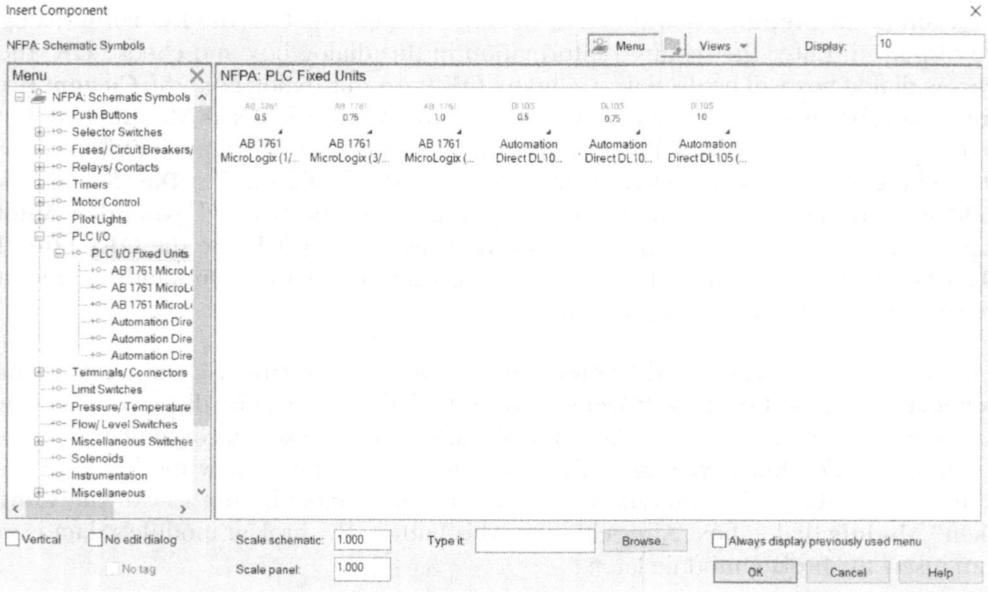

*Figure 10-7 The **Insert Component** dialog box*

In this dialog box, the nonparametric PLC modules are displayed in the **NFPA: PLC Fixed Units** area. Now, select the required module from the **NFPA : PLC Fixed Units** area; the **Insert Component** dialog box will disappear and you will be prompted to specify the insertion point for the PLC module. Next, specify the insertion point; the **Edit PLC Module** dialog box will be displayed. Enter the required values in this dialog box and choose the **OK** button; the complete PLC module will be inserted into the drawing. Figure 10-8 shows a complete nonparametric PLC module.

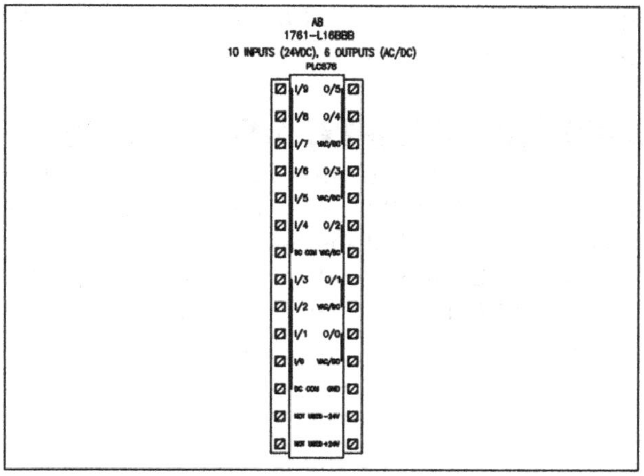

Figure 10-8 *A complete nonparametric PLC module*

EDITING A PLC MODULE

Command: AEEDITCOMPONENT

You can edit a PLC module inserted into the drawing. To do so, choose the **Edit** tool from the **Edit Components** panel of the **Schematic** tab; you will be prompted to select a component. Select the required PLC module from the drawing; the **Edit PLC Module** dialog box will be displayed, as shown in Figure 10-9. In the **Edit PLC Module** dialog box, you can edit the I/O point description, installation/location codes, catalog data, and so on.

Figure 10-9 *The* ***Edit PLC Module*** *dialog box*

After specifying the required options in the **Edit PLC Module** dialog box, choose the **OK** button from this dialog box to save the changes made.

INSERTING INDIVIDUAL PLC I/O POINTS

You can insert individual PLC I/O points into a drawing. These PLC I/O points are parent or child schematic symbols. To insert individual PLC I/O points into a drawing, choose the **Icon Menu** tool from the **Insert Components** panel of the **Schematic** tab; the **Insert Component** dialog box will be displayed. Select the **PLC I/O** icon displayed in the **NFPA: Schematic Symbols** area of this dialog box; the **NFPA: PLC I/O** area will be displayed, as shown in Figure 10-10. This area displays the parent and child symbols of a PLC module.

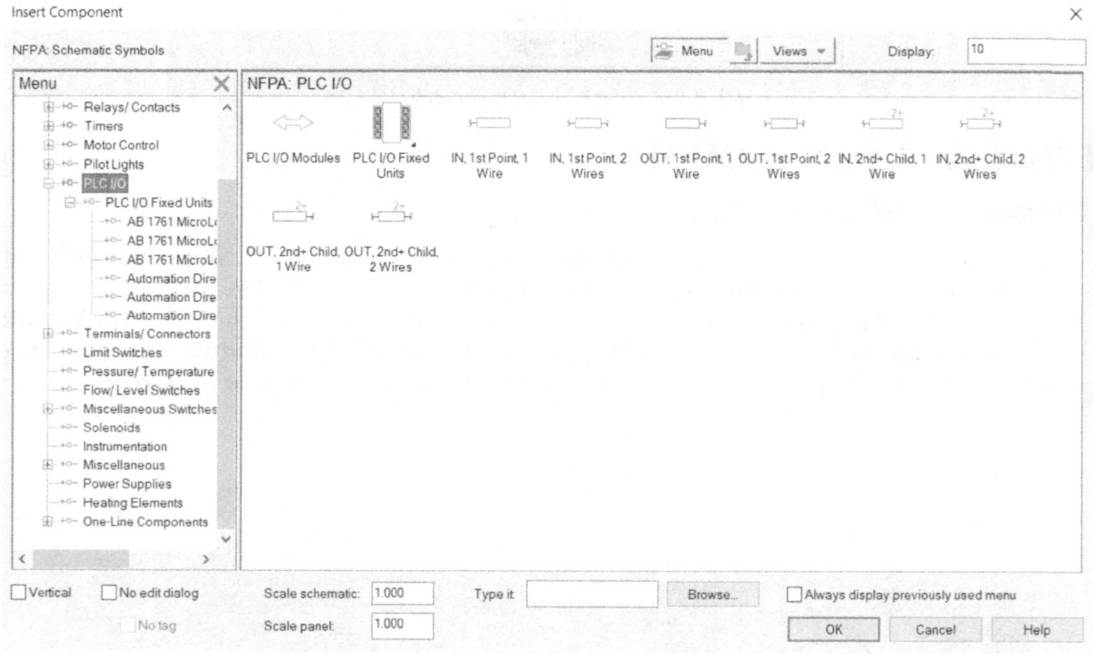

Figure 10-10 *The* ***NFPA: PLC I/O*** *area of the* ***Insert Component*** *dialog box*

The symbols with prefix 1st Point are the parent symbols and with 2nd+ are the child symbols. Select the desired symbol from the **NFPA: PLC I/O** area of the **Insert Component** dialog box; you will be prompted to specify the insertion point. Specify the insertion point; the **Edit PLC I/O Point** dialog box will be displayed, as shown in Figure 10-11. In this dialog box, specify the I/O address, catalog information, and so on for a parent symbol. After specifying the required options in the **Edit PLC I/O Point** dialog box, choose the **OK** button; the PLC symbol will be inserted into a drawing.

Note

If you insert a child PLC symbol into a drawing, then the ***Manufacturer****,* ***Catalog****, and* ***Assembly*** *edit boxes will not be activated in the* ***Edit PLC I/O Point*** *dialog box.*

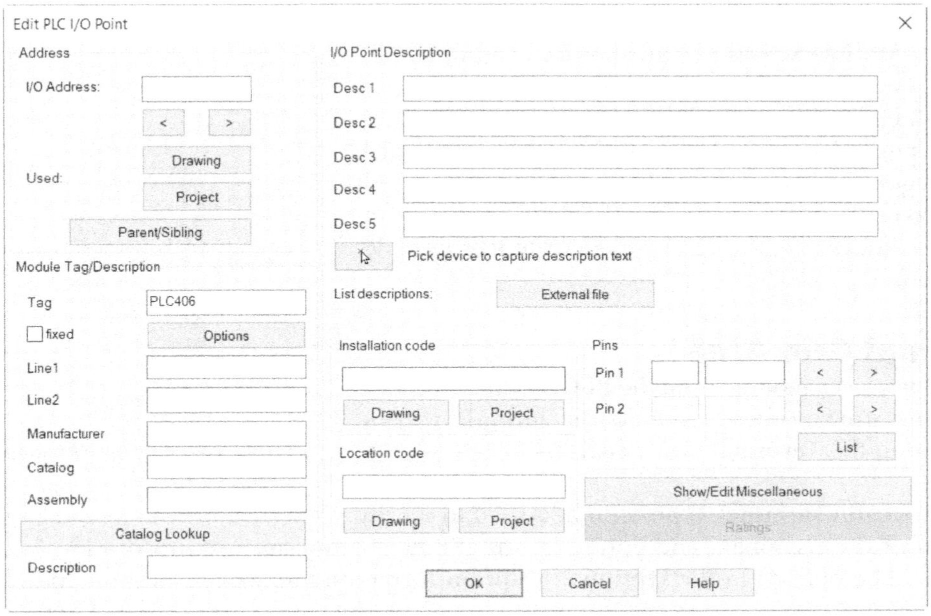

*Figure 10-11 The **Edit PLC I/O Point** dialog box*

CREATING AND MODIFYING PARAMETRIC PLC MODULES

Command: AEPLCDB

You can create as well as modify parametric PLC modules by using the **PLC Database File Editor** tool. This tool is also used to edit and modify the PLC database file *ACE_PLC.MDB*. To create a PLC module, choose the **PLC Database File Editor** tool from the **Other Tools** panel of the **Schematic** tab; the **PLC Database File Editor** dialog box will be displayed, refer to Figure 10-12. The options in the **PLC Database File Editor** dialog box are discussed next.

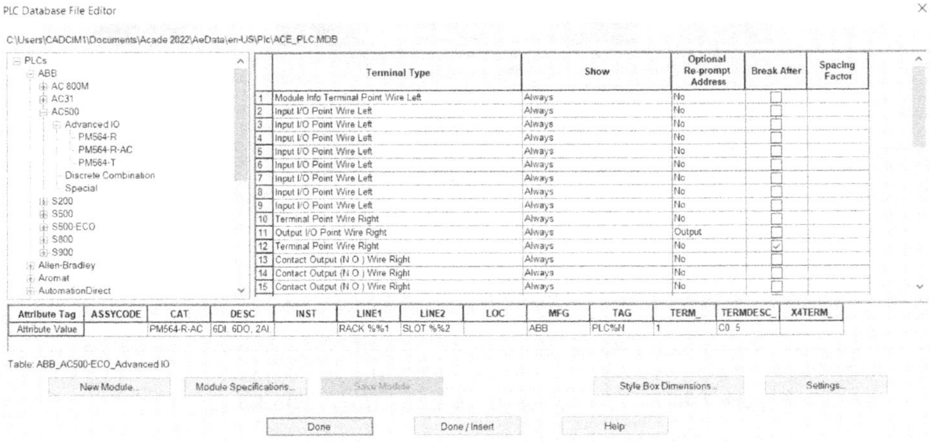

*Figure 10-12 The **PLC Database File Editor** dialog box*

PLC Module Selection List

The PLC Module Selection List, shown in Figure 10-13, displays the expanded tree view of PLC data files present in the PLC database. The list consists of manufacturer, series, type, and part number categories. If you right-click on a category shown in Figure 10-13, a shortcut menu will be displayed. Using the options in the shortcut menu, you can cut, copy, delete, paste, rename, create new module, and so on.

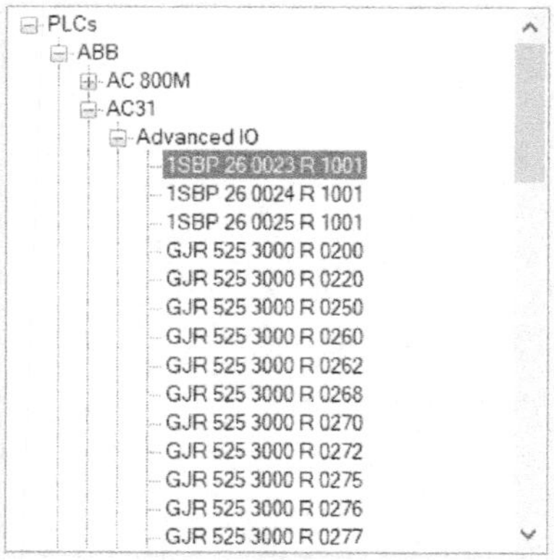

Figure 10-13 The PLC Module Selection List

Terminal Grid Area

When you select a module from the PLC Module Selection List, the information related to the module will be displayed in the **Terminal Grid** area.

In this area, the **Terminal Type** column is used to specify the type of the terminal. The options in the **Show** column are used to display the unused terminal. The options in the **Optional Re-prompt** column are used to prompt you to specify new beginning address for the I/O points of a PLC module. Select the **Break After** check box to break a module with the corresponding terminal type selected in the **Terminal Grid** area.

The **Spacing Factor** column is used to override the current rung spacing for I/O point with the specified spacing factor. The **Edit Terminal** option is used to edit the selected terminal. To do so, right-click in the **Terminal Grid** area; a shortcut menu will be displayed. Choose the **Edit Terminal** option from the shortcut menu; the **Select Terminal Information** dialog box will be displayed, as shown in Figure 10-14. Next, select the desired category and type of a terminal from this dialog box and choose **OK**; the selected category and its type will be displayed in the selected cell or column in the **Terminal Grid** area of the **PLC Database File Editor** dialog box.

*Figure 10-14 The **Select Terminal Information** dialog box*

Terminal Attributes Area

The **Terminal Attributes** area displays the attribute values associated with the selected terminal.

New Module

The **New Module** button is used to create a new PLC module. To do so, choose the **New Module** button from the **PLC Database File Editor** dialog box; the **New Module** dialog box will be displayed, as shown in Figure 10-15.

***Figure 10-15** The **New Module** dialog box*

After specifying the required options in the **New Module** dialog box, choose the **OK** button in this dialog box; a new module will be created and added to the PLC Module Selection List.

The **Module Specifications** button is used to edit the specifications of an existing PLC module.

The **Style Box Dimensions** button is used to define the dimensions of a module box for a specific graphic style. The **Settings** button is used to add or edit the symbols available for building a module. To add or edit the symbols, choose the **Settings** button from the **PLC Database File Editor** dialog box; the **Terminal Block Settings** dialog box will be displayed, as shown in Figure 10-16. In this dialog box, the **Block File Name** column displays the block file name of a terminal of a PLC module. Enter the file name of the terminal block in the required cell of the **Block File Name** column by clicking on it. Alternatively, select the block file name cell and then choose the **Browse** button from the **Terminal Block Settings** dialog box; the **Select File** dialog box will be displayed. Select the drawing file and choose the **Open** button; the file name will be displayed in the selected cell of the **Block File Name** column.

The **Category** column is used to define the category for the terminal of a PLC module. When you click on any cell in the **Category** column, the cell will change into a drop-down list. You can specify the required category in the respective cell of the **Category** column manually or select a category from the drop-down list. By default, the **Top Terminal**, **Top Input**, **Top Output**, **Input**, **Terminal**, and **Output** options are available in the drop-down list.

After specifying the required options in the **Terminal Block Settings** dialog box, choose the **OK** button in this dialog box to return to the **PLC Database File Editor** dialog box.

Note

*When you right-click on any of the columns in the **Terminal Block Settings** dialog box, a shortcut menu will be displayed. This shortcut menu has options such as **New Row**, **Delete Row**, and so on.*

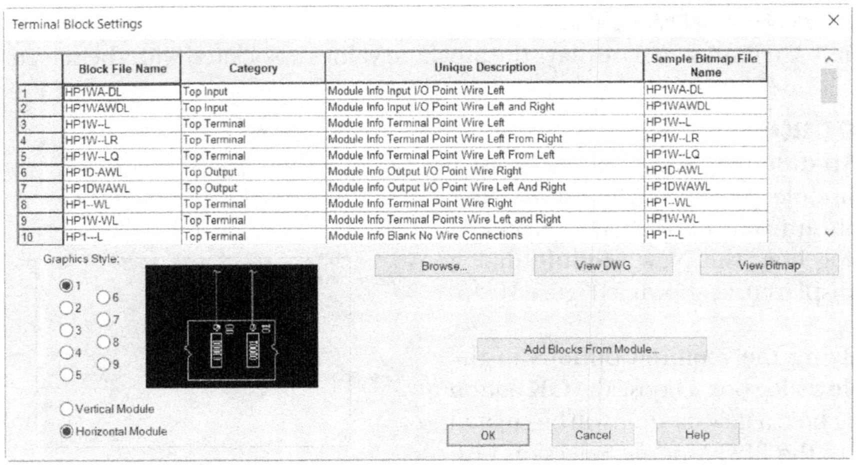

*Figure 10-16 The **Terminal Block Settings** dialog box*

After specifying the options in the **PLC Database File Editor** dialog box, choose the **Done** button to save the changes made and close the dialog box.

Now, to insert the PLC, choose the **Done / Insert** button from the **PLC Database File Editor** dialog box; the **PLC Parametric Selection** dialog box will be displayed. Select the required module and then choose the **OK** button from the **PLC Parametric Selection** dialog box. Next, follow the steps described earlier in the chapter to insert the PLC module.

CREATING PLC I/O WIRING DIAGRAMS

Command: AESS2PLC

The **PLC I/O Utility** tool is used to create a set of PLC I/O wiring diagrams directly from the spreadsheet data. Using this tool, you can generate multiple drawings quickly. To create a drawing using spreadsheet data, choose the **PLC I/O Utility** tool from the **Import** panel of the **Import/Export Data** tab; the **Select PLC I/O Spreadsheet Output File** dialog box will be displayed, as shown in Figure 10-17. The PLC I/O spreadsheet output file will be available in the *.xls*, *.csv*, and *.mdb* formats. Select the required file from this dialog box and choose the **Open** button; the **Spreadsheet to PLC I/O Utility** dialog box will be displayed, as shown in Figure 10-18.

Choose the **Start** button for creating a set of PLC I/O drawings based on the information carried in the selected PLC spreadsheet. Once the drawings are created, the **Spreadsheet to PLC I/O Utility** dialog box will be displayed again.

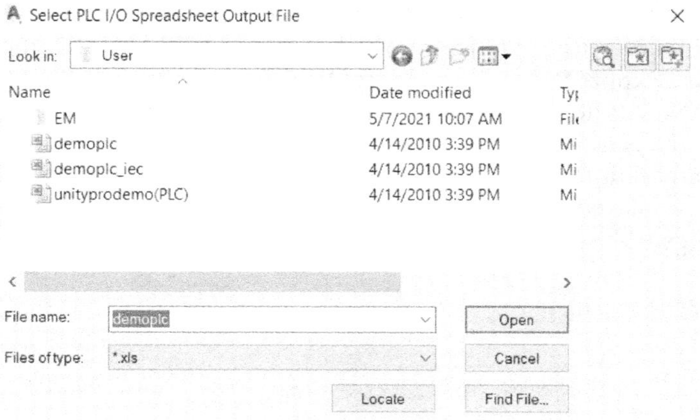

Figure 10-17 *The* ***Select PLC I/O Spreadsheet Output File*** *dialog box*

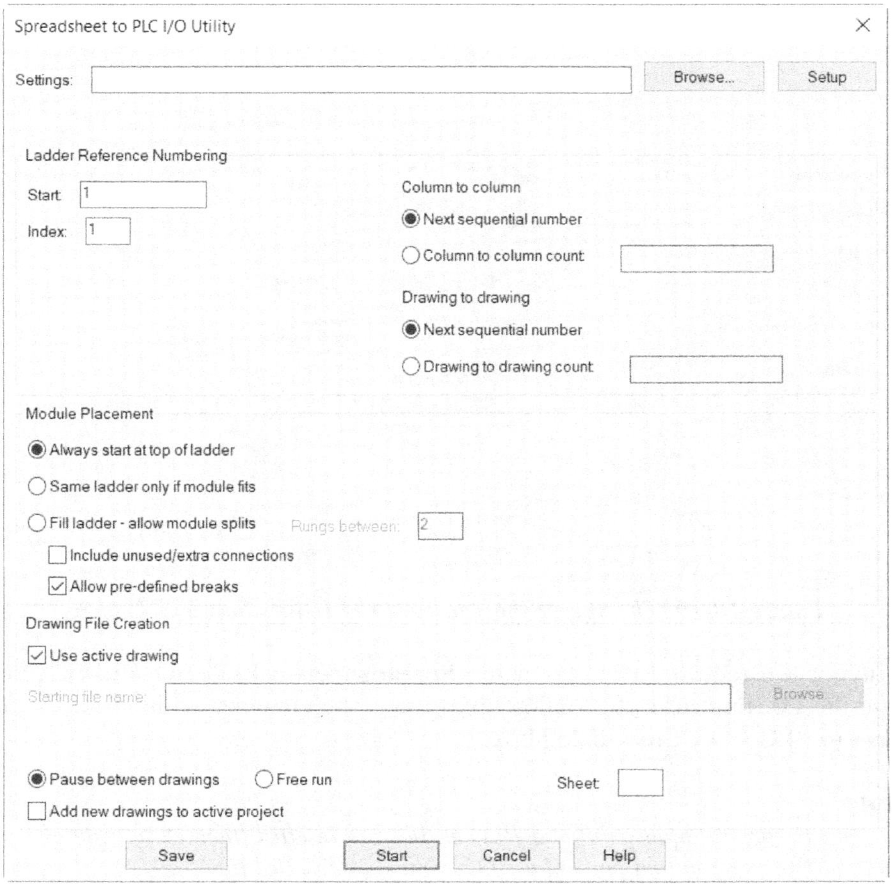

Figure 10-18 *The* ***Spreadsheet to PLC I/O Utility*** *dialog box*

Modifying and Saving the New Setting Configuration

To modify and save a new setting configuration, choose the **PLC I/O Utility** tool from the **Import** panel of the **Import/Export Data** tab; the **Select PLC I/O Spreadsheet Output File** dialog box will be displayed. Next, select the required spreadsheet file from this dialog box and choose the **Open** button; the **Spreadsheet to PLC I/O Utility** dialog box will be displayed. Choose the **Setup** button adjacent to the **Settings** edit box; the **Spreadsheet to PLC I/O Utility Setup** dialog box will be displayed, as shown in Figure 10-19. In this dialog box, you can define the settings for drawings.

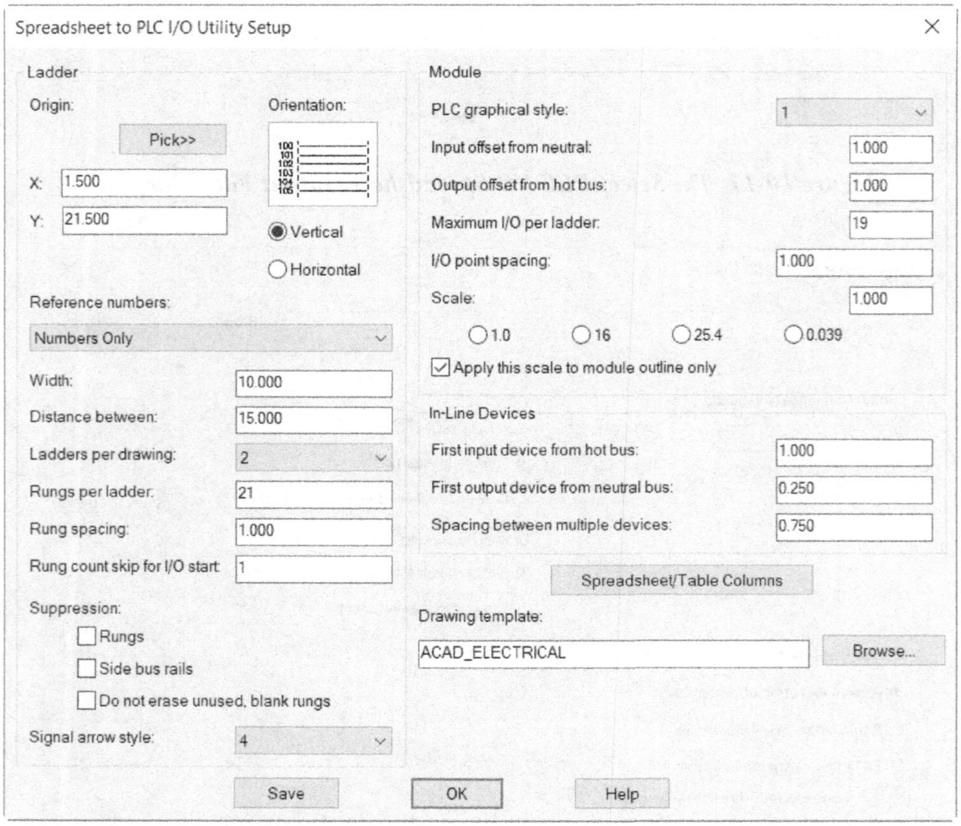

***Figure 10-19** The **Spreadsheet to PLC I/O Utility Setup** dialog box*

After specifying the required options in the **Spreadsheet to PLC I/O Utility Setup** dialog box, choose the **OK** button in this dialog box to save the changes made and to return to the **Spreadsheet to PLC I/O Utility** dialog box.

Note

The devices for an input point of a PLC module are inserted from left to right or top to bottom and the devices for output point of a PLC module are inserted from right to left or bottom to top.

MAPPING THE SPREADSHEET INFORMATION

To map and review the spreadsheet information to the attributes of a PLC module, choose the **Spreadsheet/Table Columns** button from the **Spreadsheet to PLC I/O Utility Setup** dialog box; the **Spreadsheet to PLC I/O Drawing Generator** dialog box will be displayed, as shown in Figure 10-20.

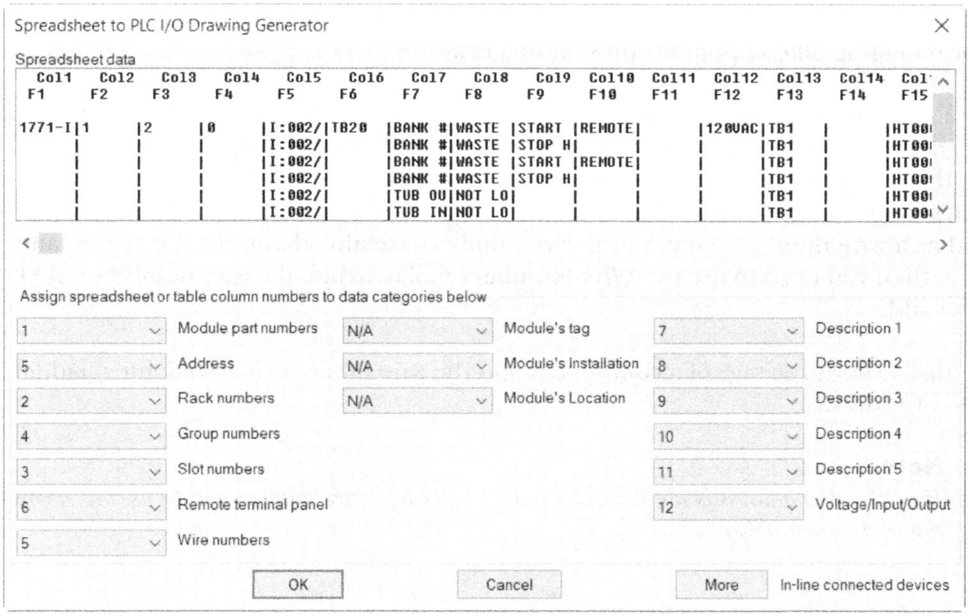

Figure 10-20 *The* ***Spreadsheet to PLC I/O Drawing Generator*** *dialog box*

After specifying the required options in the **Spreadsheet to PLC I/O Drawing Generator** dialog box, choose the **OK** button from this dialog box to save the changes made and to return to the **Spreadsheet to PLC I/O Utility Setup** dialog box.

TAGGING BASED ON PLC I/O ADDRESS

The tags of a component and wire numbers can use PLC address instead of ladder reference number when inserted into a ladder. You can tag a component based on PLC address and then insert it into a drawing. Also, you can insert a wire number, which will use a PLC address. To use PLC address as the tags of a component and wire numbers, you need to set options in the **Drawing Properties** dialog box.

Right-click on the active drawing of an active project in the **PROJECT MANAGER**; a shortcut menu will be displayed. Choose **Properties > Drawing Properties** from the shortcut menu; the **Drawing Properties** dialog box will be displayed. Next, choose the **Components** tab from the **Drawing Properties** dialog box and select the **Search for PLC I/O address on insert** check box, as shown in Figure 10-21. Next, choose the **OK** button to close this dialog box.

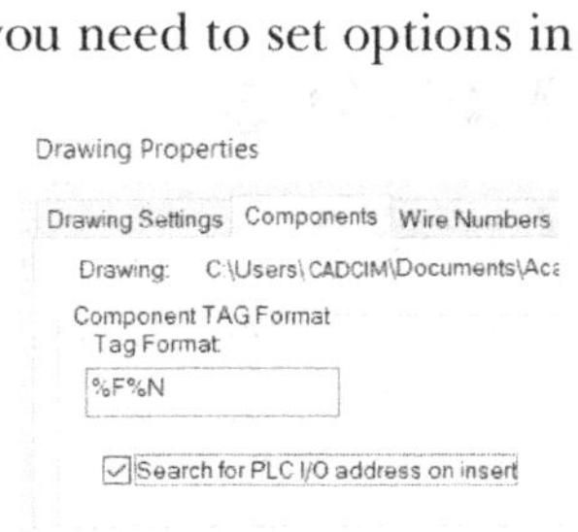

Figure 10-21 *Partial view of the* ***Components*** *tab of the* ***Drawing Properties*** *dialog box*

This check box is used to search for a PLC I/O address when you insert a component and then the tag of the inserted component will be based on the PLC I/O address. You can also tag individual components based on PLC I/O address while inserting or editing the component. To do so, choose the **Use PLC Address** button from the **Insert/Edit Component** dialog box; the I/O address of PLC attached to the component will be displayed in the edit box of the **Component Tag** area in the **Insert/Edit Component** dialog box. Note that if the components are already inserted into the ladder of a drawing, then you need to use the **Retag Component** tool for retagging all components present in a drawing.

To use PLC address as wire numbers, you need to select the **Search for PLC I/O address on insert** check box from the **Wire Numbers** tab of the **Drawing Properties** dialog box. Next, choose the **OK** button in this dialog box to exit. Now, if you insert wire numbers in the drawing, first the software will search the I/O address of a PLC. If the I/O address is found, the same will be used as a wire number. Note that if wire numbers are already inserted into the ladder of a drawing, then you need to use the **Wire Numbers** tool to assign the wire numbers based on the PLC I/O address.

Figure 10-22 shows the tags of a component and the wire numbers inserted into a ladder based on PLC I/O address.

Note

If a PLC I/O address is not found then the ladder reference number will be used as a component tag and wire number.

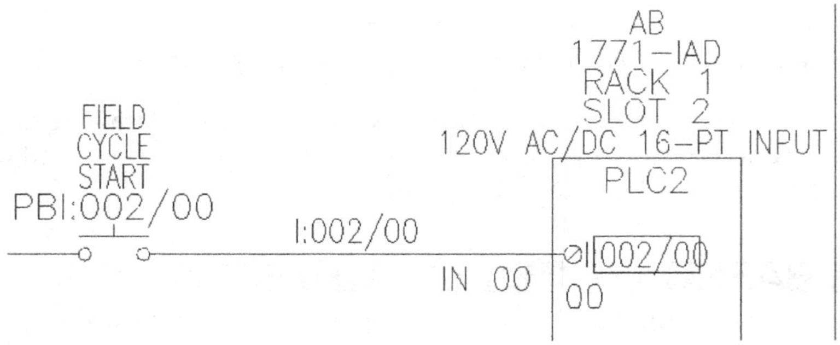

Figure 10-22 *Component tags and wire numbers inserted based on the PLC I/O address*

TUTORIALS

Tutorial 1

In this tutorial, you will first insert a ladder and then a parametric PLC module into the ladder. Also, you will add empty spaces in the PLC module and then break the module to continue inserting it a few rungs later. Next, you will continue inserting the remaining part of the module into a different ladder. **(Expected time: 20 min)**

The following steps are required to complete this tutorial:

a. Create a new drawing.
b. Insert a ladder into the drawing.
c. Insert the parametric PLC module into the drawing, add empty spaces, and break the module.
d. Save the drawing file.

Creating a New Drawing

1. Create a new drawing with the name *C10_tut01* in the **CADCIM** project. Select the **ACAD_ELECTRICAL** template and enter **Parametric PLC** as description while creating the drawing.

2. Choose the **OK** button from the **Create New Drawing** dialog box; *C10_tut01.dwg* is created in the **CADCIM** project and displayed at the bottom of the drawing list in the **CADCIM** project. Next, move *C10_tut01.dwg* to the *TUTORIALS* subfolder of the **CADCIM** project.

Inserting a Ladder into the Drawing

1. Choose the **Insert Ladder** tool from **Schematic > Insert Wires/Wire Numbers > Insert Ladder** drop-down; the **Insert Ladder** dialog box is displayed.

2. Set the following parameters in the **Insert Ladder** dialog box:

Width: **10.000**	Spacing: **1.000**
1st Reference: **200**	Rungs: **10**
1 Phase: Select this radio button	**Yes**: Select this radio button

3. Choose the **OK** button from the **Insert Ladder** dialog box; you are prompted to specify the start position of the first rung. Enter **4,19** at the Command prompt and press ENTER; a ladder is inserted into the drawing.

Inserting the Parametric PLC Module into the Ladder

1. Choose the **Insert PLC (Parametric)** tool from **Schematic > Insert Components > Insert PLC** drop-down; the **PLC Parametric Selection** dialog box is displayed.

2. In the Manufacturer Catalog tree, click on the (+) sign beside **Allen-Bradley**; a list of series is displayed.

3. Click on the (+) sign beside the **1756** series; the **1756** series is expanded.

4. Double-click on the **Discrete Input** type from the Manufacturer Catalog tree; the list is expanded and the various modules of the discrete input type are displayed in the Manufacturer Catalog tree.

5. Select the **1756-IA16** module from the lower part of the **PLC Parametric Selection** dialog box.

6. From the **Graphics Style** area in the **PLC Parametric Selection** dialog box, make sure the **1** radio button is selected.

7. Choose the **OK** button from the **PLC Parametric Selection** dialog box; you are prompted to specify the insertion point for the PLC module.

8. Enter **11,19** at the Command prompt and press ENTER; the **Module Layout** dialog box is displayed.

9. In this dialog box, select the **Allow spacers/breaks** radio button and choose the **OK** button; the **I/O Point** dialog box is displayed.

10. In the **I/O Point** dialog box, enter **1** in the **RACK** edit box and **2** in the **SLOT** edit box.

11. Choose the **OK** button; the **I/O Address** dialog box is displayed. In this dialog box, select **1:12/00** from the **Quick picks** drop-down list; **1:12/00** is displayed in the **Beginning address** edit box.

12. Choose **OK** from the **I/O Address** dialog box; the **Custom Breaks/Spacing** dialog box is displayed.

13. Choose **Add Spacer** from this dialog box; a spacer is added to the selected PLC module.

14. Choose the **Insert Next I/O Point** button; an I/O point is inserted into the module.

15. Choose the **Insert Next I/O Point** button again; an I/O point is inserted into the module.

16. Choose the **Add Spacer** button; one more spacer is added to the PLC module.

17. Choose the **Insert Next I/O Point** button again; an I/O point is inserted into the selected module.

18. Next, choose the **Break Module Now** button; the inserted symbols are combined into a single module block and you are prompted to specify an insertion point for the rest of the PLC module.

19. Next, press ESC; the **Data Saved** message box is displayed. Choose the **OK** button from the message box; the I/O points of a Parametric PLC module are inserted into the ladder, as shown in Figure 10-23.

20. Choose the **Insert Ladder** tool from **Schematic > Insert Wires/Wire Numbers > Insert Ladder** drop-down; the **Insert Ladder** dialog box is displayed.

21. Set the following parameters in the **Insert Ladder** dialog box:

Width: **10.000**
Spacing: **1.000**
1st Reference: **210**
Rungs: **18**
1 Phase: Select this radio button
Yes: Select this radio button

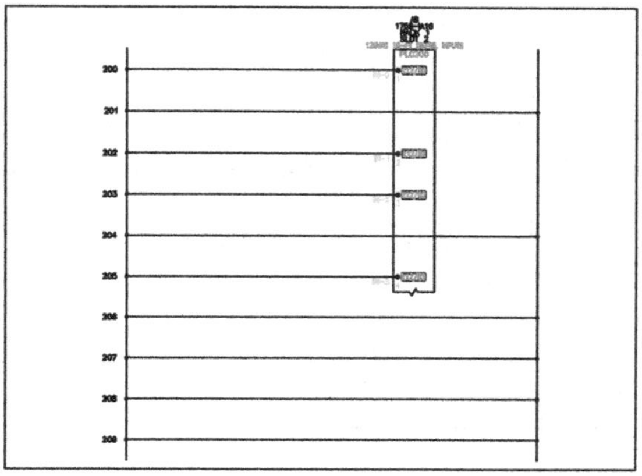

Figure 10-23 *A PLC module inserted into the ladder*

22. Choose the **OK** button from the **Insert Ladder** dialog box; you are prompted to specify the start position of first rung.

23. Enter **18,19** at the Command prompt and press ENTER; a ladder is inserted into the drawing, as shown in Figure 10-24.

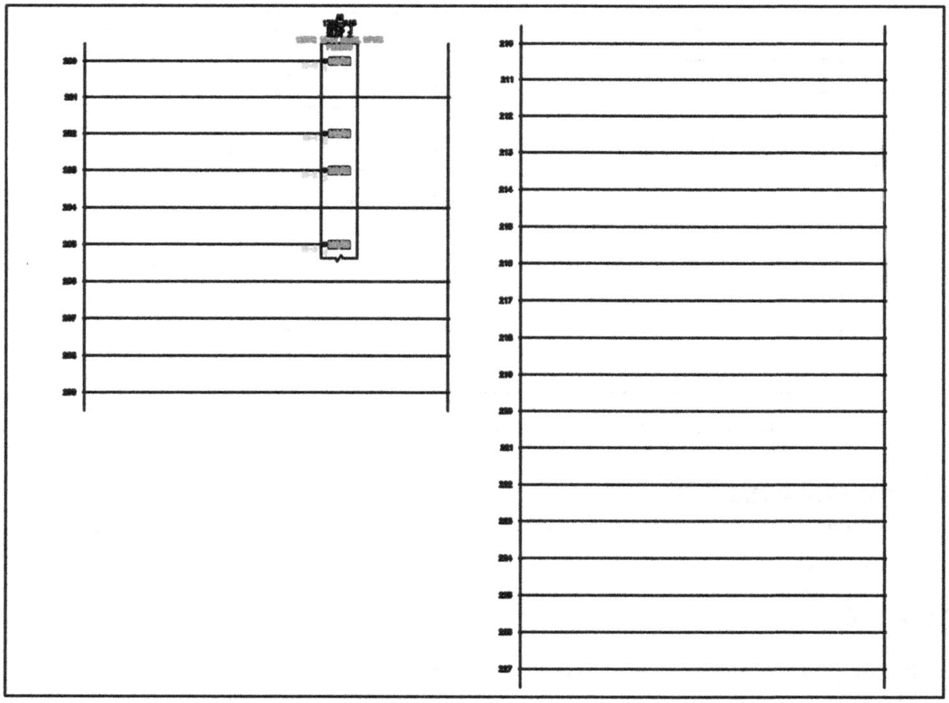

Figure 10-24 *A ladder with 18 rungs inserted into the drawing*

24. Choose the **Insert PLC (Parametric)** tool from **Schematic > Insert Components > Insert PLC** drop-down; the **Continue "Broken" Module** dialog box is displayed.

25. Choose the **Continue Module** button from the **Continue "Broken" Module** dialog box; you are prompted to specify an insertion point for the PLC module.

26. Enter **25,18** at the Command prompt and press ENTER; a PLC module is placed on the right of the ladder at rung 211 and the **Module Layout** dialog box is displayed simultaneously.

27. Select the **Insert all** radio button from the **I/O points** area of the **Module Layout** dialog box and choose **OK**; the **I/O Addressing** dialog box is displayed.

28. In this dialog box, choose the **Decimal** button; the remainder of the PLC module is inserted into the ladder, as shown in Figure 10-25.

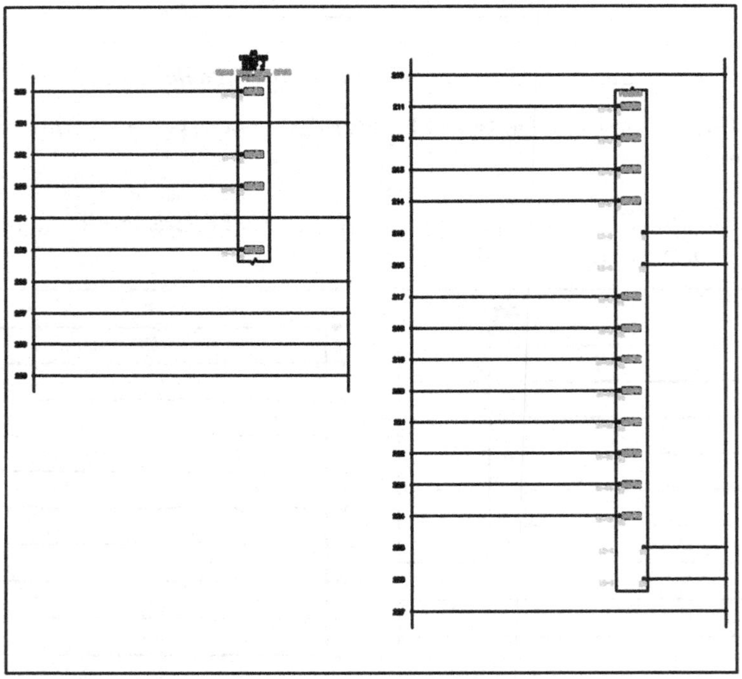

Figure 10-25 *The remaining PLC module inserted into the ladder*

Saving the Drawing File

1. Choose **Save** from the **Application Menu** to save the drawing file *C10_tut01.dwg*.

Tutorial 2

In this tutorial, you will insert a ladder into a drawing and then create a new PLC module using the **PLC Database File Editor** tool. Next, you will insert this new module into the ladder.

(Expected time: 30 min)

The following steps are required to complete this tutorial:

a. Create a new drawing.
b. Insert a ladder into the drawing.

c. Create and insert the PLC module using the **PLC Database File Editor** tool.
d. Save the drawing file.

Creating a New Drawing

1. Create a new drawing with the name *C10_tut02* in the **CADCIM** project. Select the **ACAD_ELECTRICAL** template and enter **Parametric PLC** as description while creating the drawing. Next, move *C10_tut02.dwg* to the *TUTORIALS* subfolder of the **CADCIM** project.

Inserting a Ladder into the Drawing

1. Choose the **Insert Ladder** tool from **Schematic > Insert Wires/Wire Numbers > Insert Ladder** drop-down; the **Insert Ladder** dialog box is displayed.

2. Set the following parameters in the **Insert Ladder** dialog box:

Width: **10.000**	Spacing: **1.000**
1st Reference: **1**	Rungs: **10**
1 Phase: Select this radio button	**Yes**: Select this radio button

3. Choose the **OK** button from the **Insert Ladder** dialog box; you are prompted to specify the start position of the first rung. Enter **10,19** at the Command prompt and press ENTER; a ladder is inserted into the drawing.

Creating and Inserting a New PLC Module

1. Choose the **PLC Database File Editor** tool from the **Other Tools** panel of the **Schematic** tab; the **PLC Database File Editor** dialog box is displayed.

2. Choose the **New Module** button from the **PLC Database File Editor** dialog box; the **New Module** dialog box is displayed.

3. Set the following parameters in the **New Module** dialog box:

Manufacturer: **Automation Direct**	Series: **DL205**
Series Type: **Discrete Input**	Code: **3456**
Description: **16-Point Input**	Module Type: **DC**
Base Addressing: **Prompt**	Rating: **4-20 mA**
Terminals: **16**	

4. Choose the **Module Prompts** button; the **Prompts at Module Insertion Time** dialog box is displayed.

5. Choose **%%1** from the **Prompt Number** area and enter **RACK** in the **New Prompt Text** edit box. Next, choose the **Change** button on the right of the **New Prompt Text** edit box; RACK is displayed in the **Prompt Text** area.

6. Choose **%%2** from the **Prompt Number** area and enter **SLOT** in the **New Prompt Text** edit box. Next, choose the **Change** button; SLOT is displayed in the **Prompt Text** area.

7. Choose the **OK** button from the **Prompts at Module Insertion Time** dialog box to save the changes and exit the dialog box.

8. Choose the **OK** button from the **New Module** dialog box to return to the **PLC Database File Editor** dialog box.

9. In the **Terminal Type** column, right-click on the first cell; a shortcut menu is displayed.

10. In the shortcut menu, choose the **Edit Terminal** option; the **Select Terminal Information** dialog box is displayed.

11. In this dialog box, select the **Top Input** category from the **Category** area and then choose the **Module info Input I/O Point Wire Left** type from the **Types for Category Top Input** area.

12. Choose the **OK** button from the **Select Terminal Information** dialog box; the **Module info Input I/O Point Wire Left** type is displayed in the first cell of the **Terminal Type** column.

13. Click on cell 1 of the **Show** column; the cell changes into a drop-down list. Select the **When Excluding Unused** option from the drop-down list.

14. Clear the **Break After** check box of the cell 1 if selected.

15. Right-click on cell 2 of the **Terminal Type** column; a shortcut menu is displayed. Choose the **Edit Terminal** option from the shortcut menu; the **Select Terminal Information** dialog box is displayed.

16. Select the **Top Terminal** category from the **Category** area. Next, choose the **Module Info Terminal Point Wire Left** type from the **Types for Category Top Terminal** area.

17. Choose the **OK** button; the **Module Info Terminal Point Wire Left** type is displayed in cell 2 of the **Terminal Type** column.

18. Click on cell 2 of the **Show** column; the cell changes into a drop-down list. Select **When Including Unused** from the drop-down list.

19. Similarly, right-click on cell 3 and choose the **Edit Terminal** option from the shortcut menu; the **Select Terminal Information** dialog box is displayed.

20. Select **Input** from the **Category** area and choose the **Input I/O Point Wire Left** category from the **Types for Category Input** area.

21. Choose **OK**; the **Input I/O Point Wire Left** is displayed in cell 3 of the **Terminal Type** column. Next, click on cell 3 of the **Show** column; the cell changes into a drop-down list. Now, select the **When Including Unused** option from the drop-down list.

22. Next, click on cell 4 of the **Terminal Type** column and then drag the cursor upto cell 10 of the **Terminal Type** column.

23. Right-click on cell 4; a shortcut menu is displayed. Choose the **Edit Terminal** option from the shortcut menu; the **Select Terminal Information** dialog box is displayed.

24. In this dialog box, select **Input** from the **Category** area and choose the **Input I/O Point Wire Left** category from the **Types for Category Input** area.

25. Choose **OK**; the **Input I/O Point Wire Left** option is displayed in cell 4 to cell 10 of the **Terminal Type** column.

26. Click on cell 4 of the **Show** column and select the **Always** option from the drop-down list. Similarly, select the **Always** option from cell 5 to cell 10 in the **Show** column, if it is not already selected.

27. Right-click on cell 11; a shortcut menu is displayed. Choose the **Edit Terminal** option from the shortcut menu; the **Select Terminal Information** dialog box is displayed.

28. In this dialog box, select **Terminal** from the **Category** area and choose the **Terminal Point Wire Rightx** type from the **Types for Category Terminal** area.

29. Choose **OK**; the **Terminal Point Wire Rightx** type is displayed in cell 11 of the **Terminal Type** column.

30. Choose the **Save Module** button from the **PLC Database File Editor** dialog box. Next, choose the **Done / Insert** button from the **PLC Database File Editor** dialog box; the **PLC Parametric Selection** dialog box is displayed.

31. Select **Automation Direct** from the Manufacturer Catalog tree of the **PLC Parametric Selection** dialog box and then click on it; a tree view of series is displayed. Next, select 3456 module from the lower part of the **PLC Parametric Selection** dialog box.

32. Choose the **OK** button from the **PLC Parametric Selection** dialog box; you are prompted to specify an insertion point for the 3456 module.

33. Enter **18,19** at the Command prompt and press ENTER; the **Module Layout** dialog box is displayed.

    The **Insert all** radio button in the **I/O points** area of the **Module Layout** dialog box is selected by default.

34. Choose the **OK** button from the **Module Layout** dialog box; the **I/O Point** dialog box is displayed. In this dialog box, enter **1** in the **RACK** edit box and **2** in the **SLOT** edit box and then choose the **OK** button; the **I/O Address** dialog box is displayed.

35. Select **1:12/00** from the **Quick picks** drop-down list; **1:12/00** is displayed in the **Beginning address** edit box.

36. Choose the **OK** button from the **I/O Address** dialog box; the Automation Direct-3456 PLC module is inserted into the ladder, as shown in Figure 10-26.

Saving the Drawing File

1. Choose **Save** from the **Application Menu** to save the drawing file *C10_tut02.dwg*.

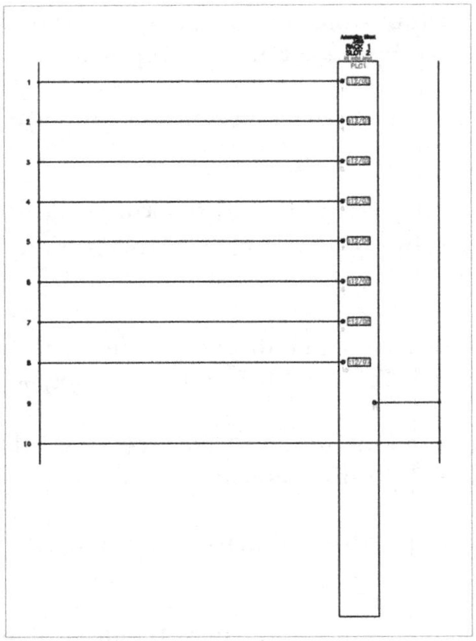

Figure 10-26 *The parametric PLC module inserted into the ladder*

Tutorial 3

In this tutorial, you will generate PLC drawings using the **PLC I/O Utility** tool. Also, you will change the default settings and save the settings in a *.wdi* file. **(Expected time: 20 min)**

The following steps are required to complete this tutorial:

a. Create a new drawing.
b. Generate PLC drawings and save the settings.

Creating a New Drawing

1. Create a new drawing with the name *C10_tut03* in the **CADCIM** project. Select the **ACAD_ELECTRICAL** template and enter **Parametric PLC** as description while creating the drawing and move *C10_tut03.dwg* to the *TUTORIALS* subfolder of the **CADCIM** project.

Generating PLC Drawings and Saving the Settings

1. Choose the **PLC I/O Utility** tool from the **Import** panel of the **Import/Export Data** tab; the **Select PLC I/O Spreadsheet Output File** dialog box is displayed. 

2. In this dialog box, select the *demoplc.xls* file and choose the **Open** button; the **Spreadsheet to PLC I/O Utility** dialog box is displayed.

3. Choose the **Setup** button; the **Spreadsheet to PLC I/O Utility Setup** dialog box is displayed.

4. Set the following parameters in the **Ladder** area of the **Spreadsheet to PLC I/O Utility Setup** dialog box.

   X : **2** Y: **21**

   Select **3** from the **PLC graphical style** drop-down list from the **Module** area.

5. Choose the **Save** button from the **Spreadsheet to PLC I/O Utility Setup** dialog box; the **Save Settings** dialog box is displayed.

6. Enter **PLC_Settings** in the **File name** edit box and select the **CADCIM** project folder (**Documents > Acade 2022 > AeData > Proj > CADCIM**) from the **Save in** drop-down list. Next, choose the **Save** button; the **Spreadsheet to PLC I/O Utility** dialog box is displayed again.

7. Set the following parameters in the **Spreadsheet to PLC I/O Utility** dialog box:

   Ladder Reference Numbering area
   Start: **100**

   Drawing File Creation area
   Select the **Use active drawing** check box, if it is not already selected.
   Make sure the **Free run** radio button is selected.
   Make sure the **Add new drawings to active project** check box is selected.

8. Choose the **Start** button in the **Spreadsheet to PLC I/O Utility** dialog box; the drawings are created. You will notice that the following three drawings are created and added to your **CADCIM** project: *C10_tut03.dwg*, *C10_tut04.dwg*, and *C10_tut05.dwg*.

Self-Evaluation Test

Answer the following questions and then compare them to those given at the end of this chapter:

1. The _________ tool is used to insert a complete PLC module into the drawing.

2. The _________ tool is used to start the automatic drawing creation process.

3. You can tag components based on the PLC address by selecting the _________ check box in the _________ tab of the **Drawing Properties** dialog box.

4. A PLC module cannot be inserted as a single block. (T/F)

5. PLC modules can be built with a variety of graphical symbols. (T/F)

6. The **PLC Database File Editor** tool is used only to create PLC modules. (T/F)

Review Questions

Answer the following questions:

1. Which of the following commands is used to insert parametric PLC modules?

   (a) **AEPLC** (b) **AEPLCDB**
   (c) **AESS2PLC** (d) **AEPLCP**

2. Which of the following tools is used to copy an entire module into a new module?

   (a) **Spreadsheet to PLC I/O Utility** (b) **Insert PLC (Full Units)**
   (c) **PLC Database File Editor** (d) **Surfer**

3. You can insert PLC I/O points as independent symbols or as a complete PLC module into a drawing. (T/F)

4. While inserting a parametric PLC module, you can break a module and restart it at a different location. (T/F)

5. The **Spreadsheet to PLC I/O Utility** tool is used to read the Excel spreadsheet file only. (T/F)

EXERCISES

Exercise 1

Create a new drawing named *C10_exer01.dwg* in the **NEW_PROJECT** project and then insert a ladder into the drawing with the following specifications: Width = 6, Spacing =1, 1st Reference =1, and Rungs =12. Next, use the **Insert PLC (Parametric)** tool to insert PLC module with the following specifications: HONEYWELL, Discrete Input, 621-3550, 16 point input module. Next, break the module at rung 11 and then insert a new ladder with following specifications: Width = 6, Spacing =1, 1st Reference =1, Rungs = 12. Continue the previous PLC module by using the **Insert PLC (Parametric)** tool again. Figure 10-27 shows the complete parametric PLC module. **(Expected time: 20 min)**

Exercise 2

Create a new drawing named *C10_exer02.dwg* in the **NEW_PROJECT** project and then insert a ladder into it. Also, insert a PLC module [L16-AWA 10in/out AC-DC/115AC-DC (3/4" spacing)] into a ladder using the **Insert PLC (Full Units)** tool. Figure 10-28 shows the values to be used for inserting the PLC module. Figure 10-29 shows the complete PLC module.

(Expected time: 15 min)

Hint: Accept the default values in the **Edit PLC Module** dialog box.

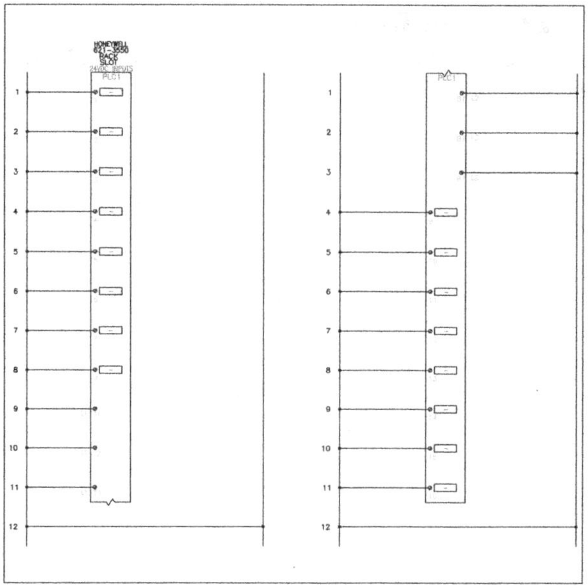

Figure 10-27 *Complete parametric PLC module*

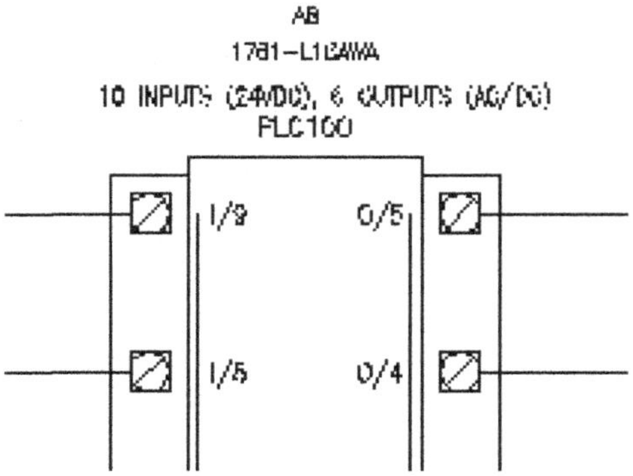

Figure 10-28 *Values for inserting the PLC module*

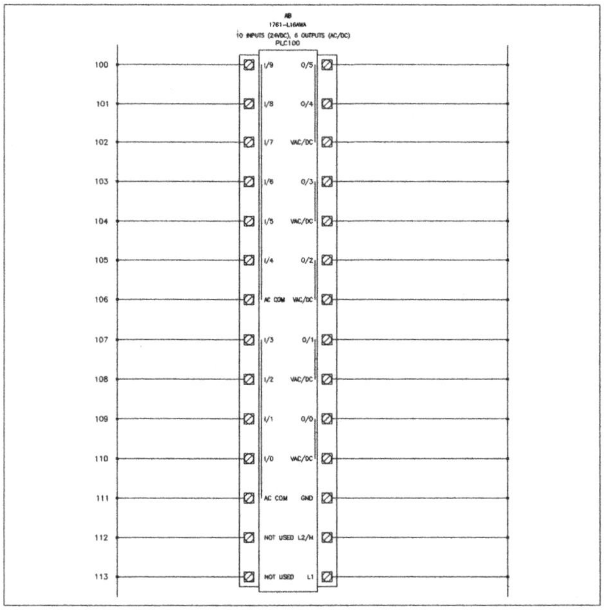

Figure 10-29 *Complete PLC module*

Answers to Self-Evaluation Test

1. **Insert PLC (Full Units)**, **2.** **Spreadsheet to PLC I/O Utility**, **3.** **Search for PLC I/O address on insert**, **Components**, **4.** F, **5.** T, **6.** F

Chapter 11

Terminals

Learning Objectives

After completing this chapter, you will be able to:

- *Insert and edit terminal symbols*
- *Insert terminals from the schematic list and the panel list*
- *Insert terminals manually*
- *Select, create, edit, and insert terminal strips*
- *Configure the settings of a terminal strip table*
- *Generate a terminal strip table*
- *Edit and create the terminal properties database*
- *Resequence terminal numbers*
- *Copy the properties of a terminal block*
- *Insert and edit jumpers*

INTRODUCTION

The terminals are used to connect wires together or to plug them into a component. In this chapter, you will learn to insert terminal symbols manually from the schematic list, or from the panel list. Also, you will come to know how to insert, edit, and create terminal strips. You will be able to insert jumpers, configure the settings, and generate terminal strip table. Further, you will learn about editing and creating a terminal properties database.

INSERTING TERMINAL SYMBOLS

Command: AECOMPONENT

Terminal symbols are used to represent wire connection points on the schematic drawings. Terminal symbols are schematic symbols that are linked by user-defined codes. To insert terminal symbol into a drawing, choose the **Icon Menu** tool from the **Insert Components** panel of the **Schematic** tab; the **Insert Component** dialog box will be displayed. Select the **Terminals/Connectors** icon from the **NFPA: Schematic Symbols** area of the **Insert Component** dialog box; the **NFPA: Terminals and Connectors** area will be displayed, as shown in Figure 11-1.

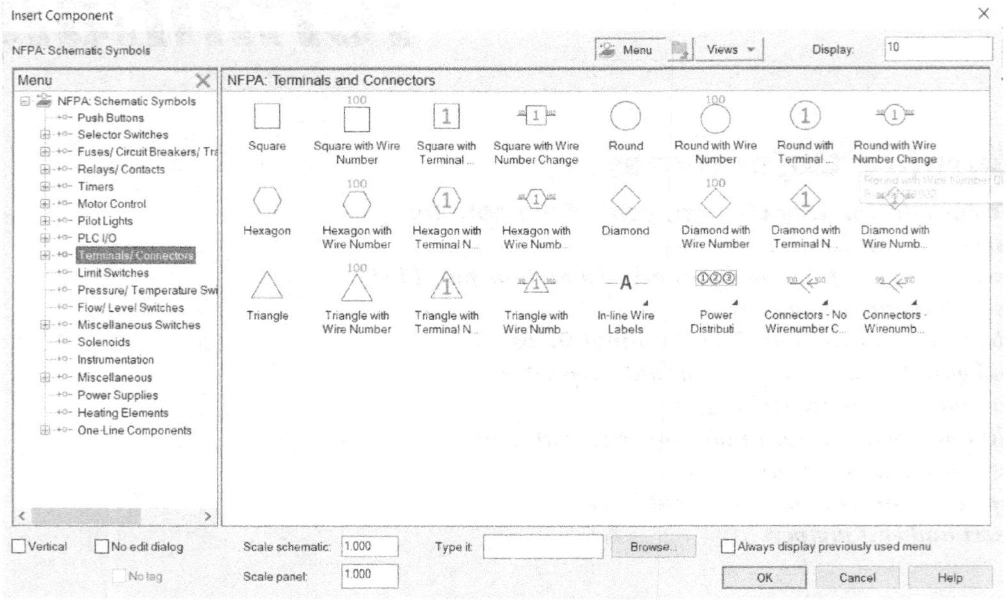

Figure 11-1 The ***Insert Component*** *dialog box showing the* ***NFPA: Terminals and Connectors*** *area*

To insert a terminal symbol into a drawing, select the required terminal from the **NFPA: Terminals and Connectors** area of the **Insert Component** dialog box; the symbol attached to the cursor will be displayed and you will be prompted to specify the insertion point for the terminal symbol. Specify the insertion point; the **Insert/Edit Terminal Symbol** dialog box will be displayed. Enter the required information in this dialog box and choose **OK**; the terminal symbol will be inserted into the drawing.

Note

*If you insert a non-intelligent terminal, the **Insert / Edit Terminal Symbol** dialog box will not be displayed.*

Annotating and Editing Terminal Symbols

Command: AEEDITCOMPONENT

You can annotate and edit terminal symbols. To do so, choose the **Edit** tool from the **Edit Components** panel of the **Schematic** tab; you will be prompted to select the component. Select the terminal that you need to edit; the **Insert / Edit Terminal Symbol** dialog box will be displayed, as shown in Figure 11-2. Different options and areas in this dialog box are discussed next.

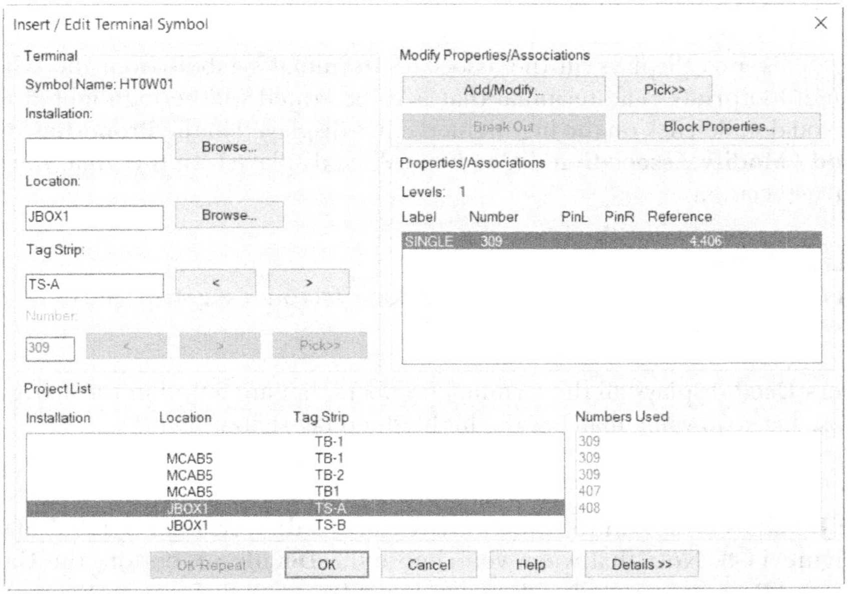

*Figure 11-2 The **Insert / Edit Terminal Symbol** dialog box*

Terminal Area

The options in the **Terminal** area are used to specify the installation code, location code, tag strip value, and terminal number. The symbol name is displayed above the **Installation** edit box in this area.

Enter the Tag ID for the terminal in the **Tag Strip** edit box. Choose the **<** or **>** buttons to decrease or increase the last digit of the Tag ID value. Note that if the terminal has already a Tag ID assigned to it, then it will appear in the **Tag Strip** edit box.

Enter the terminal number in the **Number** edit box. Choose the **<** or **>** buttons to decrease or increase the specified terminal number. Alternatively, choose the **Pick>>** button to select the text or attribute from the active drawing to use as terminal number. The **Number** edit box will not be available if the terminal symbol does not have a TERM01 attribute or the terminal number and the wire number are same.

Modify Properties/Associations Area

The options in the **Modify Properties/Associations** area are used to modify the properties and associations of the terminal symbol, associate multiple terminal symbols, associate schematic terminal symbols and their panel terminal footprint, define block properties of a terminal, and so on. If the attributes are present in the terminal block, then the information about the terminal associations will be stored in it. Otherwise, the information will be stored in Xdata. The **Modify Properties/Associations** area is used to manage the linked terminals. Note that the options in this area will not be available, if you insert one-line terminal symbol; if you insert a terminal symbol that is not a part of the active project; and if you are inserting terminals using the **Insert Terminal (Panel List)** and **Terminal Strip Editor** tools. Also, note that the **Block Properties** button will not be activated if the active drawing is not from the active project.

Properties/Associations Area

The **Properties/Associations** area displays the associations of the terminal that you are editing or inserting. This area displays all the associated terminal symbols from the schematic and terminal panel footprints. The terminal that is to be edited will be highlighted in this area. Note that if you double-click on the highlighted entry displayed in the **Properties/Associations** area, the **Add / Modify Association** dialog box will be displayed. In this area, you can modify the terminal association.

Project List Area

The **Project List** area displays the installation and location codes, tag strip, and terminal numbers that have been used in the active project.

The **Numbers Used** displays all the terminal numbers that are found in the active drawing or project, whose tag strip value matches the highlighted tag strip value.

Details

Choose the **Details >>** button to expand the **Insert / Edit Terminal Symbol** dialog box, as shown in Figure 11-3. Note that when you choose the **Details>>** button, the **Catalog Data**, **Description**, and **Ratings** areas will be displayed in the **Insert / Edit Terminal Symbol** dialog box.

Enter the required information in the **Insert / Edit Terminal Symbol** dialog box and choose the **OK** button; the information of the terminal will be updated on the terminal symbol.

Note

1. When you insert a terminal symbol, the values of the previously inserted terminal symbol will be displayed in the ***Insert / Edit Terminal Symbol*** *dialog box.*

2. You can also insert multiple terminals together using the ***Multiple Insert (Icon Menu)*** *tool and* ***Multiple Insert (Pick Master)*** *tool.*

***Figure 11-3** The expanded **Insert / Edit Terminal Symbol** dialog box*

INSERTING TERMINAL FROM THE SCHEMATIC LIST

Command: AEPANELTERMINALSCH

The **Insert Terminal (Schematic List)** tool is used to insert panel terminals from the schematic terminal list. Choose this tool from the **Insert Terminals** drop-down in the **Terminal Footprints** panel of the **Panel** tab, as shown in Figure 11-4; the **Schematic Terminals List --> Panel Layout Insert** dialog box will be displayed, as shown in Figure 11-5.

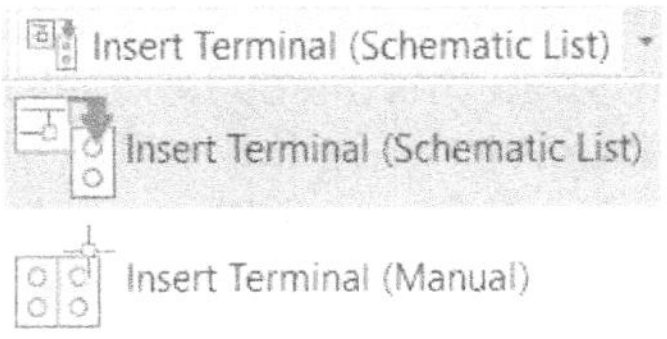

***Figure 11-4** The **Insert Terminals** drop-down*

The **Project** radio button is selected by default. Choose the **OK** button in the **Schematic Terminals List --> Panel Layout Insert** dialog box; the **Select Drawings to Process** dialog box will be displayed. Select the required files and then choose the **Process** button; the drawings from the top list of the **Select Drawings to Process** dialog box will be transferred to the bottom list for processing. Next, choose the **OK** button in the **Select Drawings to Process** dialog box;

the **Schematic Terminals (active project)** dialog box will be displayed. Next, select the terminal from the **Schematic Terminals (active project)** dialog box; the **Insert** button will be activated. Note that the **Insert** button will be activated only if the selected terminal is not inserted into a drawing and if it has catalog data. Choose the **Insert** button; you will be prompted to specify the location point for the terminal. Specify the location point; you will be prompted to specify the rotation of the terminal. Move the cursor in the horizontal or vertical direction to place the terminal horizontally or vertically, respectively. Next, press ENTER; the **Panel Layout - Terminal Insert/Edit** dialog box will be displayed. Enter the required information in this dialog box and choose the **OK** button; the terminal will be inserted in the drawing and the **Schematic Terminals (active project)** dialog box will be displayed again.

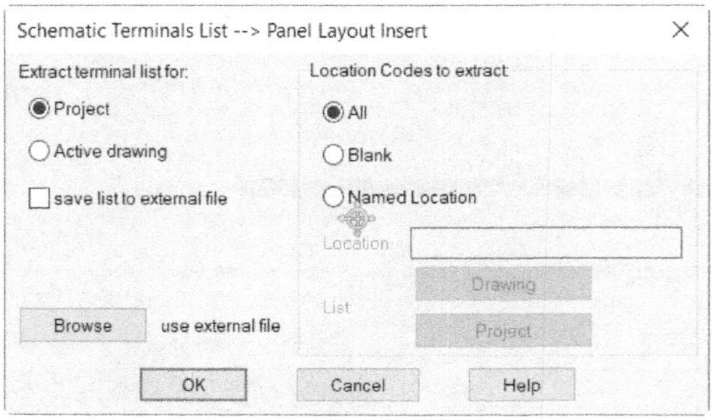

***Figure 11-5** The **Schematic Terminals List --> Panel Layout Insert** dialog box*

Now, if you choose the **Pick File** button in the **Schematic Terminals (active project)** dialog box, the **Schematic Terminals List --> Panel Layout Insert** dialog box will be displayed again. Select the **Active drawing** radio button and choose the **OK** button in this dialog box; the **Schematic Terminals (active drawing)** dialog box will be displayed. Next, select the terminal from the list displayed and then insert the terminal symbol as discussed earlier.

If you insert panel terminals from the schematic terminal list, wire numbers will also be displayed with these panel terminals in the drawing. To further understand this, activate the **NFPADEMO** project. Next, open the *demo009.dwg* file from this project. Now, choose the **Insert Terminal (Schematic List)** tool from the **Insert Terminals** drop-down in the **Terminal Footprints** panel of the **Panel** tab; the **Schematic Terminals List --> Panel Layout Insert** dialog box will be displayed. In this dialog box, select the **Project** radio button and choose **OK**; the **Select Drawings to Process** dialog box will be displayed. Choose **Do All** and then the **OK** button in this dialog box; the **Schematic Terminals (active project)** dialog box will be displayed. In this dialog box, select the first terminal in the list and enter **3** in the **Footprint Scale** button. Next, choose the **Insert** button; you will be prompted to specify the location for the selected terminal. Next, specify the location, as shown in Figure 11-6 and then press ENTER twice; the **Panel Layout Insert/Edit** dialog box will be displayed. Choose the **OK** button in this dialog box. Figure 11-7 shows the wire number displayed along with the inserted panel terminal in the *demo009.dwg* file.

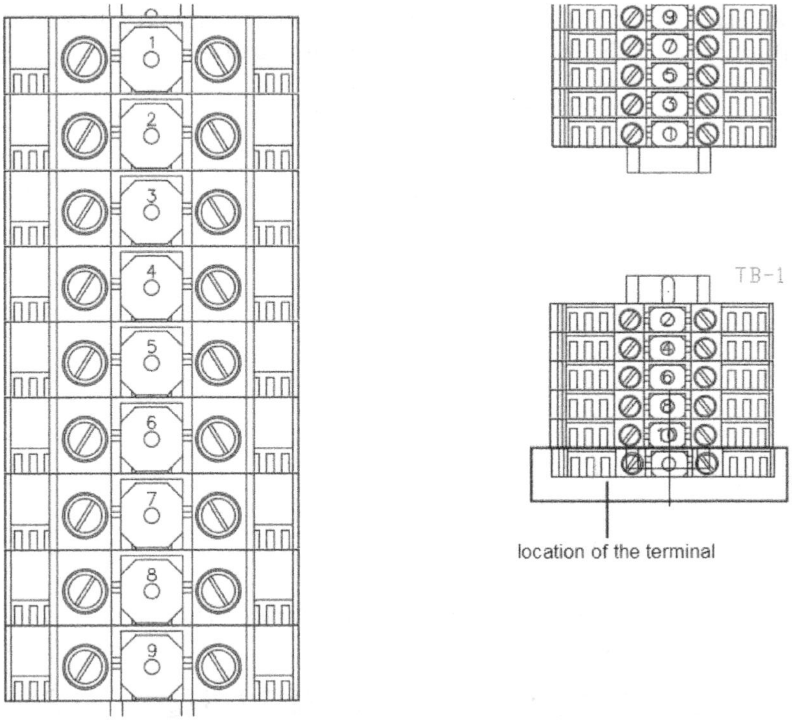

Figure 11-6 *Specifying the location of the terminal*

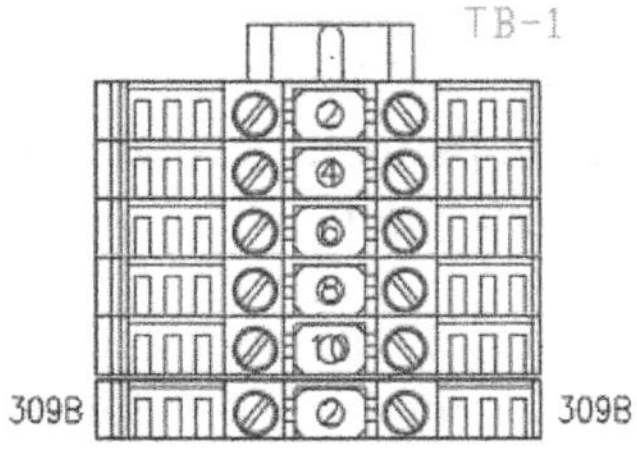

Figure 11-7 *Wire number displayed*

INSERTING TERMINALS MANUALLY

Command: AEPANELTERMINAL

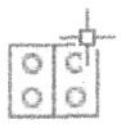

The **Insert Terminal (Manual)** tool is used to insert panel terminal footprints manually. To do so, choose the **Insert Terminal (Manual)** tool from the **Terminal Footprints** panel of the **Panel** tab; the **Insert Panel Terminal Footprint - Manual** dialog box will be displayed, as shown in Figure 11-8. The options in the **Insert Panel Terminal Footprint - Manual** dialog box are the same as that of the **Insert Component Footprint - Manual** dialog box that have been discussed in Chapter 8. The only difference is that in case of the **Insert Panel Terminal Footprint - Manual** dialog box, the **Terminal number preference** area is available.

Next, select the required terminal from the **Manual terminal selection or creation** area and specify its position; the **Panel Layout - Terminal Insert/Edit** dialog box will be displayed. Enter the required information in this dialog box and choose **OK**; a terminal will be inserted into the drawing and information in the form of Xdata will be assigned to it automatically.

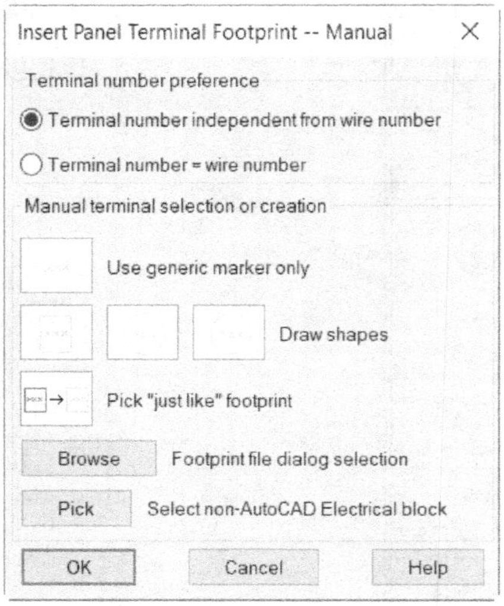

Figure 11-8 *The **Insert Panel Terminal Footprint - Manual** dialog box*

INSERTING TERMINALS FROM THE PANEL LIST

Command: AETERMINALPNL

The **Terminal (Panel List)** tool is used to insert a schematic terminal into a drawing using the panel list. Choose the **Terminal (Panel List)** tool from the **Insert Component** panel of the **Schematic** tab; the **Panel Terminal List --> Schematic Terminals Insert** dialog box will be displayed, as shown in Figure 11-9.

The **Project** radio button is selected by default. Choose the **OK** button in the **Panel Terminal List -> Schematic Terminals Insert** dialog box; the **Select Drawings to Process** dialog box will be displayed. Select the required drawings and choose the **Process** button; the drawings will be transferred from the top list to the bottom list. Next, choose the **OK** button in the **Select Drawings to Process** dialog box; the **Panel Terminals** dialog box will be displayed, as shown in Figure 11-10. The options in the **Panel Terminals** dialog box are same as discussed for the **Schematic Terminals** dialog box. The only difference in these two dialog boxes is that the **Panel Terminals** dialog box has the **Last symbol used** area.

Next, select the panel terminal reference from the **Panel Terminals** dialog box; the **Insert** button, the **Catalog Check** button, the **Scale** edit box, and the **Vertical** check box will be activated. Choose the **Insert** button; the **Insert** dialog box will be displayed. Next, select the terminal from this dialog box; the **OK** button will be activated. Choose the **OK** button; you will be prompted to specify the insertion point. Specify the insertion point; the **Insert / Edit Terminal Symbol** dialog box will be displayed. Specify the required information in this dialog box. Choose the **OK** button; the terminal will be inserted in the drawing and the **Panel Terminals** dialog box will appear on the screen again. Choose the **Close** button to exit from this dialog box.

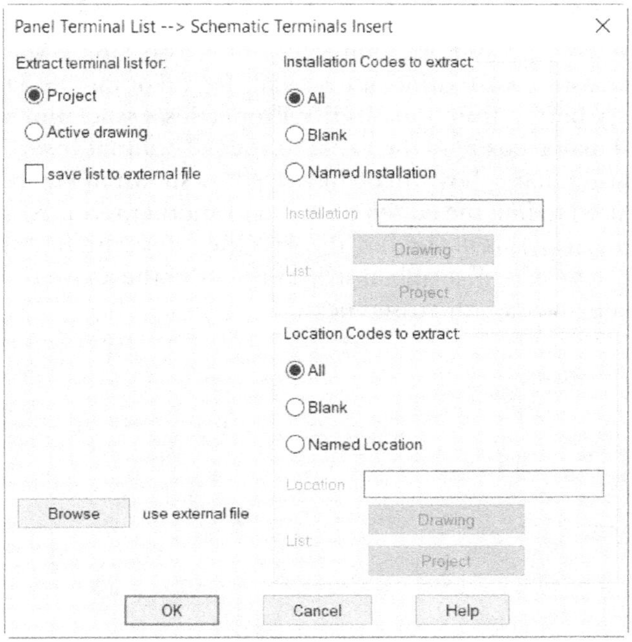

*Figure 11-9 The **Panel Terminal List --> Schematic Terminals Insert** dialog box*

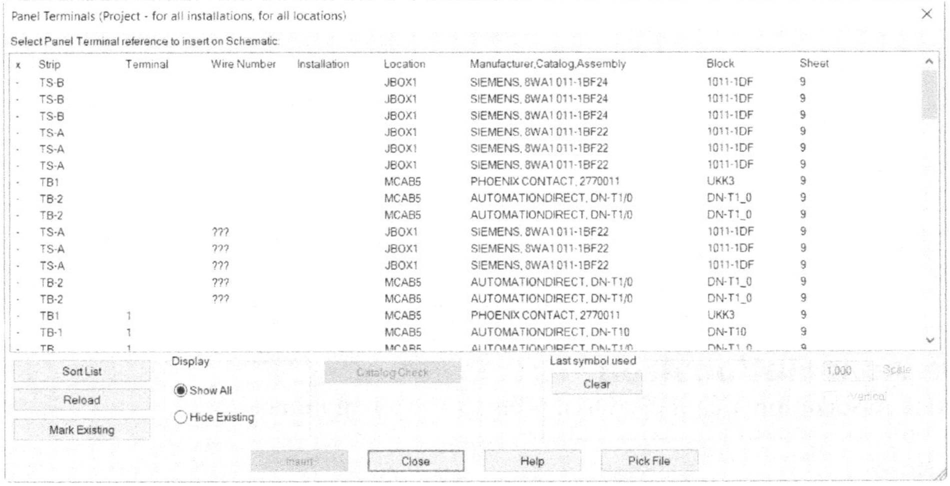

*Figure 11-10 The **Panel Terminals** dialog box*

Note

*1. The **Insert** dialog box will be displayed only when you are inserting the terminal symbol into the drawing for the first time.*

*2. If you select multiple terminals from the **Panel Terminals** dialog box and choose the **Insert** button, the **Spacing for Insertion** dialog box will be displayed. The options in this dialog box have already been discussed in the earlier chapters.*

ADDING AND MODIFYING ASSOCIATIONS

In AutoCAD Electrical, you can add or modify a terminal association. You can also associate a terminal symbol to an existing association or terminal. To add or modify a terminal association, choose the **Add/Modify** button from the **Modify Properties/Associations** area of the **Insert / Edit Terminal Symbol** dialog box or of the **Panel Layout - Terminal Insert/Edit** dialog box; the **Add / Modify Association** dialog box will be displayed, as shown in Figure 11-11. The options in this dialog box are used to link the edited terminal to another terminal symbol for creating a multiple level terminal. Also, this dialog box displays the information of the active association of the edited terminal as well as all terminal strips present in the active project. Different areas and options in this dialog box are discussed next.

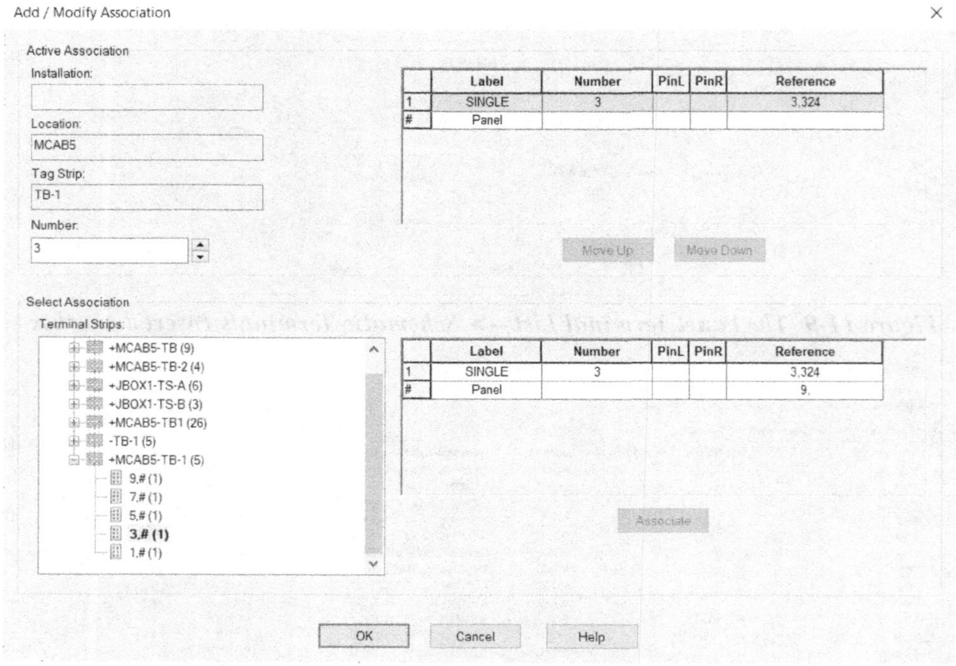

***Figure 11-11** The **Add / Modify Association** dialog box*

Active Association Area

The **Active Association** area is used to modify the terminal number.

Note

*1. You cannot edit installation, location, and tag strip values in the **Add/Modify Association** dialog box.*

*2. If the terminal to be edited is from a drawing that is not part of the active project, the **Add/Modify** and **Block Properties** buttons in the **Insert / Edit Terminal Symbol** will not be activated.*

Number

The **Number** edit box displays the terminal number. Note that this edit box will not be available if you are editing a panel terminal.

Active Association Grid

This grid displays the information of all the terminal symbols associated with the edited terminal. Note that if the schematic terminal is associated to a panel footprint, you will find a panel row added to the Active Association Grid. You cannot move this panel row up and down.

Select Association Area

In the **Select Association** area, you can select the terminal for association.

TERMINAL BLOCK PROPERTIES

When you choose the **Block Properties** button from the **Insert / Edit Terminal Symbol** dialog box or from the **Panel Layout - Terminal Insert/Edit** dialog box, the **Terminal Block Properties** dialog box will be displayed. This dialog box is used to define and manage terminal block properties. This dialog box is also used to maintain the number of levels of the terminal being edited. Note that this button is disabled if the active drawing is not part of the active project.

After specifying the required options in the **Terminal Block Properties** dialog box, choose the **OK** button to exit.

SELECTING, CREATING, EDITING, AND INSERTING TERMINAL STRIPS

Command: AETSE

The **Editor** tool is used to search the project file for terminal symbols and groups them by the values for tag strip, installation, and location. Using this tool, you can select, create, edit, and then insert a terminal strip into a drawing. Choose the **Editor** tool from the **Terminal Footprints** panel of the **Panel** tab; the **Terminal Strip Selection** dialog box will be displayed, refer to Figure 11-12.

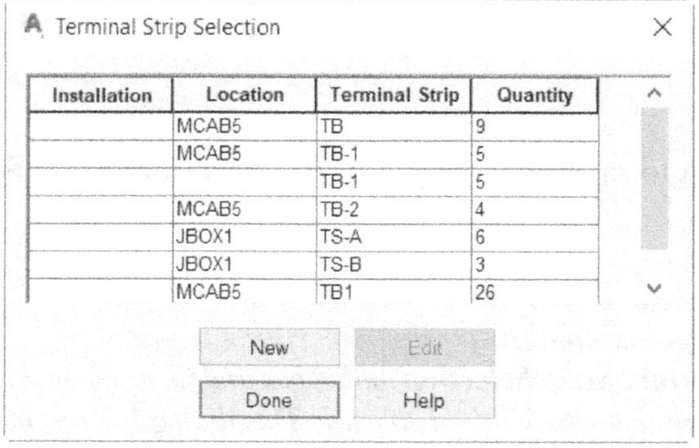

Figure 11-12 *The **Terminal Strip Selection** dialog box*

Note

*If the current drawing is not in the active project, the **Current Drawing is Not in the Active Project** message box will be displayed. Choose the **Yes** button in this message box to continue working in the same drawing.*

The **Terminal Strip Selection** dialog box displays the terminals that are present within the active project. This dialog box consists of the installation codes, location codes, terminal strip values, and the number of terminals associated with each terminal strip. The different options in this dialog box are discussed next.

New

The **New** button is used to create a new terminal strip. Choose the **New** button; the **Terminal Strip Definition** dialog box will be displayed, as shown in Figure 11-13.

Specify the location code and the installation code for the new terminal in the respective edit boxes. Alternatively, you can choose the **Browse** button to specify the same.

Specify the tag name of the strip for the new terminal strip in the **Terminal Strip** edit box. Also, note that you cannot have duplicate terminal strip names in the active project.

Specify the number of blocks in the terminal strip in the **Number of Terminal Blocks** edit box.

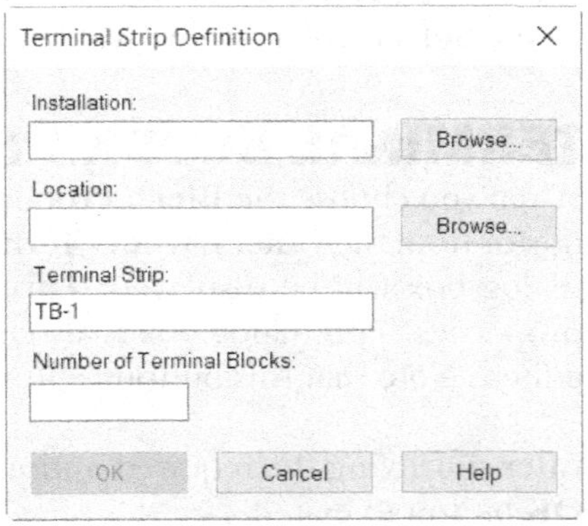

*Figure 11-13 The **Terminal Strip Definition** dialog box*

Enter the values in the **Terminal Strip** and **Number of Terminal Blocks** edit boxes; the **OK** button of the **Terminal Strip Definition** dialog box will be activated. Choose the **OK** button; the **Terminal strip editor** dialog box will be displayed.

Edit

The **Edit** button will be activated only after selecting the terminal strip from the **Terminal Strip Selection** dialog box. This button is used to edit a terminal strip. Select the terminal strip from the **Terminal Strip Selection** dialog box and then choose the **Edit** button; the **Terminal Strip Editor** dialog box will be displayed. This dialog box displays the information of the selected terminal.

Done

Choose the **Done** button to save the changes in the **Terminal Strip Selection** dialog box and exit from this dialog box.

Note

*You can create a new terminal strip using an existing terminal strip. To do so, select the terminal strip from the **Terminal Strip Selection** dialog box and then choose the **New** button; the **Terminal Strip Definition** dialog box will be displayed. This dialog box displays the values of the existing terminal strip. Next, change the values according to your requirement and then choose the **OK** button to create a new terminal strip.*

*If you leave the **Installation** and **Location** edit boxes blank, only the terminal strip with the strip tag name will be created.*

Editing the Terminal Strip

To edit the terminal strip, choose the **Editor** tool from the **Terminal Footprints** panel of the **Panel** tab; the **Terminal Strip Selection** dialog box will be displayed. Next, select the terminal strip; the **Edit** button will be available. Choose the **Edit** button; the **Terminal Strip Editor** dialog box will be displayed, refer to Figure 11-14. In this dialog box, you can make changes to the terminals before inserting the strip into the drawing. The **Terminal Strip Editor** dialog box consists of the **Terminal Strip**, **Catalog Code Assignment**, **Cable Information**, and **Layout Preview** tabs. These tabs are discussed next.

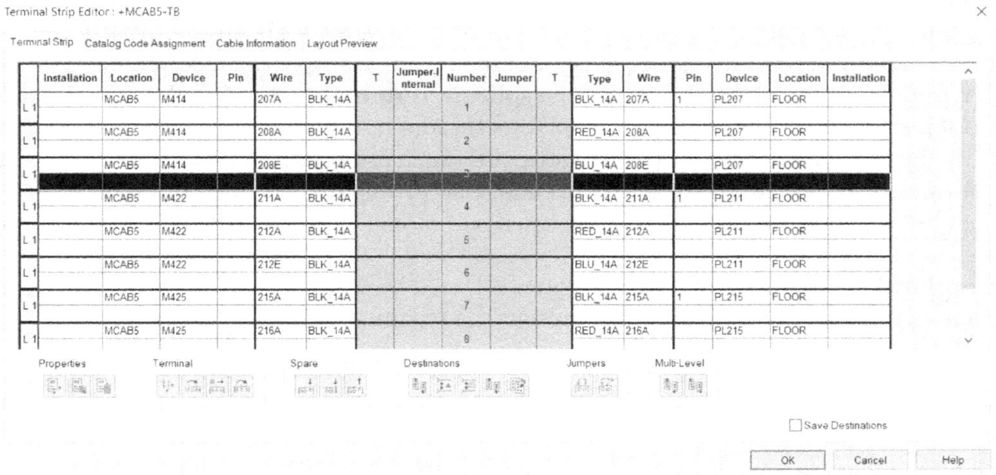

Figure 11-14 The ***Terminal Strip Editor*** *dialog box*

Terminal Strip Tab

The **Terminal Strip** tab is chosen by default, refer to Figure 11-14. This tab displays the information of each terminal. In this tab, you can modify, copy, and paste terminal block properties, edit, reassign, renumber, and move terminals, destination settings, assign jumpers, associate terminals, and so on. Also, this tab displays the information of each terminal. These terminals can be sorted by selecting the column headings. The first click on the column heading sorts the column in ascending order and if you click again on the column heading, it will be sorted in the descending order. This sorting criterion applies to all tabs of this dialog box, so you do not have to sort again if you switch between the tabs such as **Catalog Code Assignment**, **Cable Information**, and **Layout Preview**. In this dialog box, the terminals will be displayed at the center of the list box with associated catalog information and the destinations on both sides. The terminals in this dialog box are separated by thick bold lines. Also, the column that is in the extreme left side of the grid indicates different levels for the multiple level terminal for example, L1, L2, L3, L4 and so on. This tab also consists of **Properties**, **Terminal**, **Spare**, **Destinations**, **Jumpers**, and **Multi-Level** areas. These areas consist of various buttons. These buttons are used to reorder terminals, change information of a terminal, insert spare terminals, and so on.

Catalog Code Assignment Tab

The options in this tab are used to modify the catalog data of the terminals. All the areas in this tab are the same as those in the **Terminal Strip** tab except that of the **Catalog** area. The options in the **Catalog** area are used to assign, delete, copy, and paste the catalog number to the selected terminal.

Cable Information Tab

In the **Cable Information** tab, the information of the cable associated to the terminals in the terminal strip is displayed. This tab displays terminal information at the center of the list, cable name, wire conductor information, and device destination information on both sides.

Layout Preview Tab

The **Layout Preview** tab is shown in Figure 11-15. In this tab, you can see the preview of the terminal strip and select the required format of the terminal strip to be inserted into a drawing. The **Layout Preview** tab consists of the **Graphical Terminal Strip**, **Tabular Terminal Strip (Table Object)**, and **Jumper Chart (Table Object)** radio buttons. These radio buttons are used to specify the type of terminal strip to be generated such as graphical, tabular, or jumper chart. The options in the **Layout Preview** tab will change according to the radio button that you select. Different areas and options in this tab are discussed next.

The **Graphical Terminal Strip** radio button is selected by default, refer to Figure 11-15, and is used to insert the AutoCAD blocks to represent the terminal strip.

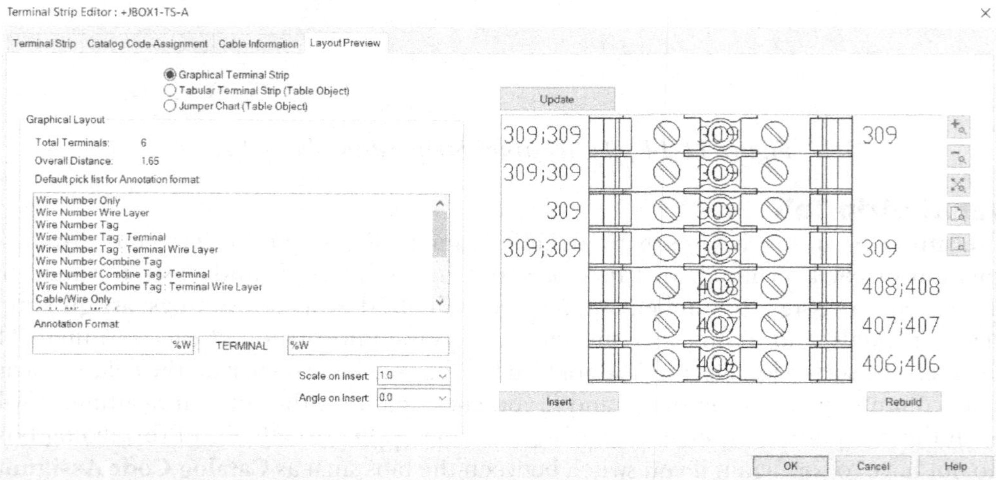

*Figure 11-15 The **Terminal Strip Editor** dialog box showing the **Layout Preview** tab*

After specifying the required options in the **Terminal Strip Editor** dialog box, choose the **OK** button; the **Graphical Terminal Strip Insert/Update Required** message box will be displayed, prompting you to rebuild or insert the updated terminal strip. Choose the desired option from this message box; you will return to the **Terminal Strip Selection** dialog box. Choose the **Done** button to exit from this dialog box.

Defining the Settings of the Terminal Strip Table

You can define the settings of the terminal strip table in the **Terminal Strip Table Settings** dialog box. To do so, select the **Tabular Terminal Strip (Table Object)** radio button from the **Layout Preview** tab of the **Terminal Strip Editor** dialog box. Next, choose the **Settings** button from the **Tabular Layout** area; the **Terminal Strip Table Settings** dialog box will be displayed, as shown in Figure 11-16.

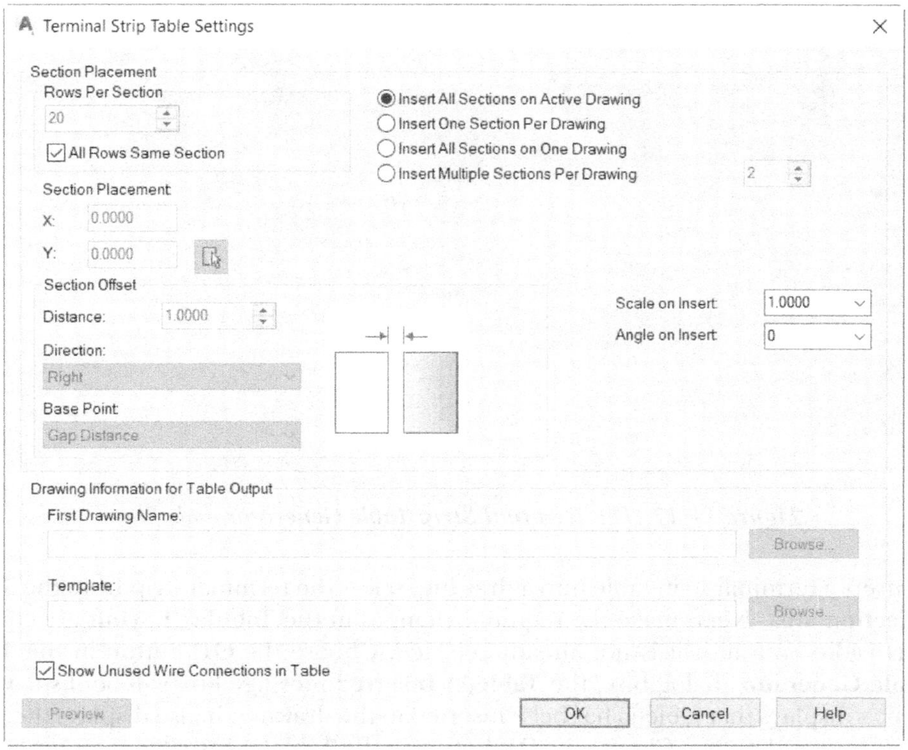

***Figure 11-16** The **Terminal Strip Table Settings** dialog box*

After specifying the required settings of the table in the **Terminal Strip Table Settings** dialog box, choose the **OK** button to save the changes and exit from this dialog box.

GENERATING THE TERMINAL STRIP TABLE

Command: AETSEGENERATOR

The **Terminal Strip Table Generator** tool is used to create drawings along with a terminal strip table. This tool is also used to insert one or more than one terminal strip in the form of a table in a drawing. This tool is also used to rebuild, refresh the existing terminal strip tables present in an active project, and add the newly created drawings to the active project. To insert a tabular terminal strip, choose the **Table Generator** tool from the **Terminal Footprints** panel of the **Panel** tab; the **Terminal Strip Table Generator** dialog box will be displayed, as shown in Figure 11-17. You can insert a terminal strip as a single object or split it into multiple table objects. The new drawings will be created as needed and will automatically be added to the active project. Note that the terminal strip that you select will start in a new drawing.

If you click on any of the column headers in the **Terminal Strip Selection** area, the terminal strips in this area will be sorted accordingly.

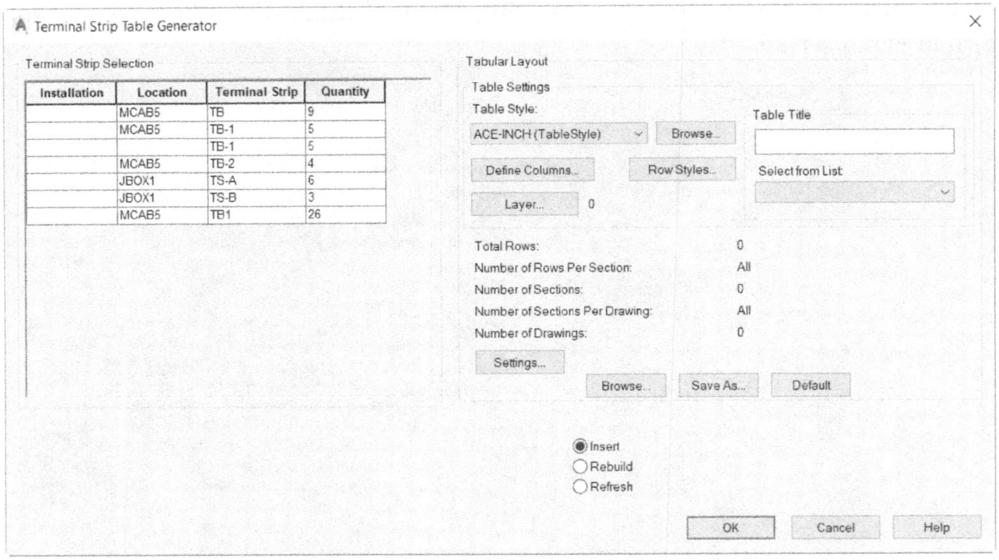

***Figure 11-17** The **Terminal Strip Table Generator** dialog box*

Now, to insert a terminal strip table into a drawing, select the terminal strip from the **Terminal Strip Selection** area. Next, make the required changes in the **Tabular Layout** area and select the **Insert** radio button, if it is not already selected. Choose the **OK** button in the **Terminal Strip Table Generator** dialog box; the **Table(s) Inserted** message box will be displayed. This message box displays that table(s) has been inserted in the drawing. It also displays the path and location of the new drawing. Choose the **OK** button in the **Table(s) Inserted** message box to exit from the dialog box. You will notice that new drawing is added at the end of the active project.

Note

*You can also select multiple terminal strips from the **Terminal Strip Selection** area by using the SHIFT or CTRL key or by clicking and dragging the mouse.*

EDITING THE TERMINAL PROPERTIES DATABASE TABLE

Command: AECATALOGOPEN

The **Catalog Browser** tool is used to edit the existing entry or create a new entry in the terminal category list of the catalog database. To do so, choose the **Catalog Browser** tool from the **Icon Menu** drop-down in the **Insert Components** panel of the **Schematic** tab, refer to Figure 11-18; the **Catalog Browser** dialog box will be displayed. In this dialog box, select the **TRMS (Terminals)** category from the **Category** drop-down list. Next, delete the content from the **Search** field and choose the **Search** button; the catalog database of all the terminals (for all the manufacturers) will be displayed. Scroll to the right in the Database grid. You will notice that there are some columns available in it. Now, choose the **Edit** button located below the **Search** field and again scroll to the right in the Database grid. On doing so, some more columns are added to this grid, namely **LEVELS**, **LEVELDESCRIPTION**, **TPINL**, **TPINR**, **WIRESPERCONNECTION**, **INTERNALJUMPER**, and so on, refer to Figure 11-19.

To edit the entry of a terminal, you need to click in the respective cell of the column and make the necessary changes. Also, you can right-click in the cell of the column and use the options in the shortcut menu displayed to edit it. Some of the major columns are discussed next.

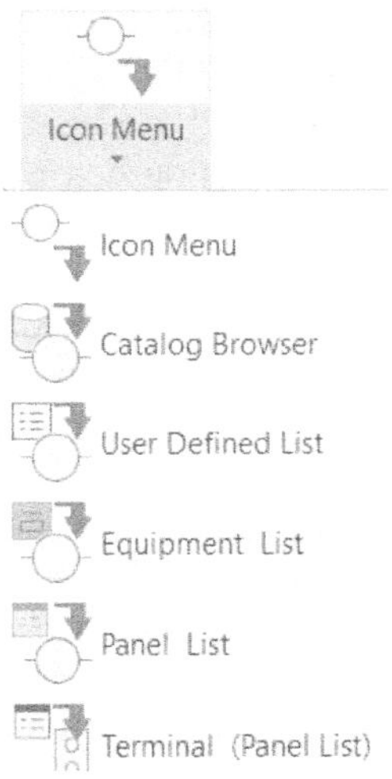

*Figure 11-18 The **Icon Menu** drop-down*

If you double-click on a cell of the **SYMBOL2D** column, it will be converted into a field consisting of an edit box and a Browse button, as shown in Figure 11-20. You can enter the name of the 2D symbol in this edit box. If you click on the Browse button, the **2D Symbol** dialog box will be displayed. You can enter names for the 2D symbols in the edit boxes available in this dialog box. If you choose the **Icon menu** button from this dialog box, the **Insert Component** dialog box will be displayed. Now, you can select a symbol from this dialog box to be used as a 2D symbol for the corresponding terminal entry. Also, you can enter comments for this symbol in the **Comment** edit boxes. Note that a row is added at the bottom of the Database grid when you enter symbol name in the last row of this dialog box.

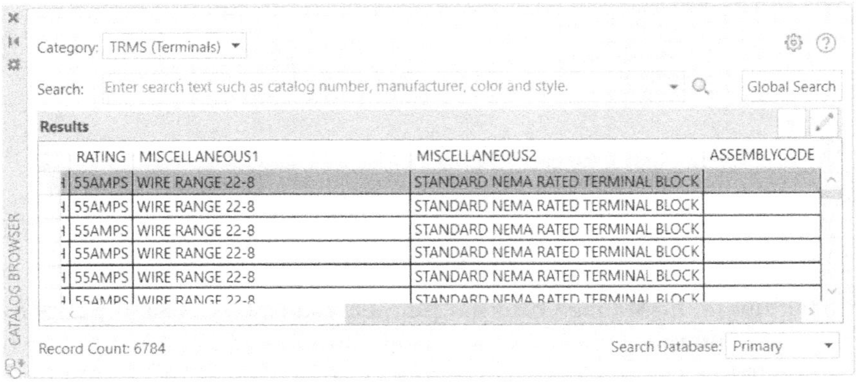

*Figure 11-19 The columns added to the **Catalog Browser** dialog box*

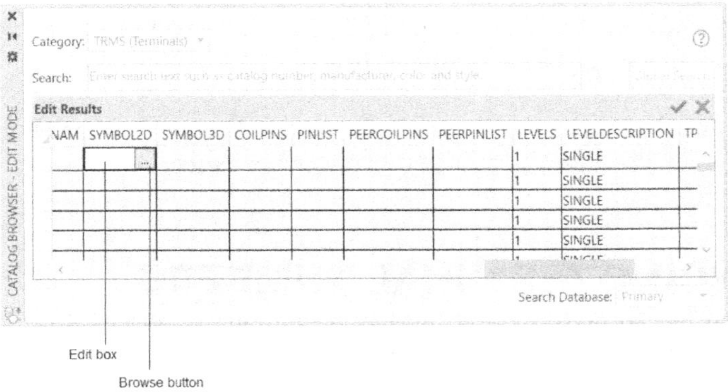

Figure 11-20 The field with an edit box and the Browse button

If you double-click on the cell of the **LEVELS**, **LEVELDESRIPTION**, **TPINL**, **TPINR**, or **WIRESPERCONNECTION** column, it will be converted into a field consisting of an edit box and a Browse button, refer to Figure 11-20. Choose the Browse button; the **Terminal Block Properties** dialog box will be displayed.

Edit the values in this dialog box and choose the **OK** button; the values in the corresponding column of the **Catalog Browser** dialog box will change accordingly. Note that the number of rows in the **Terminal Block Properties** dialog box is equal to the number of levels in the corresponding terminal block.

If you double-click on the cell of the **INTERNALJUMPER** column, it will be converted into a field consisting of an edit box and a Browse button. Click on the Browse button; the **Manage Internal Jumpers** dialog box will be displayed, as shown in Figure 11-21. Note that the Browse button for this field will be activated only if the number of levels is more than the one in the corresponding **LEVEL** column.

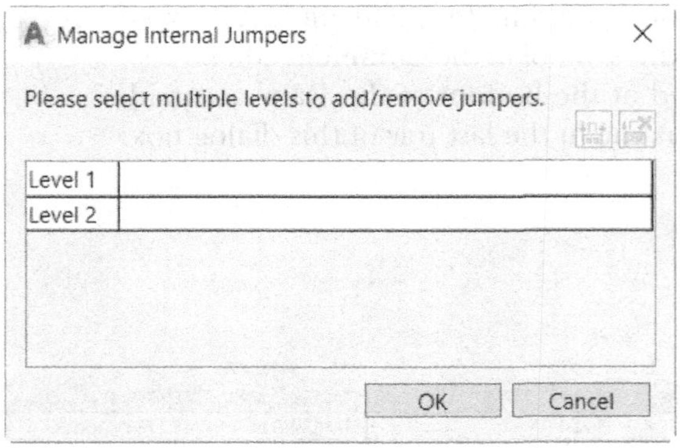

***Figure 11-21** The **Manage Internal Jumpers** dialog box*

There are two buttons in the **Manage Internal Jumpers** dialog box: **Assign Jumper** and **Delete Jumper**. The **Assign Jumper** button will be activated when you select more than one row. If the jumper is assigned between any two levels of the terminal block, the **Delete Jumper** button will be activated.

To create a new entry in the Terminal category list, scroll down in the Database grid and then add details in the blank row available at the bottom.

Once you edit the entry or create a new entry, choose the **Accept Changes** button located below the **Search** field to save the changes made. If you want to cancel the changes made, choose the **Cancel Changes** button.

RESEQUENCING TERMINAL NUMBERS

AutoCAD Electrical has some tools that help in resequencing terminal numbers in drawings. However, these tools do not resequence the terminals that use the wire number as a terminal number. The terminal strips can be resequenced by two methods and are discussed next.

Note
The panel terminals cannot be renumbered using these tools. To resequence a terminal strip that has panel terminals, use the ***Terminal Strip editor*** *tool.*

First Method

Command: AETERMRENUMPICK

To resequence the terminals by selecting them directly from the drawing, choose **Components > Terminals > Terminal Strip Utilities > Terminal Renumber (Pick Mode)** tool from the menu bar; you will be prompted to enter the start number. Enter the start number for the terminal number and then press ENTER. Alternatively, you can directly press ENTER to accept the default value. Next, you will be prompted to select the terminal symbol. Select the terminal symbols one by one from the drawing; the terminal number will get renumbered automatically and increment with each pick. Press ENTER to exit the command.

Second Method

Command: AETERMRENUM

To resequence the terminal numbers across the entire project, choose **Components > Terminals > Terminal Strip Utilities > Terminal Renumber (Project-Wide)** tool from the menu bar; the **Project-wide Schematic Terminal Renumber** dialog box will be displayed,. The options in this dialog box are discussed next.

The **Tag-ID** edit box is used to specify the Tag-ID of terminal that you need to resequence. Enter the terminal strip Tag-ID in the edit box in the **Tag-ID** area. Alternatively, you can choose the **Drawing** or **Project** buttons to select the Tag-ID of the terminal strip to be renumbered from the active drawing or active project. Note that only the specified Tag-ID will be renumbered.

Select the **Include Installation/Location in terminal strip Tag-ID match** check box to include the installation and location codes in the terminal strip for the Tag-ID match. On selecting this check box, the **Installation code** and **Location code** areas will be activated. If you want to filter the search, enter the installation code and the location code in the **Installation code** and **Location code** areas, respectively. Alternatively, choose the **Drawing** or **Project** button to search for the installation and location codes in the active drawing or in the active project.

Note that the terminals will be renumbered only if they match with the specified Tag-ID, Installation code, and location code.

Next, enter the beginning number of terminal in the **Starting Terminal Number** edit box. After specifying the required options in the **Project-wide Schematic Terminal Renumber** dialog box, choose the **OK** button in this dialog box; the **Select Drawings to Process** dialog box will be displayed. Next, select the drawings that you need to process and choose the **Process** button; the selected drawings will be moved from the top list to the bottom list. Next, choose the **OK** button; the terminal numbers will be resequenced.

COPYING TERMINAL BLOCK PROPERTIES

Command: AECOPYTERMINALPROP

The **Copy Terminal Block Properties** tool is used to copy terminal properties of one terminal symbol and paste them into another terminal symbol. To do so, choose the **Copy Terminal Block Properties** tool from the **Edit Components** panel of the **Schematic** tab; you will be prompted to select the master terminal. Select the master terminal to copy the terminal properties from; you will be prompted to select the target terminal(s). Select the terminal(s) to apply the terminal properties and press ENTER; the terminal properties will be applied to the target terminals. Press ENTER to exit the command. Now, if you want to view whether the terminal properties of the master terminal have been applied to the target terminal, right-click on it and choose the **Edit Component** option; the **Insert / Edit Terminal Symbol** dialog box will be displayed. In this dialog box, choose the **Block Properties** button to view the terminal properties of the target terminal.

EDITING JUMPERS

Command: AEJUMPER

The **Edit Jumper** tool is used to edit or delete jumper information on a terminal. This tool is also used to jumper two or more terminals together. To edit or delete a jumper on a terminal, choose the arrow on the right of the **Edit Components** panel of the **Schematic** tab; a flyout will be displayed. Choose the **Edit Jumper** tool from the flyout; you will be prompted to select a terminal symbol. Select the terminal symbol. The prompt sequence for this command is given next.

Command: **AEJUMPER** Enter
Select terminal <Browse>: *Select terminal symbol.*
Select jumpered terminals [Browse/Edit/Show] <Edit>: *Enter any of the options or ESC to exit the command.*

The **AEJUMPER** command has the following three options.

Browse **Edit** **Show**

These options are discussed next.

The Browse Option

If you enter **B** at the Command prompt, the **Select Terminals To Jumper** dialog box will be displayed, as shown in Figure 11-22. This dialog box displays a list of the terminals present within an active project in the **Schematic Terminals** area. In this dialog box, you can add terminals to the current jumper and view jumpers or terminals at the lower part of the dialog box by choosing the **View** button.

The Edit Option

When you enter **E** at the Command prompt "*Select jumpered terminals [Browse/Edit/Show] <Edit>*", the **Edit Terminal Jumpers** dialog box will be displayed.

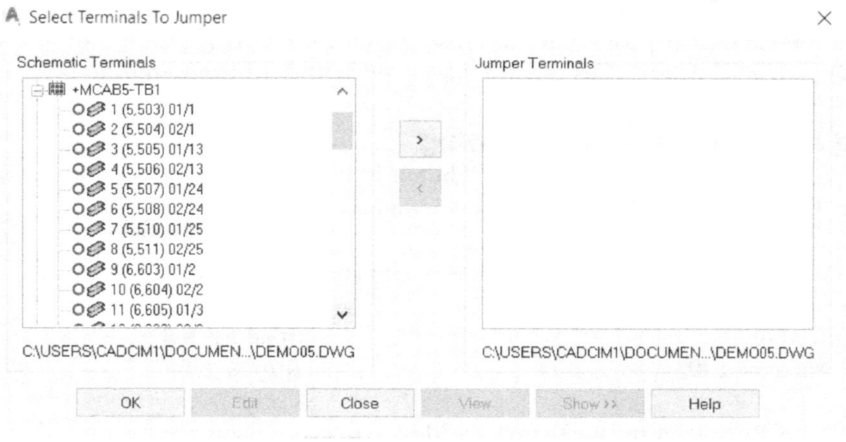

Figure 11-22 The ***Select Terminals To Jumper*** *dialog box*

The Show Option

When you enter **S** at the Command prompt "*Select jumpered terminals [Browse/Edit/Show] <Edit>*", a red dashed line which represents the jumper (temporary line), is drawn between the primary and secondary terminals within the same drawing, as shown in Figure 11-23. Choose **View > Redraw** from the menu bar to remove this red dashed line (temporary line).

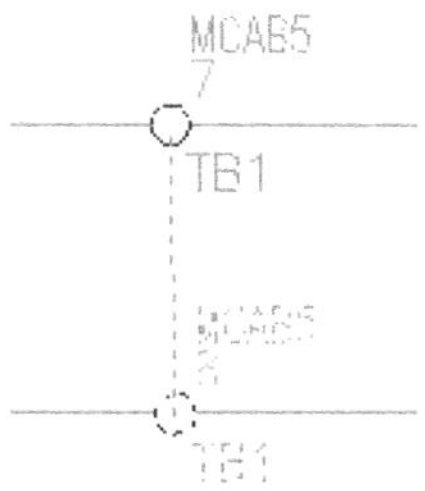

Figure 11-23 Displaying jumper between terminals

TUTORIALS

Tutorial 1

In this tutorial, you will create a ladder and then insert a terminal symbol in it. Also, you will enter required information in the **Insert / Edit Terminal Symbol** dialog box.

(Expected time: 10 min)

The following steps are required to complete this tutorial:

a. Create a new drawing.
b. Insert a ladder in the drawing.
c. Insert a terminal symbol into the drawing and enter the required information.
d. Save the drawing.

Creating a New Drawing

1. Activate the **CADCIM** project. Next, click on the *TUTORIALS* subfolder and then create the *C11_tut01.dwg* drawing, as already discussed in the previous chapters.

Inserting a Ladder into the Drawing

1. Choose the **Insert Ladder** tool from **Schematic > Insert Wires/Wire Numbers > Insert Ladder** drop-down; the **Insert Ladder** dialog box is displayed.

2. Set the following values in the edit boxes of the **Insert Ladder** dialog box:

   Width = **5.000** Spacing = **1**
   1st Reference = **100** Rungs = **5**

3. Choose the **OK** button in the **Insert Ladder** dialog box; you are prompted to specify the start position of the first rung.

4. Enter **9,11** at the Command prompt and press ENTER; the ladder is inserted in the drawing.

Inserting Terminal Symbol and Entering Information

1. Choose the **Icon Menu** tool from **Schematic > Insert Components > Icon Menu** drop-down; the **Insert Component** dialog box is displayed. 

2. Select the **Terminals/Connectors** icon from the **NFPA: Schematic Symbols** area; the **NFPA: Terminals and Connectors** area is displayed.

3. Select **Square with Terminal Number** located on the third column and first row of the **NFPA: Terminals and Connectors** area; you are prompted to specify the insertion point.

4. Enter **12.5,11** at the Command prompt and press ENTER; the **Insert / Edit Terminal Symbol** dialog box is displayed.

5. Enter **MACHINE** in the **Location** edit box, **TB1** in the **Tag Strip** edit box, and **1** in the **Number** edit box in the **Terminals** area.

6. Choose the **Details >>** button to expand the **Insert / Edit Terminal Symbol** dialog box if the dialog box is not already expanded.

7. Choose the **Catalog Lookup** button from the **Catalog Data** area; the **Catalog Browser** dialog box is displayed.

8. In this dialog box, enter **1492-WFB4** in the **Search** field and choose the **Search** button.

9. Select **1492-WFB4** catalog value from the list displayed in the **Catalog Browser** dialog box.

10. Choose the **OK** button in the **Catalog Browser** dialog box; the catalog values are displayed in the **Manufacturer** and **Catalog** edit boxes of the **Catalog Data** area.

11. Choose the **Block Properties** button in the **Modify Properties/Associations** area of the **Insert/Edit Terminal Symbol** dialog box; the **Terminal Block Properties** dialog box is displayed.

    By default, 1 is displayed in the **Levels** edit box, SINGLE is displayed under the **Level Description** column, and 2 is displayed under the **Wires Per Connection** column.

12. Enter **1** and **2** in the row under the **Pin Left** and **Pin Right** columns, respectively.

13. Choose the **OK** button from the **Terminal Block Properties** dialog box; the values you entered in this dialog box are displayed in the **Properties/Associations** area.

14. Choose the **OK** button from the **Insert / Edit Terminal Symbol** dialog box; the terminal symbol is inserted in the drawing, as shown in Figure 11-24.

Note

*If the **Assign Symbol To Catalog Number** message box is displayed on choosing the **OK** button from the **Insert / Edit Terminal Symbol** dialog box, choose the **Map symbol to catalog number** option from this message box.*

Saving the Drawing File

1. Choose **Save** from the **Application Menu** to save the drawing file *C11_tut01.dwg*.

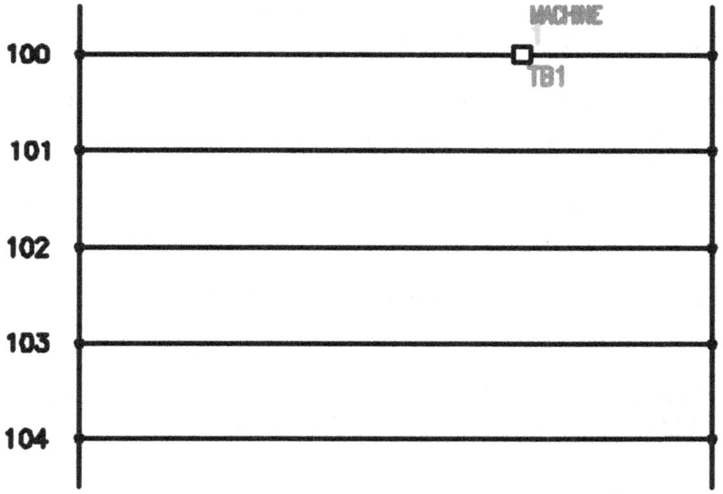

Figure 11-24 *Terminal symbol inserted in the ladder*

Tutorial 2

In this tutorial, you will insert DIN Rail in a drawing and then insert the terminal strip by extracting the terminal information from *Demo04.dwg* that you added to the **CADCIM** project in Tutorial 3 of Chapter 2. Also, you will add information to the terminal strip.

(Expected time: 25 min)

The following steps are required to complete this tutorial:

a. Create a new drawing.
b. Insert a DIN Rail into the drawing.
c. Insert a terminal strip into the drawing and add information to it.
d. Generate the terminal strip table.
e. Save the drawing.

Creating a New Drawing

1. Activate the **CADCIM** project. Next, click on the *TUTORIALS* subfolder and then create the *C11_tut02.dwg* drawing as already discussed in the previous chapters.

Inserting the DIN Rail into the Drawing

1. Choose the **Icon Menu** tool from **Panel > Insert Component Footprints > Icon Menu** drop-down; the **Alert** message box is displayed.

2. Choose the **OK** button in this message box; the non-visible 'WD_PNLM' block is inserted into the drawing and the **Insert Footprint** dialog box is displayed.

3. Choose the **DIN Rail** icon from the **Panel Layout Symbols** area of the **Insert Footprint** dialog box; the **Din Rail** dialog box is displayed.

4. Select **AB, 199-DR1, Symmetrical Rail 35mm x 7.5mm 1m Ion** from the drop-down list in the **Rail Type** area, if it is not already selected.

5. Choose the **Pick Rail Information >>** button from the **Origin and Length** area; you are prompted to specify the insertion point for the rail.

6. Enter **13,17** at the Command prompt and press ENTER; you are prompted to specify the end point of rail.

7. Enter **13,7** at the Command prompt and press ENTER; the **Din Rail** dialog box is displayed again.

8. Select the **Vertical** radio button from the **Orientation** area and the **NC holes** radio button from the **Panel Mounting** area, if not already selected.

9. Make sure 1.0 is displayed in the **Scale** edit box.

10. Choose the **OK** button in the **Din Rail** dialog box; the **Panel Layout - Component Insert/Edit** dialog box is displayed.

11. Choose the **OK** button in this dialog box; DIN Rail is inserted into the drawing, as shown in Figure 11-25.

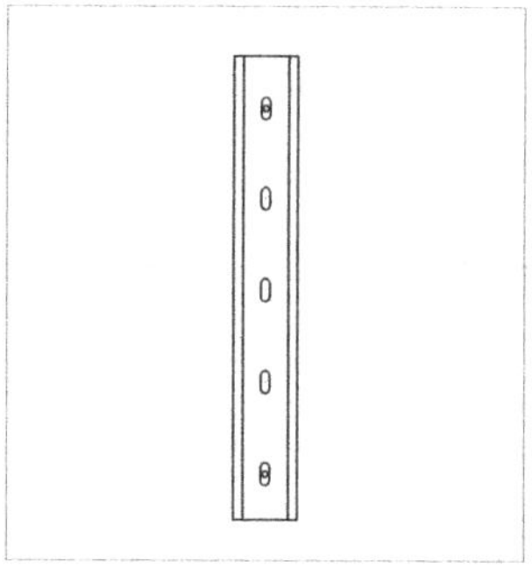

Figure 11-25 DIN Rail inserted in the drawing

Inserting a Terminal Strip into the Drawing and Adding Information to it

1. Choose the **Editor** tool from the **Terminal Footprints** panel of the **Panel** tab; the **Terminal Strip Selection** dialog box is displayed.

> **Note**
> *If the **QSAVE** message box is displayed after choosing the **Editor** tool from the **Terminal Footprints** panel of the **Panel** tab, choose the **OK** button in the **QSAVE** message box to display the **Terminal Strip Selection** dialog box.*

2. Select the **TS-A** terminal strip from this dialog box, refer to Figure 11-26.

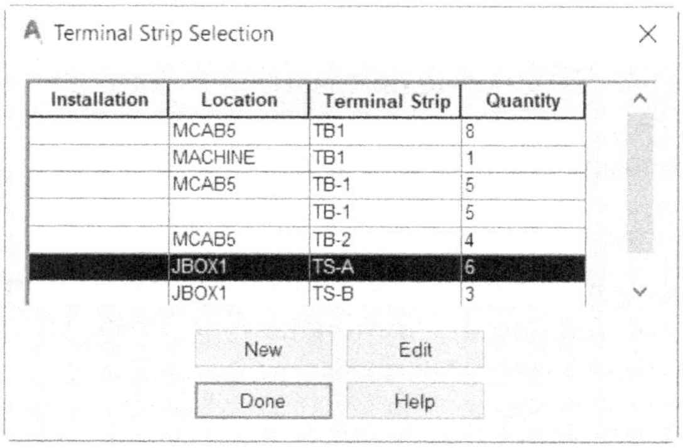

*Figure 11-26 The **Terminal Strip Selection** dialog box*

3. Choose the **Edit** button; the **Terminal Strip Editor** dialog box is displayed.

> **Note**
> *After choosing the **Edit** button, if the **Defined Terminal Wiring Constraints Exceeded** message box is displayed, choose the **OK** button in this message box.*

4. Choose the **Catalog Code Assignment** tab from the **Terminal Strip Editor** dialog box and then select all the rows in this tab.

5. Choose the **Assign Catalog Number** button from the **Catalog** area; the **Catalog Browser** dialog box is displayed.

6. Enter **DN-T10** in the **Search** field of this dialog box and choose the **Search** button.

7. Next, select **DN-T10, GRAY, 18-10AWG, 50pcs/ft (166/m), TERMINAL BLOCK** from this dialog box.

8. Choose the **OK** button from this dialog box; the catalog data is added to the terminal strip.

9. Select the last row in the terminal list displayed in the **Catalog Code Assignment** tab and choose the **Insert Spare Terminal** button from the **Spare** area; the **Insert Spare Terminal** dialog box is displayed.

10. Enter **2** in the **Quantity** edit box and then choose the **Insert Below** button; the spare terminals are inserted below the last row of the terminal list in the **Catalog Code Assignment** tab.

11. Next, choose the **Layout Preview** tab and then select the **Graphical Terminal Strip** radio button from the **Terminal Strip Editor** dialog box, if not already selected.

12. Next, select the **Wire Number Tag** option from the **Default pick list for Annotation format** list box.

13. Make sure 1.0 is selected in the **Scale on Insert** drop-down list.

14. Make sure 0.0 is selected in the **Angle on Insert** drop-down list.

15. Choose the **Update** button; a preview of the terminal strip is displayed in the preview window.

16. Choose the **Insert** button and enter **13,16** at the Command prompt and then press ENTER; the terminal strip is inserted into the drawing and the **Terminal Strip Editor** dialog box is displayed again.

17. Choose the **OK** button in the **Terminal Strip Editor** dialog box to return to the **Terminal Strip Selection** dialog box.

18. Choose the **Done** button in the **Terminal Strip Selection** dialog box to exit the dialog box. Figure 11-27 shows the terminal strip inserted in a DIN Rail. Figure 11-28 shows the zoomed view of the terminal strip.

Note

*If the **QSAVE** message box is displayed, choose the **OK** button in this dialog box to save the drawing.*

Generating the Terminal Strip Table

1. Choose the **Table Generator** tool from the **Terminal Footprints** panel of the **Panel** tab; the **Terminal Strip Table Generator** dialog box is displayed.

The **Terminal Strip Selection** area in this dialog box displays the terminal strips used in the active Project.

2. Select the terminal strip named **TS-A** from the **Terminal Strip Selection** area.

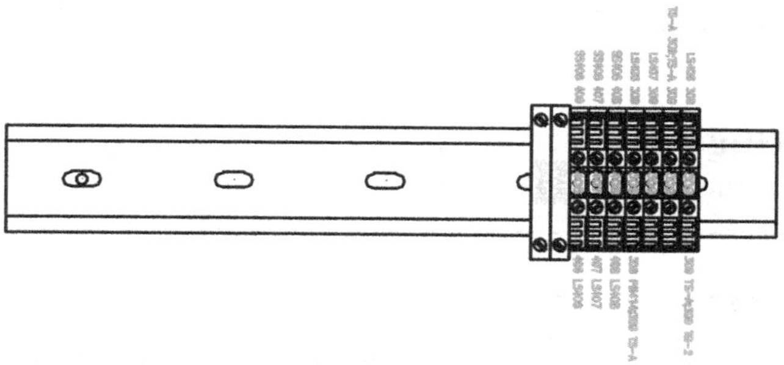

Figure 11-27 *Terminal strip inserted in a DIN Rail*

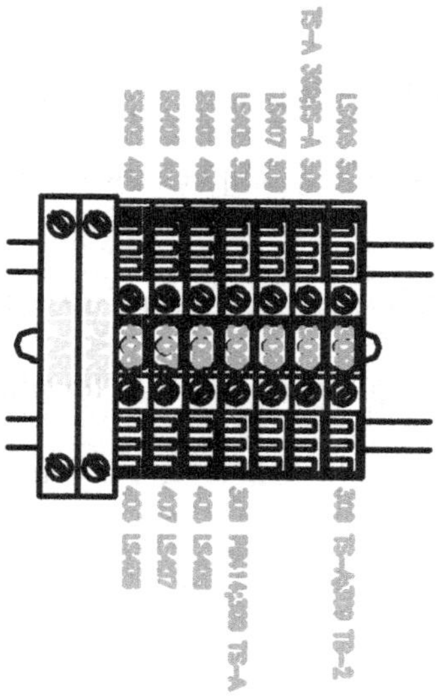

Figure 11-28 *The zoomed view of the terminal strip*

3. Choose the **Settings** button from the **Tabular Layout** area; the **Terminal Strip Table Settings** dialog box is displayed.

4. Enter **1.5** in the **X** edit box and **10** in the **Y** edit box of the **Section Placement** area.

5. Choose the **Browse** button located next to the **First Drawing Name** edit box; the **First Drawing Name** dialog box is displayed. In this dialog box, browse to the location where *C11_tut02* is saved, select **C11_tut02** from this dialog box and then choose **Save**; the path for the selected drawing file is displayed in the **First Drawing Name** edit box. Next, choose the **OK** button from the **Terminal Strip Table Settings** dialog box.

6. Choose **OK** from the **Terminal Strip Table Generator** dialog box; the **Table(s) Inserted** message box is displayed. Note that the terminal strip table is inserted in the new drawing *C11_tut03.dwg* as mentioned in this message box.

   The new drawing *C11_tut03.dwg* is automatically added to the active project at the bottom of the list. If it is not displayed, choose the **Refresh** button from the **PROJECT MANAGER** to display it.

Saving the Drawing File

1. Choose **Save** from the **Application Menu** to save the drawing file *C11_tut02.dwg*.

Self-Evaluation Test

Answer the following questions and then compare them to those given at the end of this chapter:

1. Which of the following dialog boxes will be displayed when you choose the **Editor** tool?

   (a) **Terminal Strip Selection** (b) **Terminal Strip Editor**
   (c) **Terminal Strip Table Generator** (d) None of these

2. The __________ tool is used to insert a jumper between the terminals that belong to different terminal strips.

3 A terminal strip table can be added to the drawing using the __________ or __________ tool.

4. The __________ tool is used to annotate and edit terminal symbols.

5. The settings of a terminal strip table can be defined in the **Terminal Strip Table Settings** dialog box. (T/F)

Review Questions

Answer the following questions:

1. Which of the following attributes on terminal symbol is used to link individual symbols to a terminal strip?

   (a) Installation attributes (b) Location attributes
   (c) Tag Strip attributes (d) All of these

2. The terminal that uses a wire number as its terminal number can be renumbered. (T/F)

3. You can generate the terminal strip information graphically but not in the table format. (T/F)

4. The **Terminal Strip Selection** dialog box displays the terminal strips present in all the projects. (T/F)

EXERCISES

Exercise 1

Create a new drawing as *C11_exer01.dwg* and then insert the ladder and terminal symbol in it, as shown in Figure 11-29. **(Expected time: 15 min)**

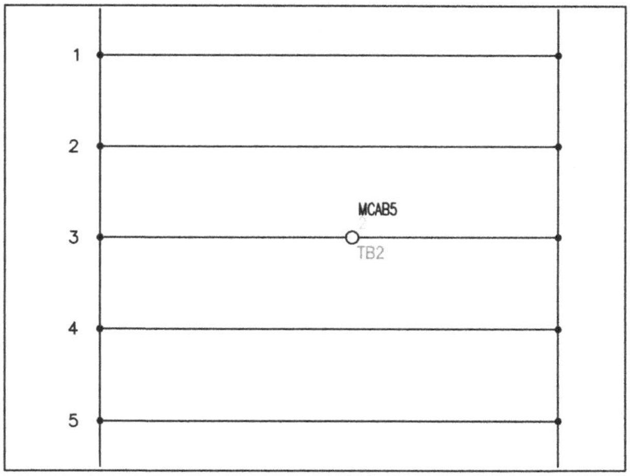

Figure 11-29 *Terminal symbol inserted in a ladder*

Exercise 2

Create a new drawing as *C11_exer02.dwg*. You will insert a DIN Rail (AB, 1492-N22) in the drawing and specify the following information in the **Din Rail** dialog box: **Length** = 4, **X** = 13, **Y** = 17, **Orientation** = vertical. Also, you will assign item number 15 to this DIN Rail. The DIN Rail that you have inserted into a drawing is shown in Figure 11-30. Next, you will insert a terminal strip by using the **Editor** tool. You will select TS-B terminal strip from the **Terminal Strip Selection** dialog box. You will also add catalog information as AUTOMATIONDIRECT, FEED-THROUGH, 20AMPS (DN-T12) to the terminal list displayed in the **Terminal Strip Editor** dialog box. You will also insert two spare terminals after the last row in the **Catalog Code Assignment** tab. The graphical terminal strip inserted into a drawing is shown in Figure 11-31. **(Expected time: 20 min)**

Hint: Select **Wire Number Tag : Terminal** from the **Default pick list for Annotation format** list box and **Scale on Insert** = 3.0.

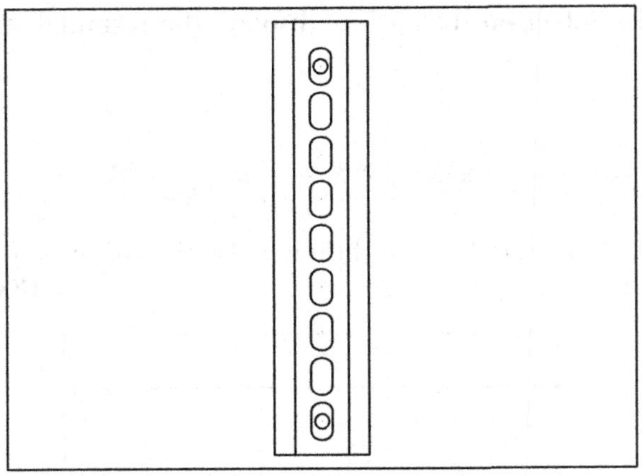

***Figure 11-30** DIN Rail inserted in a drawing*

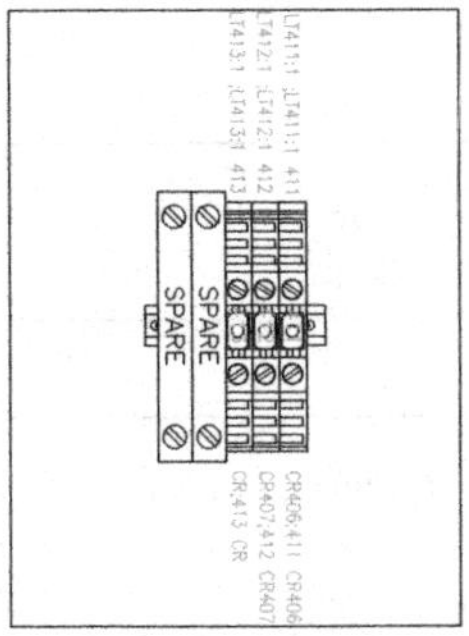

***Figure 11-31** Graphical terminal strip inserted in the drawing*

Answers to Self-Evaluation Test

1. a, **2. Edit Jumper**, **3. Terminal Strip Editor, Terminal Strip Table Generator**, **4. Edit**, **5.** T

Chapter 12

Settings, Configurations, Templates, and Plotting

Learning Objectives

After completing this chapter, you will be able to:

- *Set the project and drawing properties*
- *Understand reference files*
- *Map a title block*
- *Update a title block*
- *Create template drawings*
- *Plot Projects*
- *Understand the project task list*

INTRODUCTION

In this chapter, you will learn to set the project and drawings properties. You will also learn to use various reference files supported by AutoCAD Electrical, which will help you annotate drawings. Later in this chapter, you will learn about title blocks, creating template files, plotting drawings, and creating a project task list.

SETTING PROJECT PROPERTIES

The Project properties organize the default settings and configuration of AutoCAD Electrical. These properties are used as the default properties when a new drawing is created for a project. You can apply the project settings to the drawing settings when you create or add a drawing to a project. To invoke the **Project Properties** dialog box, right-click on the active project in the **PROJECT MANAGER**, and then choose the **Properties** option from the shortcut menu; the **Project Properties** dialog box will be displayed, as shown in Figure 12-1. In this dialog box, you can define the settings for a new project and also use these settings for new drawings. In this dialog box, you can also edit and modify the properties for project settings, components, wire numbers, cross-references, styles, and drawing format.

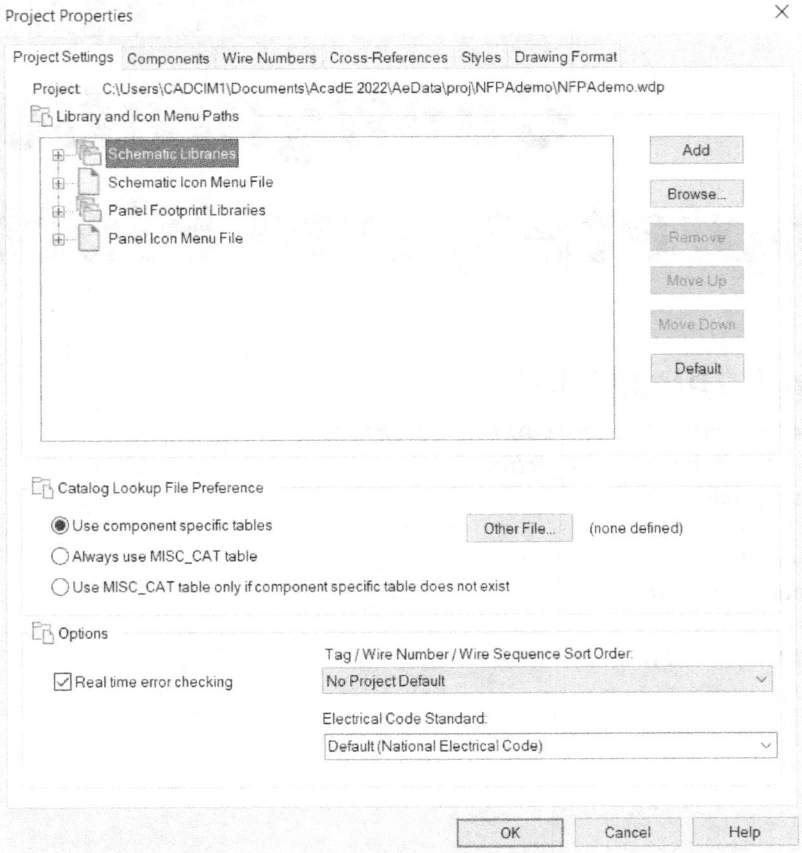

Figure 12-1 *The **Project Properties** dialog box*

The values that you define in this dialog box will be saved in the project file. The icons shown in Figure 12-2 indicate whether the settings will be applied to the project or the drawing. Different tabs and options in the **Project Properties** dialog box are discussed next.

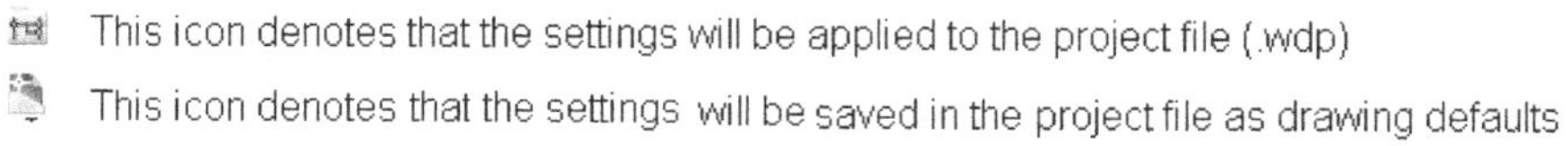

***Figure 12-2** Icons indicating the settings for the project and drawing*

Project Settings Tab

The **Project Settings** tab is chosen by default, refer to Figure 12-1. In this tab, you can modify the default settings of libraries and icon menu paths, catalog lookup files, and error checking. The information in this tab will be saved as default settings of the project in the project definition file (*.wdp*). In the **Library and Icon Menu Paths** area, you can select the schematic library, panel footprint library, and icon menu files that you want to use. In the **Catalog Lookup File Preference** area, you can control the preference for searching the bill of material catalog database. The **Options** area is used to control the real-time error checking.

Components Tab

In the **Components** tab, shown in Figure 12-3, you can modify the default settings of the components in the project. The information in this tab will be saved as default settings of the project in the project definition file (*.wdp*). Different areas in this tab are discussed next.

Component TAG Format Area

In the **Component TAG Format** area, you can specify the tag for the components in the project. The tag can be sequential or reference-based. The **Tag Format** edit box is used to specify the tag format for the new components for the entire project. By default, %F%N is displayed in this edit box. You can define a new format by specifying the replaceable parameters in this edit box. The tag format consists of two sets of information such as the family code and the alphanumeric reference number. For example, PB503 is an alphanumeric tag, where PB is the family code of the push button component and 503 is the line reference number of a component. Also, note that the %N parameter is compulsory for all tag formats.

Select the **Sequential** radio button; the edit box adjacent to this radio button will be activated. In this edit box, you can enter the starting sequential number for the components of a drawing. The **Line Reference** radio button is selected by default and is used to set the format of the tags based on their line reference. Choose the **Suffix Setup** button on the right of the **Line Reference** radio button; the **Suffix List for Reference-Based Component Tags** dialog box will be displayed. This dialog box displays the suffix list. This list is used to create unique reference-based tag when multiple components of the same family are located in the same reference location. The component tag suffix automatically gets added at the end of the tag format.

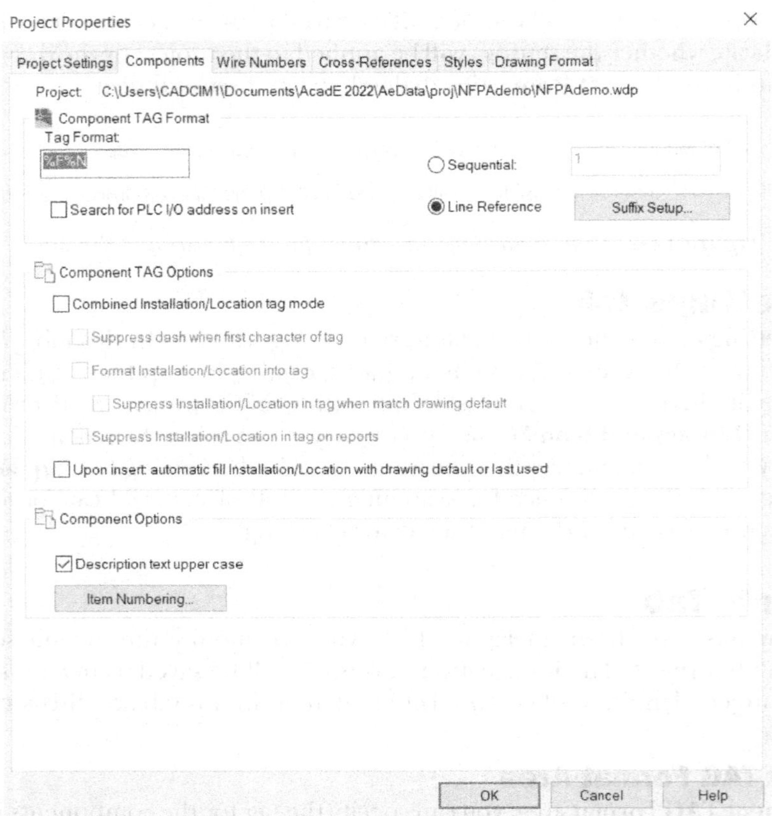

*Figure 12-3 The **Project Properties** dialog box showing the **Components** tab*

Component TAG Options Area

The options in this area are used to set different options for component tags such as the combined installation/location tag mode, suppressing dashes of tag, and so on.

Component Options Area

In the **Component Options** area, you can set the description of the text to upper case or lower case and the settings for item number of the panel component.

Wire Numbers Tab

In the **Wire Numbers** tab, shown in Figure 12-4, you can modify the default settings of the project for wire numbers. The information in this tab will be saved as default settings of the project in the project definition file (*.wdp*). Different areas and options in this tab are discussed next.

Wire Number Format Area

In the **Wire Number Format** area, you can specify the format for wire numbers. Specify the format for the new wire number in the **Format** edit box.The **Sequential** radio button is used to set the starting sequential number for the wire number. By default, the **Line Reference** radio button is selected. This radio button is used to assign wire numbers to the wires based on their line reference number.

Figure 12-4 The ***Project Properties*** *dialog box showing the* ***Wire Numbers*** *tab*

Wire Number Options Area

The options in the **Wire Number Options** area are used to specify various options for the wire number.

New Wire Number Placement Area

In the **New Wire Number Placement** area, you can specify the options for the placement of wire numbers such as above, in-line, or below the wire. Note that the options in this area will show the results only if you are inserting the new wire numbers.

Wire Type Area

In the **Wire Type** area, you can rename the column heading of the User1 to User20 fields. To do so, choose the **Rename User Columns** button in the **Wire Type** area; the **Rename User Columns** dialog box will be displayed. Enter the name for the User1 to User20 fields; these changes will be reflected in the **Set Wire Type**, **Create/Edit Wire Type**, and **Change/Convert Wire Type** dialog boxes.

Cross-References Tab

In the **Cross-References** tab, you can modify the default settings of the cross-references in the project. The new drawings that you create will be saved as default settings of the project in the project definition file (*.wdp*) for cross-referencing. The cross-reference tag indicates the location of the linked components within the project.

Styles Tab

In the **Styles** tab, you can modify the default settings for the component styles in the project. The information in this tab will be saved as default settings of the project in the project definition file (*.wdp*). Different areas and options in this tab are discussed next.

Arrow Style Area

In the **Arrow Style** area, you can specify the style of the source and destination arrows. In the **PLC Style** area, you can specify the style of the PLC modules. You can select the PLC module style from the five pre-defined styles or from the four user-defined styles. In this area, the preview of the selected PLC style will be displayed in the preview window.

Wiring Style Area

In the **Wiring Style** area, you can specify style for wires when they intersect. In this area, you can specify settings for wire crossings and wire tee connections which are used while inserting a wire.

Fan-In/Out Marker Style Area

In the **Fan-In/Out Marker Style** area, you can specify the style for the fan in/out marker symbols. You can also specify the layer of the wires going out of **Fan In/Out Source** marker and coming into the **Fan In/Out Destination** marker. By default, **1** radio button is selected. You can select the style for the fan in/out marker from the four pre-defined styles and five user-defined styles. The preview of the selected option will also be displayed in the preview window

Drawing Format Tab

In the **Drawing Format** tab, you can modify the project default settings for the drawings. Also, in this tab, you can set the defaults for ladder, format for cross-referencing, order of the tag and wire number, scale of drawing, and the layer for component. Also, the information in this tab will be saved as default settings of the project in the project definition file (*.wdp*). The different areas and options in this tab are discussed next.

Ladder Defaults Area

In the **Ladder Defaults** area, you can set the default values of orientation and spacing for the ladder. The spacing values include spacing between the rungs, width, and the multi-wire spacing.

Note

All options mentioned above, except the orientation of the ladder, can also be changed using the ***Insert Ladder*** *tool.*

Format Referencing Area

In the **Format Referencing** area, you can specify the format for the reference numbers of the ladder. The options in this area are discussed next.

The **X-Y Grid** radio button is used to insert the X-Y grid labels in the drawings. This referencing is commonly used with point-to-point wiring diagram style drawings. When you select the **X-Y Grid** radio button and choose the **Setup** button, the **X-Y Grid Setup** dialog box will be displayed. Enter the required information in this dialog box and choose the **OK** button to save the information and return to the **Project Properties** dialog box.

The **X Zones** button is used to insert the X grid labels for drawings. When you choose the **Setup** button, the **X Zones Setup** dialog box will be displayed. Next, enter the required information in the **X Zones Setup** dialog box and choose the **OK** button to return to the **Project Properties** dialog box and save the changes.

By default, the **Reference Numbers** radio button is selected. This radio button is used to change the display of reference numbers of a ladder in a drawing. Choose the **Setup** button; the **Line Reference Numbers** dialog box will be displayed. In this dialog box, you can define the style of line reference numbers of a ladder. By default, the **Numbers only** radio button is selected. Select the required radio button in this dialog box; the style of line reference numbers will be changed accordingly.

Choose the **OK** button in the **Line Reference Numbers** dialog box to save the changes made and exit the dialog box.

Tag / Wire Number / Wire Sequence Order Area

In the **Tag / Wire Number /Wire Sequence Order** area, you can set the default wire numbering, component tag, and wire sequence sort order for the active drawing. The sort order specified in this area is used for default wire sequencing of wire networks with multiple components. Your settings of sort order in this tab will override the settings that you have done in the **Project Settings** tab. Select the sort order from the **Sort Order** drop-down list. By default, the **No override** option is selected.

Scale Area

In the **Scale** area, you can specify the scale for inserting the components and the wire numbers.

Layers Area

In the **Layers** area, you can specify layers for wire numbers and component.

After specifying the required options in the **Project Properties** dialog box, choose the **OK** button to save the changes and exit the dialog box.

SETTING DRAWING PROPERTIES

Command: AEPROPERTIES

Drawing properties are similar to the project properties. Each drawing has its own settings. To define settings for a drawing, choose the **Drawing Properties** tool from the **Other Tools** panel of the **Schematic** tab; the **Drawing Properties** dialog box will be displayed, as shown in Figure 12-5. Alternatively, to display the **Drawing Properties** dialog box, right-click on the drawing file name in the **PROJECT MANAGER**; a shortcut menu will be displayed. Choose the **Properties > Drawing Properties** option from the shortcut menu. In this dialog box, you can specify the settings for the new or selected drawing.

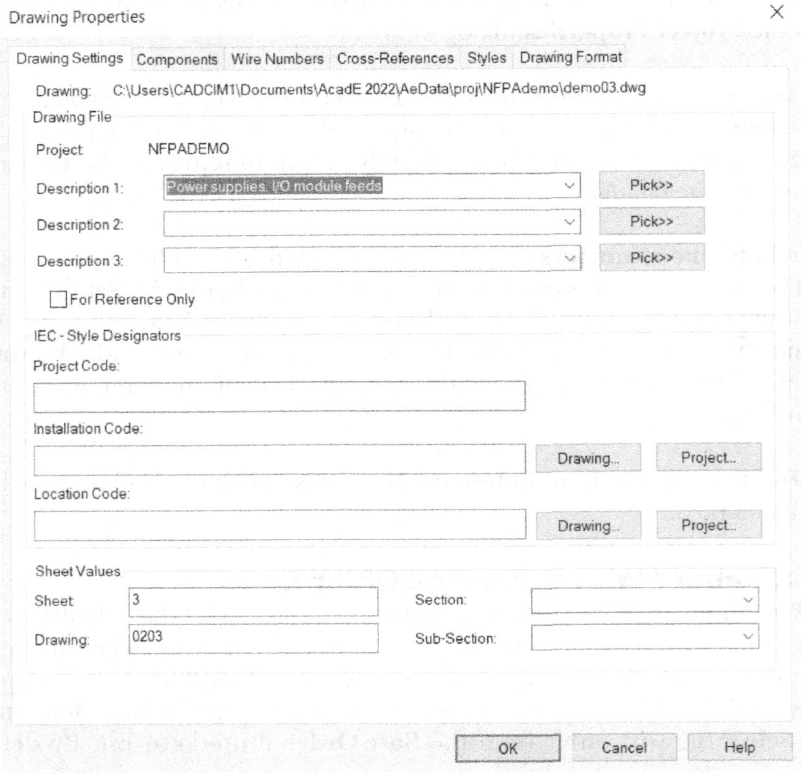

*Figure 12-5 The **Drawing Properties** dialog box*

The settings in the **Project Properties** dialog box will be overridden by the settings that you specify in the **Drawing Properties** dialog box. The settings that you specify in the **Drawing Properties** dialog box are saved in the WD_M block in the drawing file. Also, in this dialog box, you can edit and modify the properties for drawing settings, components, wire numbers, cross-references, styles, and drawing format. The options in the **Drawing Properties** dialog box are similar to that of the **Project Properties** dialog box, except in the **Drawing Settings** tab. Therefore, in the next section, only the **Drawing Settings** tab will be discussed.

Drawing Settings Tab

In the **Drawing Settings** tab, shown in Figure 12-5, you can specify and edit the information of a drawing file. Also, you can add drawing specific descriptions and sheet values. Observe that the name and location of the drawing file is displayed above the **Drawing File** area. Different areas and options in this tab are discussed next.

Drawing File Area

In the **Drawing File** area, you can edit the information of a drawing file. Also, you can add or edit the description of a drawing.

IEC - Style Designators Area

In the **IEC - Style Designators** area, you can specify the default value for the project code (%P), installation code (%I), and location code (%L). The codes that you enter in this area will be stored in the drawing on the WD_M.dwg block file attributes.

Sheet Values Area

In the **Sheet Values** area, you can set the sheet number, drawing numbers, section, and sub-section values. The values that you specify in this area will be saved in the drawing in the WD_M block file attributes.

UNDERSTANDING REFERENCE FILES

In this section, you will learn about various reference files that are supported by AutoCAD Electrical for annotating the drawings. The reference files are basically the ASCII text files, which are used for different purposes. You can edit these reference files using Notepad, Wordpad, and other word editors. These reference files can exist in multiple versions and are given as follows:

1. In the project folder, you will find the project's *.wdp* file. The name of the file will be the same as that of the project.

2. If the project-named file is not found, then AutoCAD Electrical will search the project directory for the default file.

3. If the project default file is not found, then AutoCAD Electrical will search for the default file that is present in the support directory.

By using these reference files, you can change the description of the drawing, update the title block attributes, and so on easily. Different reference files are discussed next.

Project Files (.WDP File)

The project files application is a project-based system. These files are created and maintained using the **PROJECT MANAGER**. Each project is defined by an ASCII text file and has an extension *.wdp*. The project file, also called the *.wdp* file, consists of a list of project information, default settings of a project, drawing properties, and name of the drawing files. You can create unlimited number of projects but only one project can be active at a time. Also note that if the project contains a single wiring diagram drawing, the project file is not needed.

When you create or add new drawings to a project, the project settings that are stored in the project file are referenced for maintaining the same format throughout the project drawings. A single project file can find an unlimited number of drawings located in many different directories. If you add a drawing in a project that is stored in the same directory as the project file, then only the file name will be stored in the project file (*.wdp*) but if the drawing you add belongs to a different directory than the project file, then both the file name and complete location of the drawing will be stored in the project file(*.wdp*). The default path of the project file is *C:\User\User Name\My Documents\Acade 2022\AeData\Proj\'your project'*. The symbol libraries, settings of drawing files, and other reference files are stored in the project directory and by using them you can easily change the settings and configurations for different projects.

Project Description Line Files (.WDL File)

When you right-click on a project name, a shortcut menu will be displayed. Next, choose the **Descriptions** option from the shortcut menu; the **Project Description** dialog box will be displayed. You can customize the generic LINEx label description that is displayed in the **Project Description** dialog box by creating a *.wdl* file and then changing the values in it.

To create a *.wdl* file, open a Notepad file and then enter the values that you want to display in place of Line1, Line2, and so on. Next, save the Notepad file with extension *.wdl* in the project folder wherein you have stored all drawings of the project, project file, and other reference files. The naming convention for these files is <projectname>_wdtitle.wdl or default_wdtitle.wdl. The LINEx label values are changed to match the attribute values of the drawing title block but they can be used for different purposes such as drawing descriptions, report information, and so on.

Note that it is not necessary to have entries in order and also the line number can be skipped. You can store unlimited number of lines in the *.wdl* file but the file should consist of one line per label in the format LINEx = label. After you save the *.wdl* file in the respective project folder, right-click on the name of that project in the **PROJECT MANAGER**; a shortcut menu will be displayed. Next, choose the **Descriptions** option from the shortcut menu; the **Project Description** dialog box will be displayed. You will notice that on saving the *.wdl* file in the project folder, the **Project Description** dialog box will be modified. Also, select the **in reports** check box to include the information in the reports.

Tip
It is advisable to save the project file and project drawings in the same directory.

Component Reference Files

In this topic, you will learn about various component reference files such as *.wdd*, *.loc*, *.inst* files. All these files are ASCII text files. These files save the time of the users as they can just select the values from the list, instead of entering the value every time. These files are discussed next.

.WDD File

The generic description file is *wd_desc.wdd*. This file consists of the description of the components. The *.wdd* file can be accessed by choosing the **Defaults** button in the **Description** area of the **Insert / Edit Component** dialog box. When you choose the **Defaults** button, the **Descriptions** dialog box will be displayed. Now, from the **Descriptions** dialog box, you can select the description for the component. Choose the **OK** button to display the description in the **Description** area of the **Insert / Edit Component** dialog box. The *.wdd* file can be a family- specific ASCII text file, for example LS.WDD for family 'LS' limit switch. If family-specific file is not found, then it will search for project-specific file that is *<project>.wdd*.

If both family-specific and project-specific files are not found, then AutoCAD Electrical will search for a general description file, WD_DESC.WDD file in various AutoCAD Electrical search paths and AutoCAD support paths. If the AutoCAD Electrical does not find anything, you will be prompted to browse for a *.wdd* description file. The *.wdd* file can be edited using the Notepad file or Wordpad file. To create a new *.wdd* file, you need to open a Notepad or Wordpad file and then enter the values that you want to enter in the **Descriptions** area and save the file in the respective folder with an extension *.wdd*.

.INST File

The installation codes file (default.inst) is an ASCII text file. This file contains installation codes. Now, to create the .inst file, open a Notepad or Wordpad file and then enter the installation codes that you want to use in the project and save the file in the respective folder as *Default.inst*. These installation codes will be displayed when you choose the **Project** button in the **Installation code** area of the **Insert / Edit Component** dialog box; the **All Installations - Project** dialog box will be displayed. Next, if you select the **Include external list** check box, the installation codes that you have saved in the *Default.inst* file will be displayed.

This way the *Default.inst* file is used for entering installation codes for components. This file can be named as *<projectname>.inst*. First, a file with the same path and name as the project and *.inst* extension will be searched. If the file is not found, the *Default.inst* file will be searched in the same directory as the project file and then in the subdirectory.

.LOC File

The location codes file (*Default.loc*) is an ASCII text file. This file contains location codes. Now, to create the *.loc* file, open a Notepad or Wordpad file and enter the location codes that you want to use in the project. Next, save the file in the respective folder as *Default.loc*. These location codes will be displayed when you choose the **Project** button in the **Location code** area of the **Insert / Edit Component** dialog box; the **All Locations - Project** dialog box will be displayed. Next, if you select the **Include external list** check box, the location codes that you have saved in the *Default.loc* file will be displayed. This way the *Default.loc* file is used for entering the location codes for components. Also, this file can be named as *<projectname>.loc*. First, a file with the same path and name as the project with the .loc extension will be searched. If the file is not found, then the *Default.loc* file will be searched in the same directory as the project file and then in the subdirectory.

MAPPING THE TITLE BLOCK

You can update the values of the attributes in the title block with the properties of a project or drawing. The title block of a drawing consists of a block with attributes. The attributes can be mapped to the title block in the following ways: you can create the *default.wdt* ASCII text file using any text editor such as Notepad, Wordpad or you can use **Title Block Setup** dialog box for creating the *.WDT* file and saving the mapping in an invisible WD_TB attribute on the title block. The *default.wdt* file defines each attribute's mapping.

Using this *.wdt* file, you can configure multiple blocks by separating block names with comma (,) or by creating multiple BLOCK =entries. Block names are followed by the attribute mappings.

The **Title Block Setup** tool is used to create title block mapping. Using this tool, you can pick blocks and attributes from the active drawing. To do so, choose the **Title Block Setup** tool from the **Other Tools** panel of the **Project** tab; the **Setup Title Block Update** dialog box will be displayed, as shown in Figure 12-6. Different areas and options in this dialog box are discussed next.

Note

Before updating the title blocks of a project, you must define how the project and drawing data is mapped to the matching title block attributes.

Method 1 Area

In the **Method 1** area, you can use the *.wdt* file to save the mapping information. Also, in this area, different path and location of the *.wdt* files are displayed. The methods for creating mapping information using different *.wdt* files are discussed next.

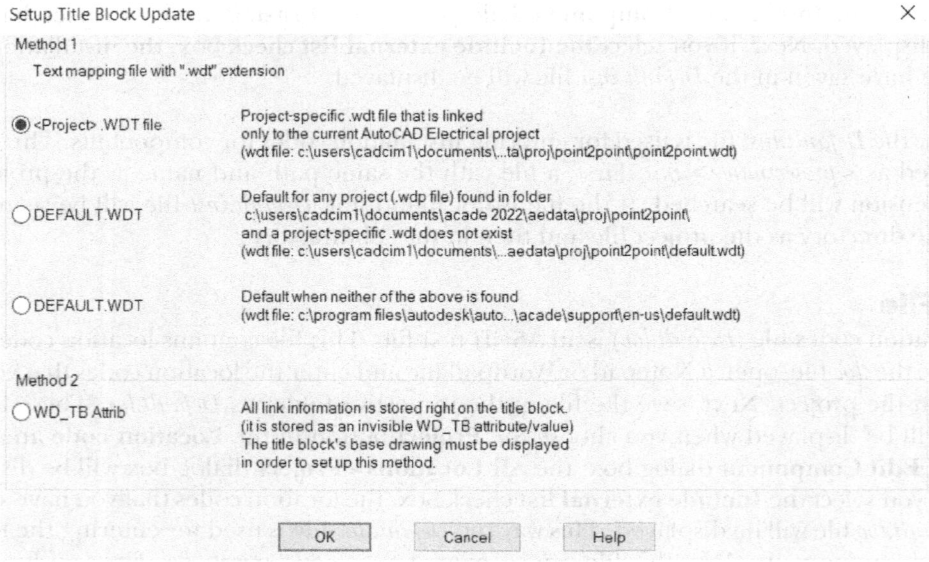

*Figure 12-6 The **Setup Title Block Update** dialog box*

Case 1

Select the **<Project>.WDT file** radio button in the **Method 1** area. Now, if the *.wdt* file already exists and you choose the **OK** button in the **Setup Title Block Update** dialog box, the **.WDT File Exists** dialog box will be displayed. The **View** button is used to view the existing *.wdt* file. The **Overwrite** button is used to overwrite the existing *.wdt* file and to create a new *.wdt* file. The **Edit** button is used to edit the existing *.wdt* file. Choose the **Cancel** button to exit the **.WDT File Exists** dialog box without saving the changes.

Case 2

Select the **<Project>.WDT file** radio button in the **Method 1** area. Now, if the *.wdt* file does not exist and you choose the **OK** button in the **Setup Title Block Update** dialog box, the **Enter Block Name** dialog box will be displayed, as shown in Figure 12-7. The options in this dialog box are discussed next.

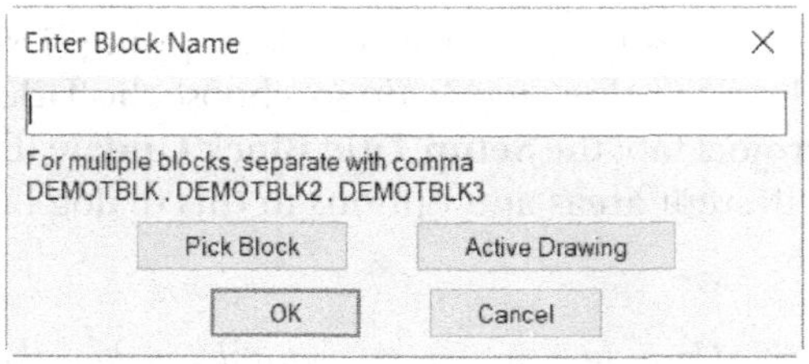

*Figure 12-7 The **Enter Block Name** dialog box*

Specify the block name in the edit box. Alternatively, you can choose the **Pick Block** button to select the block name from the drawing. Also, you can choose the **Active Drawing** button to use active drawing as the block name. If you want to enter multiple blocks in the edit box, separate block names with comma. Choose the **OK** button in the **Enter Block Name** dialog box; the **Title Block Setup** dialog box will be displayed. Specify the desired options and choose the **OK** button to save the changes and exit the **Title Block Setup** dialog box.

DEFAULT.WDT

The **DEFAULT.WDT** radio button is used when the project-specific *.wdt* file such as *<Project>. WDT* file is not found in the current project folder. When you select the **DEFAULT.WDT** radio button, AutoCAD Electrical will search for *default.wdt* file and will use it for mapping the title block. Note that the *default.wdt* file is saved in the folder of the current project.

DEFAULT.WDT

The **DEFAULT.WDT** radio button is used when the project-specific *.wdt* file *<Project>.WDT* file or *default.wdt* file is not found in the current project folder. When you select the **DEFAULT. WDT** radio button, AutoCAD Electrical will search for *default.wdt* file, which is saved at the *c:\program files\autodesk\acade 2022\support\default.wdt* location. AutoCAD Electrical will use this file for mapping the title block.

Method 2 Area

In the **Method 2** area, you can save the mapping information in the WD_TB attribute. If you use the **Method 2** area, you need to open the base drawing for the title block in order to insert the attribute. To do so, enter **Explode** at the Command prompt; you will be prompted to select objects. Next, select the title block and press ENTER; the title block will be exploded. Now, you can see the WD_TB attribute at the bottom of the title block, as shown in Figure 12-8.

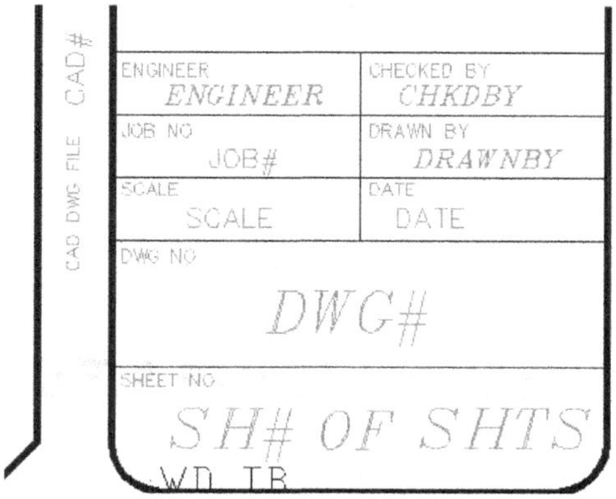

Figure 12-8 Lower right portion of the title block showing the WD_TB attribute

The WD_TB attribute helps in attribute mapping of a title block. Note that, if you double-click on the WD_TB attribute, the **Edit Attribute Definition** dialog box will be displayed. The values in the **Edit Attribute Definition** dialog box help in mapping the AutoCAD Electrical properties to the attributes in the block. Press ESC to cancel the command.

Next, choose the **Title Block Setup** tool from the **Other Tools** panel of the **Project** tab; the **Setup Title Block Update** dialog box will be displayed. In this dialog box, select the **WD_TB attribute** radio button from the **Method 2** area; the **Title Block Setup** dialog box will be displayed. Specify the desired options in the **Title Block Setup** dialog box and choose the **OK** button to save the changes made and exit the dialog box.

UPDATING TITLE BLOCKS

Command: AEUPDATETITLEBLOCK

You can update the attributes values of title blocks for the active drawing or project. To update the information of a title block, right-click on the active project; a shortcut menu will be displayed. Next, choose the **Title Block Update** option from the shortcut menu; the **Update Title Block** dialog box will be displayed, refer to Figure 12-9. Alternatively, choose the **Title Block Update** tool from the **Other Tools** panel of the **Project** tab to display the **Update Title Block** dialog box. Using this dialog box, you can update the information of the title block for an active drawing or project.

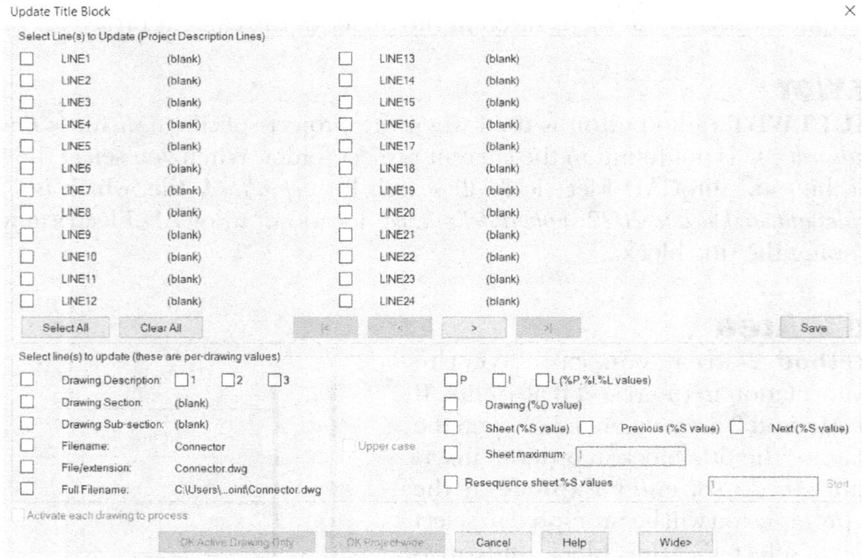

*Figure 12-9 The **Update Title Block** dialog box*

CREATING TEMPLATES

One way to customize AutoCAD Electrical is to create template drawings that contain initial drawing setup information and if desired, visible objects and text. When the user starts a new drawing, the settings associated with the template drawing are automatically loaded. If you start a new drawing from scratch, AutoCAD Electrical will load default setup values such as the WD_M block, predefined standard AutoCAD Electrical layers, and so on.

In production drawings, most of the electrical drawing setup values remain the same. For example, the company title block, layers, schematic ladders already inserted, panel settings already configured, enclosure inserted, and other drawing setup values do not change. You will save considerable time if you save these values and reload them when starting a new drawing. You can do this by creating template drawings that contain the initial drawing setup information configured according to the company specifications.

With this template, when you start a new drawing, AutoCAD Electrical does not have to pause and ask permission to insert the non-visible block. These templates consist of title block information, layer information, and so on. The extension of the template file is *.dwt*.

In order to create a template file, you can open existing drawing file or you can create a new one. Now, to create a new drawing file, choose the **New Drawing** button from the **PROJECT MANAGER**; the **Create New Drawing** dialog box will be displayed. Specify the options in this dialog box as per your requirement and then choose the **OK** button; the drawing will be created and added in the active project. Now, specify the settings for this drawing as per your requirement. For example, if you want to use this template drawing as a panel template, then you need to configure the panel settings. If you are creating a template drawing for schematic, then you need to insert ladders.

Next, choose **File > Save As** from the menu bar; the **Save Drawing As** dialog box will be displayed. Select the **AutoCAD Drawing Template (*.dwt)** option from the **Files of type** drop-down list; the **Template** folder will open automatically, where you need to save the template file. Alternatively, you can save the template file at the desired location. Next, enter the name of the template drawing in the **File name** edit box. Choose the **Save** button to save the template drawing; the **Template Options** dialog box will be displayed. After specifyng the desired information in this dialog box, the new template will be created.

PLOTTING THE PROJECT

The **Plot Project** option is used to plot full drawing set of a project or a group of drawings that you select. To plot full drawing set of a project, choose the **Publish/Plot** button from the **PROJECT MANAGER**; a flyout will be displayed, as shown in Figure 12-10. Choose the **Plot Project** option from this flyout; the **Select Drawings to Process** dialog box will be displayed. Select the drawings that you need to plot using this dialog box.

*Figure 12-10 The flyout displayed on choosing the **Publish/Plot** button*

If you choose the **Do All** button, all drawings will be moved to the bottom list of the **Select Drawings to Process** dialog box. If you choose the **Process** button, the selected drawings will be moved to the bottom list of the **Select Drawings to Process** dialog box. Choose the **OK** button from the **Select Drawings to Process** dialog box; the **Batch Plotting Options and Order** dialog box will be displayed.

After specifying the required options in the **Batch Plotting Options and Order** dialog box, choose the **OK** button from the **Order** area to plot the selected drawings in the order they appear in the project list displayed in the **PROJECT MANAGER**.

PROJECT TASK LIST

Sometimes, while editing a component, you may be prompted to update the components related to it. Also, the **Update other drawings?** message box may be displayed. You have the option to update the drawings immediately or add them to the task list. Now, if you choose the **Task** button, the components will be added to the task list for updation. Also, the information will be saved to the project task list, which can be updated later. The **Project Task List** button in the **PROJECT MANAGER** is used to execute the pending updates on the drawing files of the active project, which have been modified. Choose the **Project Task List** button; the **Task List** dialog box will be displayed. This dialog box displays information of the saved tasks.

When you select a task from the list displayed, the **Remove** and **OK** buttons will become available. Choose the **OK** button in this dialog box to perform the pending task.

TUTORIALS

Tutorial 1

In this tutorial, you will change the drawing properties for the wire numbers of a drawing and then update the wire numbers. **(Expected time: 15 min)**

The following steps are required to complete this tutorial:

a. Open the drawing *DEMO05.DWG*.
b. Save the drawing with the name *C12_tut01.dwg*.
c. Add the drawing to the **CADCIM** project list.
d. Change format for wire numbers in the **Drawing Properties** dialog box.
e. Save the drawing.

Opening and Saving the Drawing

1. Activate the **CADCIM** project and open the drawing *DEMO05.DWG* and save it with the name *C12_tut01.dwg*.

Adding the Active Drawing to the Active Project Drawing List

1. Add the drawing *C12_tut01.dwg* to the **CADCIM** project list, as discussed earlier. Next, move it to the *TUTORIALS* subfolder.

Note

*If the **Update Terminal Associations** message box is displayed, choose the **Yes** button in this message box; C12_tut01.dwg is added to the **CADCIM** project.*

2. Choose **Save** from the **Application Menu** to save the drawing.

Changing Drawing Properties

1. Right-click on the *C12_tut01.dwg* drawing file in the **PROJECT MANAGER**; a shortcut menu is displayed. Choose **Properties > Drawing Properties** from the shortcut menu; the **Drawing Properties** dialog box is displayed.

2. By default, the **Drawing Settings** tab is chosen. Now, enter **01** in the **Sheet** edit box of the **Sheet Values** area.

3. Choose the **Wire Numbers** tab and enter **W-%S%N** in the **Format** edit box of the **Wire Number Format** area.

4. Choose the **OK** button in the **Drawing Properties** dialog box to save the changes and to exit from it.

5. Choose the **Wire Numbers** tool from the **Insert Wires/Wire Numbers** panel of the **Schematic** tab; the **Sheet01 - Wire Tagging** dialog box is displayed.

6. Accept the default values and then choose the **Drawing-wide** button; wire numbers are inserted in the wires throughout the drawing, as shown in Figure 12-11.

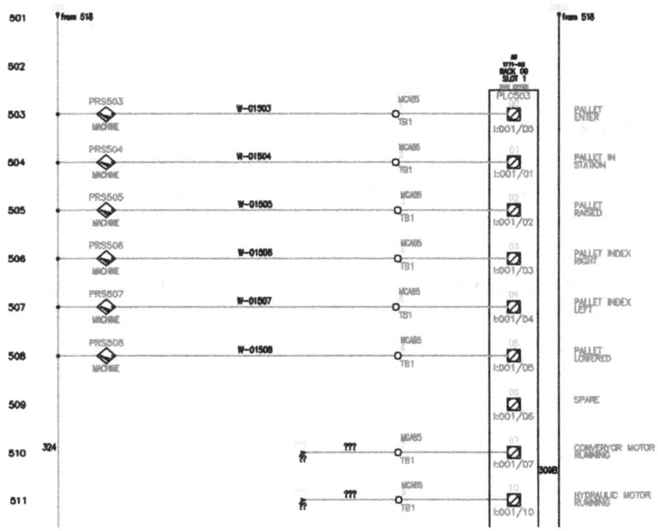

Figure 12-11 *Wire numbers inserted in the drawing after changing the drawing properties*

Saving the Drawing File

1. Choose **Save** from the **Application Menu** to save the drawing file *C12_tut01.dwg*.

Tutorial 2

In this tutorial, you will change the project properties and notice the changes taking place in the current drawing. **(Expected time: 15 min)**

The following steps are required to complete this tutorial:

a. Open and save the drawing.
b. Add drawing to the **CADCIM** project list.
c. Change the project properties.
d. Save the drawing.

Opening and Saving the Drawing

1. Activate the **CADCIM** project and then open the *C12_tut01.dwg* drawing from it.

2. Save the drawing *C12_tut01.dwg* with the name *C12_tut02.dwg*.

Adding Drawing to the CADCIM Project List

1. Add the drawing *C12_tut02.dwg* to the **CADCIM** project, as discussed earlier. Next, move it to the *TUTORIALS* subfolder.

Note

*If the **Update Terminal Associations** message box is displayed, choose the **Yes** button in it; C12_tut02.dwg is added to the TUTORIALS* subfolder of the **CADCIM** project.

2. Choose **Save** from the **Application Menu** to save the drawing.

Changing Project Properties

1. To change the project properties, right-click on the **CADCIM** project; a shortcut menu is displayed. Choose the **Properties** option from the shortcut menu; the **Project Properties** dialog box is displayed.

2. Choose the **Components** tab and select the **Combined Installation/Location tag mode** and **Format Installation/Location into tag** check boxes from the **Component TAG Options** area. Also, clear all other check boxes in this area, if they are selected.

3. Choose the **OK** button in the **Project Properties** dialog box to save the changes made and to exit it; the **IEC Tag Mode Update** dialog box is displayed, as shown in Figure 12-12.

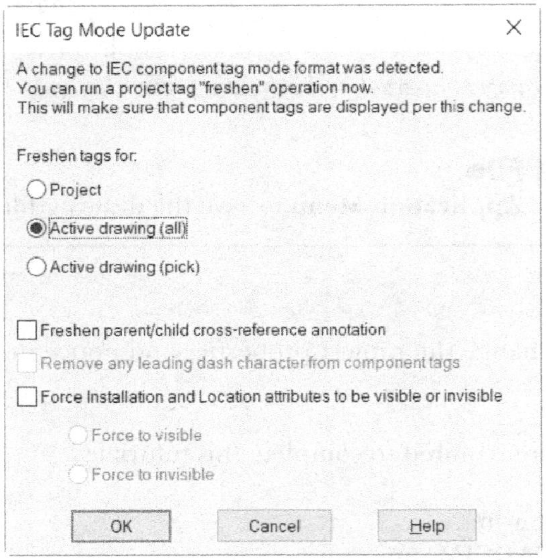

***Figure 12-12** The **IEC Tag Mode Update** dialog box*

4. Select the **Active drawing (all)** radio button and choose the **OK** button; the **QSAVE** message box is displayed.

5. Choose the **OK** button in the **QSAVE** message box and notice the changes in the drawing. Figure 12-13 shows the drawing *C12_tut02.dwg* after changing the project properties.

Saving the Drawing File

1. Choose **Save** from the **Application Menu** to save the drawing file *C12_tut02.dwg*.

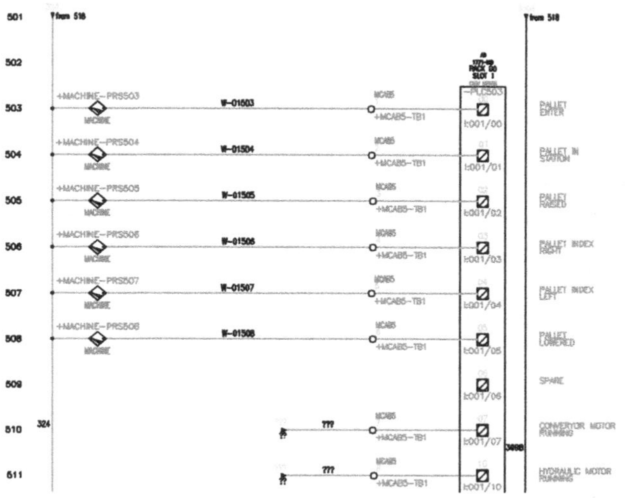

Figure 12-13 Drawing after changing the project properties

Tutorial 3

In this tutorial, you will create a new drawing, set the schematic settings, insert a ladder, and save the drawing as a template file. Finally, you will create a new drawing using the template that has been created. **(Expected time: 20 min)**

The following steps are required to complete this tutorial:

a. Create a new drawing, define the schematic settings, and insert a ladder.
b. Save the drawing as *schematic_template.dwt* template.
c. Create the new drawing using the new template file.
d. Save the drawing.

Creating a New Drawing, Defining Schematic Settings, and Inserting a Ladder

1. Activate the **CADCIM** project.

2. Create a drawing by choosing **New > Drawing** from the **Application Menu**; the **Select template** dialog box is displayed.

3. By default, the **ACAD_ELECTRICAL.dwt** is displayed in the **File name** edit box. If it is not displayed, select this template file from the **Select template** dialog box.

4. Choose the **Open** button; the drawing file is opened.

5. Choose the **Insert Ladder** tool from **Schematic > Insert Wires/Wire Numbers > Insert Ladder** drop-down; the **Alert** message box is displayed. 

6. Choose the **OK** button in the **Alert** message box to insert the invisible wd_m block; the **Insert Ladder** dialog box is displayed.

7. Set the following parameters in this dialog box:

   Width = 5 **Spacing** = 1
   1st Reference = 1 **Rungs** = 6
   Without reference numbers: Clear this check box, if it is selected.
   1 Phase: Select this radio button **Yes**: Select this radio button

8. Choose the **OK** button in the **Insert Ladder** dialog box; you are prompted to specify the start position of the first rung.

9. Enter **10, 17.5** at the Command prompt and press ENTER; the ladder is inserted in the drawing.

Saving the Template File

1. In order to save the drawing file as template, choose **Save** from the **Application Menu**; the **Save Drawing As** dialog box is displayed.

2. Select the **AutoCAD Drawing Template (.*dwt)** from the **Files of type** drop-down list and then enter **Schematic_Template** in the **File name** edit box.

3. Choose the **Save** button; the **Template Options** dialog box is displayed.

4. In this dialog box, enter **Schematic Template** in the **Description** area.

5. Choose **OK** in the **Template Options** dialog box; the drawing is saved as template drawing.

Creating a New Drawing

1. Choose the **New Drawing** button from the **PROJECT MANAGER**; the **Create New Drawing** dialog box is displayed.

2. Enter **C12_tut03** in the **Name** edit box.

3. In order to specify the template drawing in the **Template** edit box, choose the **Browse** button on the right of the **Template** edit box; the **Select template** dialog box is displayed.

4. In this dialog box, select the **Schematic_Template** and choose the **Open** button, refer to Figure 12-14; the file name and location of template drawing is displayed in the **Template** edit box of the **Create New Drawing** dialog box.

5. Choose the **OK** button in the **Create New Drawing** dialog box; the **Apply Project Defaults to Drawing Settings** message box is displayed.

6. Choose the **Yes** button in this dialog box; the *C12_tut03.dwg* file is created. Next, move it to the *TUTORIALS* subfolder.

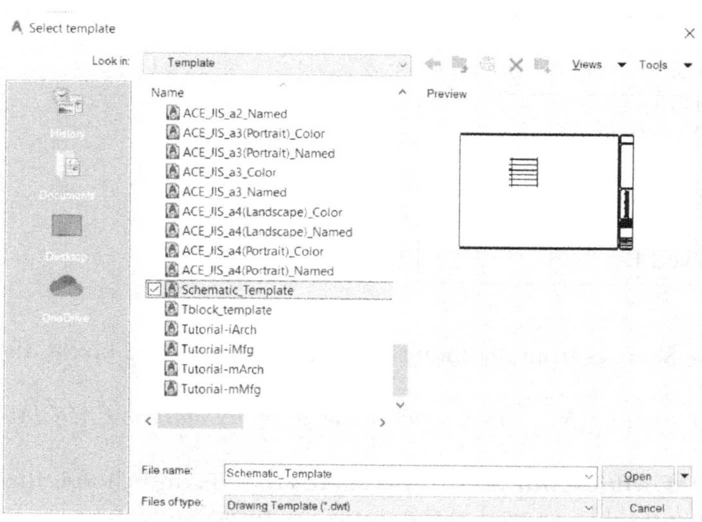

***Figure 12-14** The **Select template** dialog box*

Saving the Drawing File

1. Choose **Save** from the **Application Menu** to save the drawing file *C12_tut03.dwg*.

Tutorial 4

In this tutorial, you will create a project description line file (*.wdl*) and update the title block information of the given drawing accordingly. **(Expected time: 15 min)**

The following steps are required to complete this tutorial:

a. Open and save the drawing.
c. Add the drawing to the **CADCIM** project list.
d. Change the LINEx values.
e. Update the title block information.
f. Save the drawing.

Opening and Saving the Drawing

1. Open the *C12_tut01.dwg* drawing from the **CADCIM** project.

2. Save the drawing *C12_tut01.dwg* with the name *C12_tut04.dwg*, as discussed in the previous tutorials.

Adding the Active Drawing to the CADCIM Project List

1. Add the drawing *C12_tut04.dwg* to the **CADCIM** project drawing list, as discussed and then choose **Save** from the **Application Menu** to save the drawing.

Changing the LINEx Values

1. To change the values of the lines of the **Project Description** dialog box, open a **Notepad** file and enter the following information in it:

LINE1=Title 1:
LINE2=Title 2:
LINE3=Title 3:
LINE4=Job Number:
LINE5=Date:
LINE6=Engineer:
LINE7=Drawn By:
LINE8=Checked By:
LINE9=Scale:

2. Choose **File > Save As** from the menu bar; the **Save As** dialog box is displayed.

3. Browse to *C:\Users\User Name\Documents\Acade 2022\AeData\Proj\CADCIM*.

4. Enter **CADCIM_wdtitle.wdl** in the **File name** edit box and choose the **Save** button; the *CADCIM_wdtitle.wdl* file is saved in the respective project folder.

5. Right-click on the **CADCIM** project; a shortcut menu is displayed. Choose the **Descriptions** option from it; the **Project Description** dialog box is displayed, as shown in Figure 12-15. You will notice that in this dialog box, the line values have been changed to the specified values.

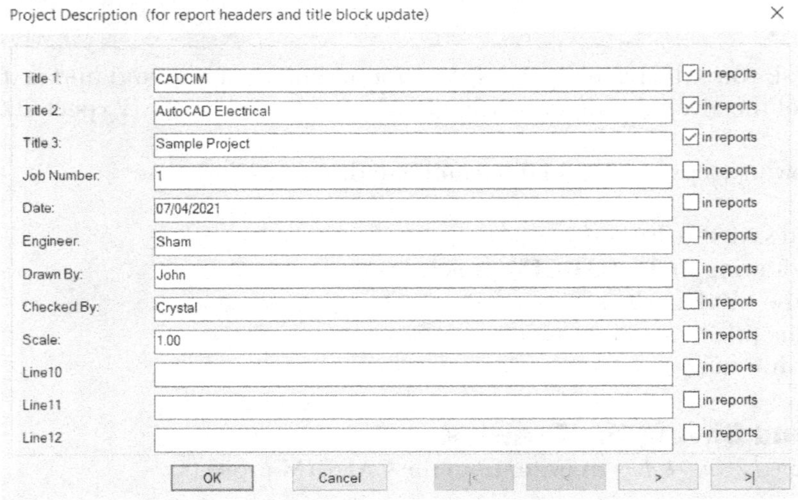

*Figure 12-15 The **Project Description** dialog box*

Note

*The **Project Description** dialog box displays the values shown in Figure 12-15 because you had specified these values in Tutorial 1 of Chapter 2. If these values are not displayed, you can enter the following information in the **Project Description** dialog box:*

LINE1= CADCIM
LINE2= AutoCAD Electrical
LINE3= Sample Project
LINE4= 1

LINE5= 07/04/2021
LINE6= Sham
LINE7= John
LINE8= Crystal
LINE9= 1.00

Updating the Title Block

1. Right-click on the **CADCIM** project; a shortcut menu is displayed. Choose the **Title Block Update** option from it; the **Update Title Block** dialog box is displayed.

2. Select the check boxes shown in Figure 12-16.

3. Choose the **OK Active Drawing Only** button; the title block information is displayed in the title block of the drawing, as shown in Figure 12-17.

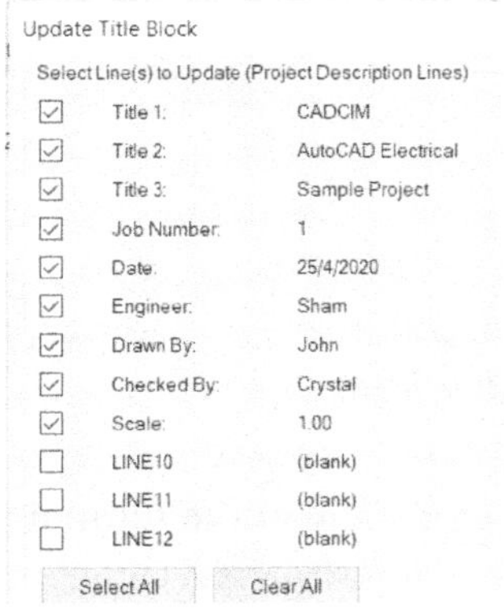

***Figure 12-16** Partial view of the **Update Title Block** dialog box*

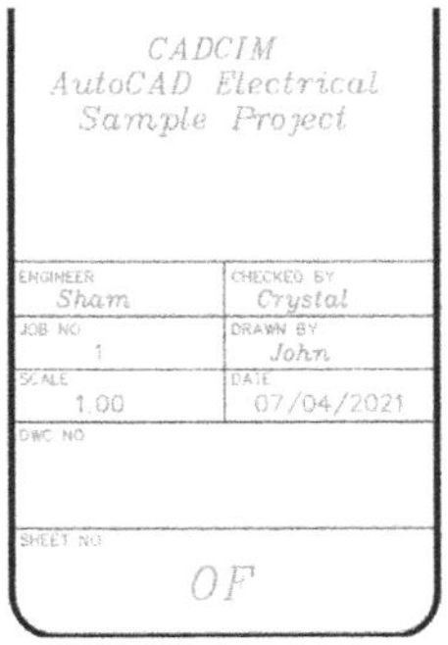

***Figure 12-17** The title block information*

Saving the Drawing File

1. Choose **File > Save** from the menu bar to save the drawing file.

Tutorial 5

In this tutorial, you will create a new block with attributes defined in it. Next, you will create a template using this block. You will also create a new drawing using this template and then set up the title block using method 1 and update the same for this drawing, refer to Figure 12-18. **(Expected time: 30 min)**

The following steps are required to complete this tutorial:

a. Create a new block with attributes.
b. Create a new template file.
c. Create a new drawing.
d. Set up the title block.

e. Update the title block.
f. Save the drawing.

Figure 12-18 Updated title block

Creating a New Block with Attributes

1. Make sure the CADCIM project is activated. Open any of the tutorial files in this project. (for ex: *c02tut01.dwg*)

2. Delete entire content, if any, as well as the title block from this drawing.

3. Choose the **Rectangle** tool from the **Draw** panel of the **Home** tab; you are prompted to specify the first corner point. Enter **F** at the Command prompt; you are prompted to specify the fillet radius. Enter **1** at the Command prompt. Next, you are prompted to specify the first corner point. Enter **0,0** at the Command prompt; you are prompted to specify the other corner point. Now, enter **30,22** at the Command Prompt; a rectangle is created with fillet corners, refer to Figure 12-19.

4. Similarly, draw two more rectangles with same fillet radius and corner points as mentioned below, refer to Figure 12-19.

	First corner point	Second corner point
Rectangle 2	30.25,22	34,10
Rectangle 3	30.25,9.75	34,0

5. Choose the **Multiline Text** tool from the **Text** drop-down in the **Annotation** panel of the **Home** tab; you are prompted to specify the first corner. Enter **32,11** at the Command Prompt and then enter **R** at the Command prompt; you are prompted to specify the rotation angle. Enter **90** at the Command prompt.

Figure 12-19 *Two more rectangles created*

6. Enter **H** at the Command prompt; you are prompted to specify height. Enter **0.6** at the Command prompt; Now, you are prompted to specify the opposite corner. Enter **33,21** at the Command Prompt; a text box appears on the screen. Enter **CADCIM Technologies** in this text box.

7. Select **CADCIM Technologies** in the text box and then select **Times New Roman** from the **Text Style** drop-down list in the **Formatting** panel of the **Home** tab. Also, choose the **B** button from the **Formatting** panel. Now, click anywhere in the viewport; the text **CADCIM Technologies** is displayed vertically. Note that you may have to vary the text height as per the dimension of the upper rectangle. Next, choose the **Move** tool from the **Modify** panel and align it at the center in the upper rectangle, refer to Figure 12-20.

8. Choose the **Line** tool from the **Draw** panel of the **Home** tab and draw seven lines in the lower rectangle, as shown in Figure 12-21.

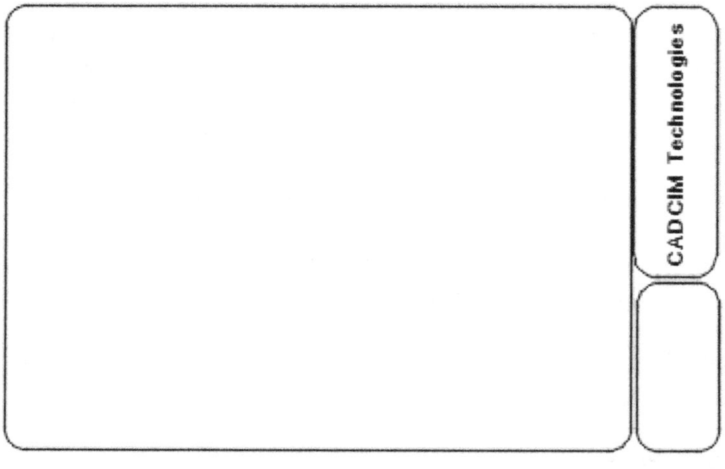

Figure 12-20 *Text aligned in the upper rectangle*

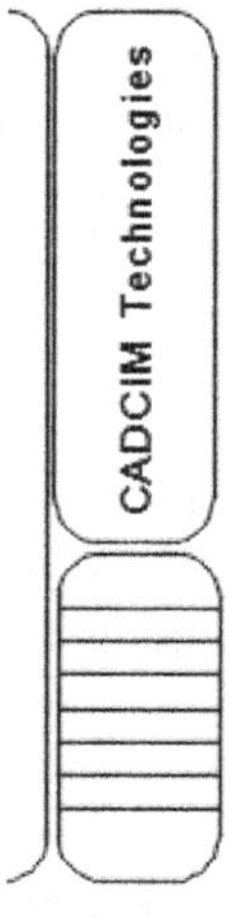

Figure 12-21 *Seven Lines drawn in the lower rectangle*

9. Choose the **Multiline Text** tool from the **Text** drop-down in the **Annotation** panel of the **Home** tab and enter **Engineer, Drawn By**, **Checked By**, **Scale**, **Drawing no**, **Sheet** in the lower rectangle, as shown in Figure 12-22.

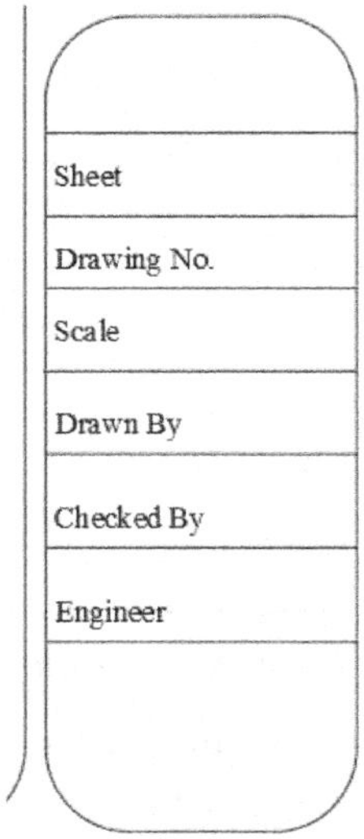

Figure 12-22 *Text entered in the lower rectangle*

Assume the text height as per the distance between the lines drawn.

Next, you need to define attributes for the block.

10. Choose the down arrow in the **Block** panel of the **Home** tab and then choose the **Define Attributes** tool; the **Attribute Definition** dialog box is displayed.

11. In this dialog box, make sure the **Invisible** check box is cleared in the **Mode** area. Enter **Engineer** in the **Tag** edit box of the **Attribute** area.

12. In the **Text Settings** area, select **ITALIC** from the **Text Style** drop-down list. Next, enter the text height as per the distance between the lines drawn in the **Text height** edit box. Also, make sure that the **Specify on-screen** check box is selected in the **Insertion Point** area and then choose **OK**; the *ENGINEER* attribute is attached to the cursor. Next, place the *ENGINEER* attribute next to the **Engineer** text, refer to Figure 12-23.

13. Similarly, add 7 more attributes to the block, as shown in Figure 12-23.

14. Choose the **Create** tool from the **Block** panel of the **Home** tab; the **Block Definition** dialog box is displayed. In this dialog box, enter **Tblock** in the **Name** edit box and choose the **Select objects** button; you are prompted to select the objects for the block. Select all the objects in the viewport (including attributes) and press ENTER. Next, choose **OK** from the dialog box; the selected objects are converted into a single block named **Tblock**.

Creating a New Template File

1. Choose **Save As** from the **Application Menu**; the **Save Drawing As** dialog box is displayed.

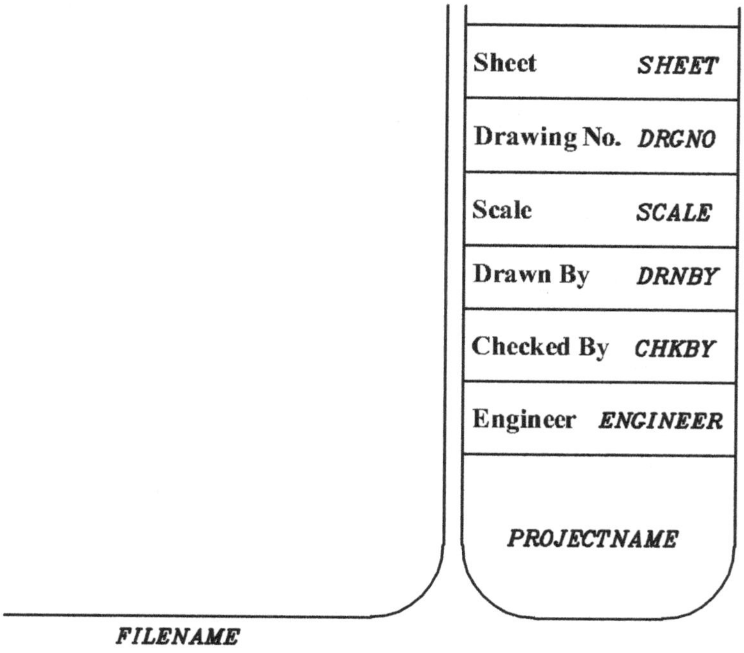

Figure 12-23 Seven more attributes added

2. Select **AutoCAD Drawing Template (*.dwt)** from the **Files of type** drop-down list and enter **Tblock_template** in the **File name** edit box. Next, choose **Save**; the **Template Options** dialog box is displayed. Enter **New CADCIM Template** in the **Description** area and choose **OK**; the drawing is saved as *TBlock_template.dwt* file.

Creating a New Drawing File

1. Choose the **New Drawing** button from the **PROJECT MANAGER**; the **Create New Drawing** dialog box is displayed. In this dialog box, enter **C12_tut05** in the **Name** edit box.

2. Choose the **Browse** button next to the **Template** edit box; the **Select Template** dialog box is displayed. In this dialog box, select **Tblock_template** and then choose **Open** to close the dialog box.

3. In the **Sheet Values** area, enter **1205** in the **Drawing** edit box and enter **5** in the **Sheet** edit box. Next, choose **OK** from the **Create New Drawing** dialog box; the *C12_tut05.dwg* is created and the **Apply Project Defaults to Drawings** message box is displayed. Choose **Yes** from this message box; the **AutoCAD Message** message box is displayed, as shown in Figure 12-24. Choose **OK** in this message box.

Setting Up the Title Block

1. Choose the **Title Block Setup** tool from the **Other Tools** panel of the **Project** tab; the **Setup Title Block Update** dialog box is displayed. In this dialog box, make sure the **<Project>.WDT file** radio button is selected and then choose **OK**; the **Enter Block Name** dialog box is displayed.

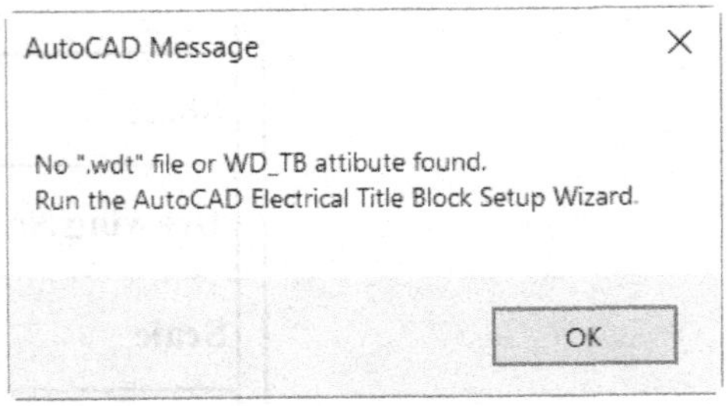

***Figure 12-24** The **AutoCAD Message** message box*

2. In this dialog box, choose the **Pick Block** button; the **Enter Block Name** dialog box disappears temporarily and you are prompted to select objects.

3. Select the title block in the drawing and press ENTER; the **Enter Block Name** dialog box is displayed again and Tblock is displayed in the edit box, as shown in Figure 12-25. Next, choose **OK** in this dialog box; the **Title Block Setup** dialog box is displayed, as shown in Figure 12-26.

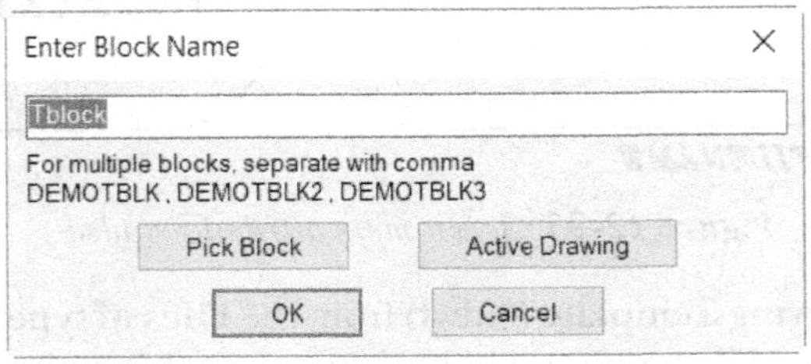

***Figure 12-25** The Tblock displayed in the edit box of the **Enter Block Name** dialog box*

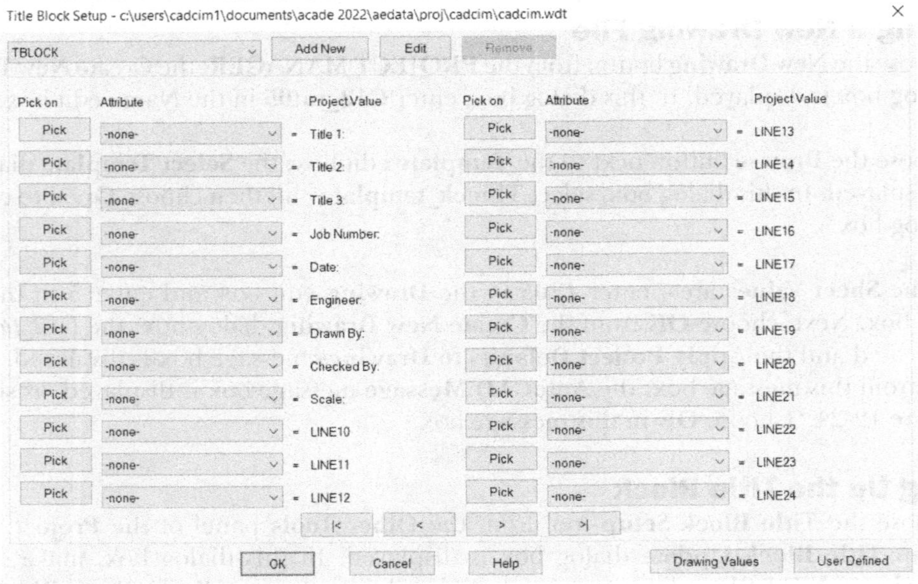

***Figure 12-26** The **Title Block Setup** dialog box*

4. In this dialog box, select **PROJECTNAME** from the **Attribute** drop-down list located at the left of the text **Line 1**, as shown in Figure 12-27.

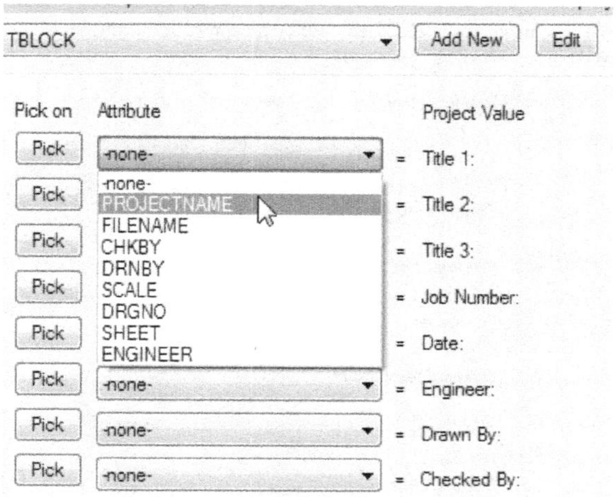

Figure 12-27 *Selecting* ***PROJECTNAME*** *from the* ***Attribute*** *drop-down list*

5. Similarly, select **ENGINEER** from the **Attribute** drop-down list located at the left of the text **Line 6**, select **DRNBY** from the **Attribute** drop-down list located at the left of the text **Line 7**, select **CHKBY** from the **Attribute** drop-down list located at the left of the text **Line 8**, and select **SCALE** from the **Attribute** drop-down list located at the left of the text **Line 9**, as shown in Figure 12-28.

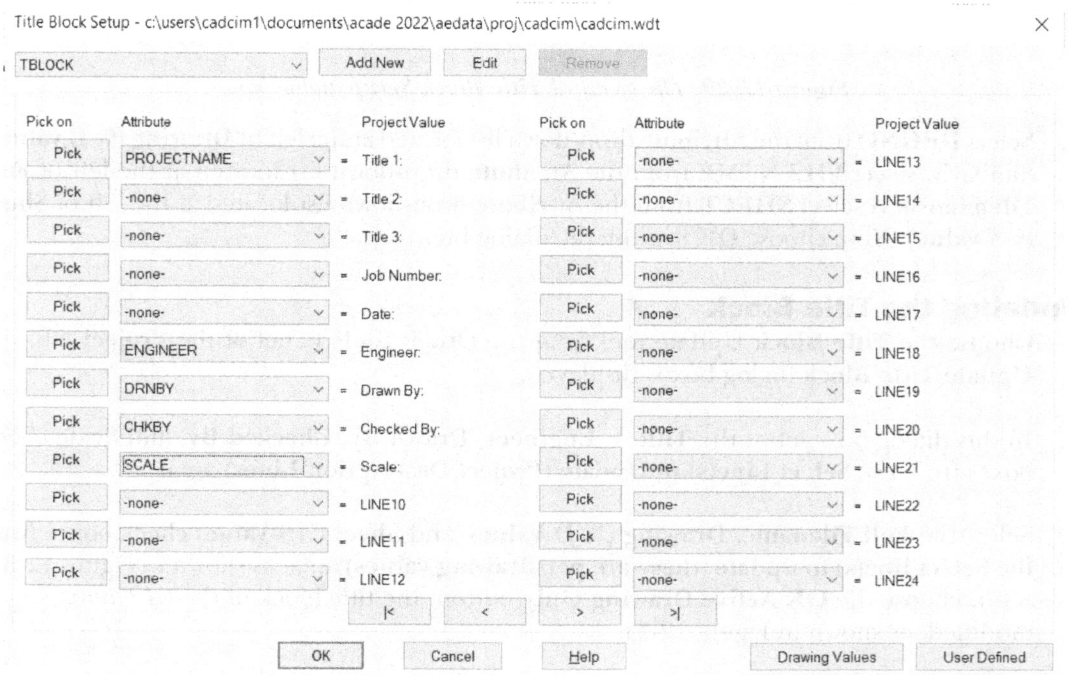

Figure 12-28 *Attributes selected from the* ***Attribute*** *drop-down list*

6. Choose the **Drawing Values** button; the modified **Title Block Setup** dialog box is modified, as shown in Figure 12-29.

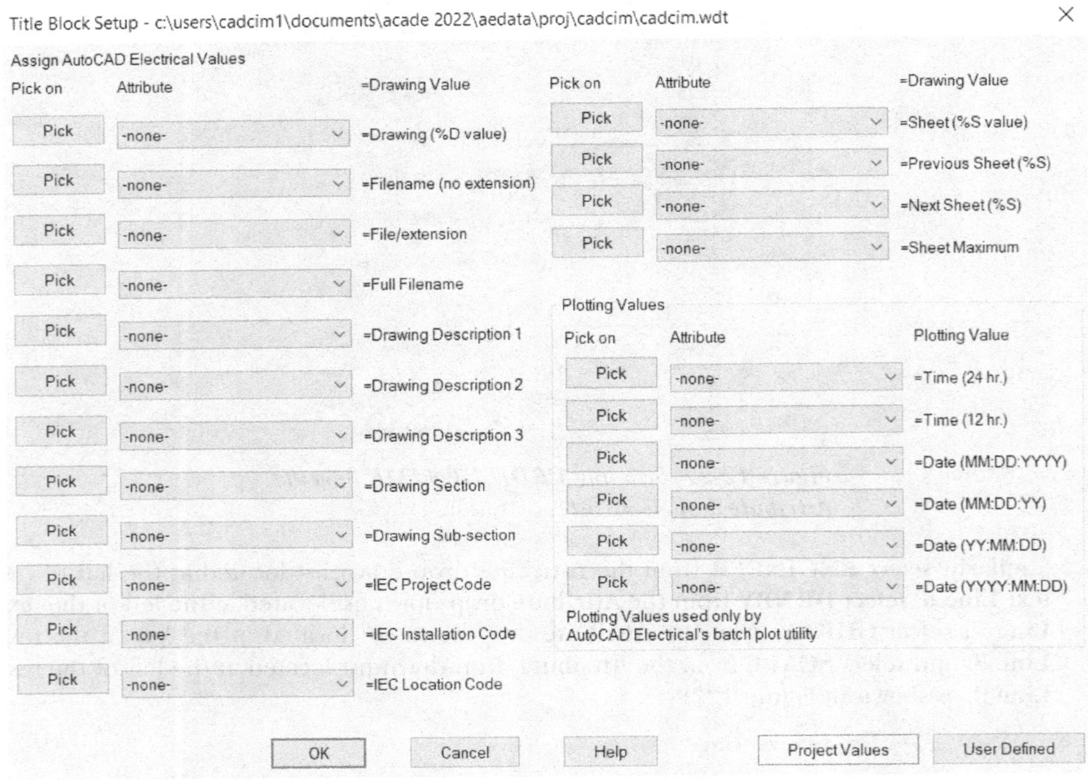

Figure 12-29 The modified ***Title Block Setup*** *dialog box*

7. Select **DRGNO** from the **Attribute** drop-down list located at the left of **Drawing (% D value)**. Similarly, select **FILENAME** from the **Attribute** drop-down list located at the left of **Full Filename** and select **SHEET** from the **Attribute** drop-down list located at the left of **Sheet (%S value)**. Now, choose **OK** to close the dialog box.

Updating the Title Block

1. Choose the **Title Block Update** tool from the **Other Tools** panel of the **Project** tab; the **Update Title Block** dialog box is displayed.

2. In this dialog box, select the **Title 1**, **Engineer**, **Drawn By**, **Checked By**, and **Scale** check boxes from the **Select Line(s) to Update (Project Description Lines)** area.

3. Select the **Full Filename**, **Drawing (% D Value)**, and **Sheet (%S Value)** check boxes from the **Select line(s) to update (these are per-drawing values)** area, as shown in Figure 12-30. Next, choose the **OK Active Drawing Only** button; the title block of the *C12_tut05.dwg* is modified, as shown in Figure 12-31.

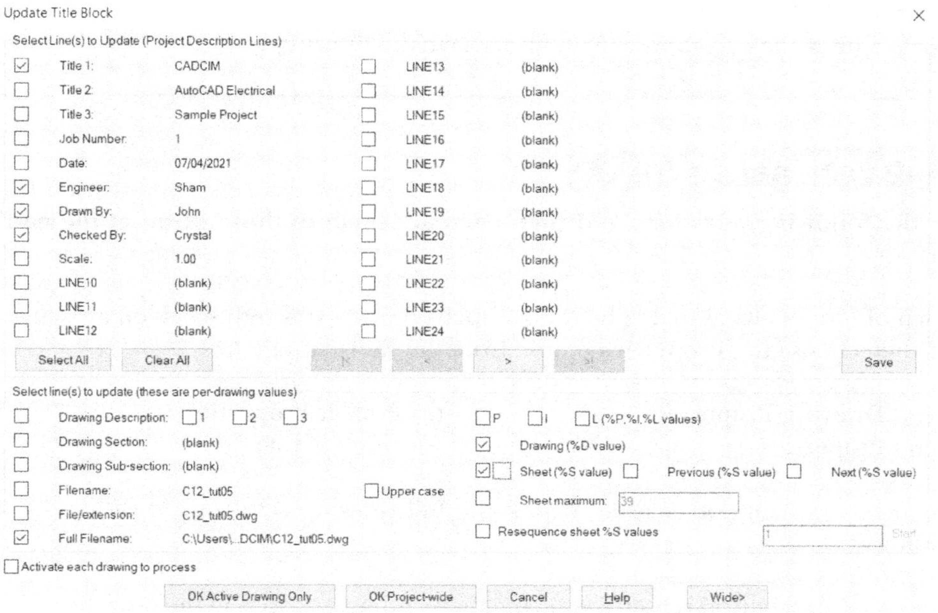

*Figure 12-30 The **Update Title Block** dialog box*

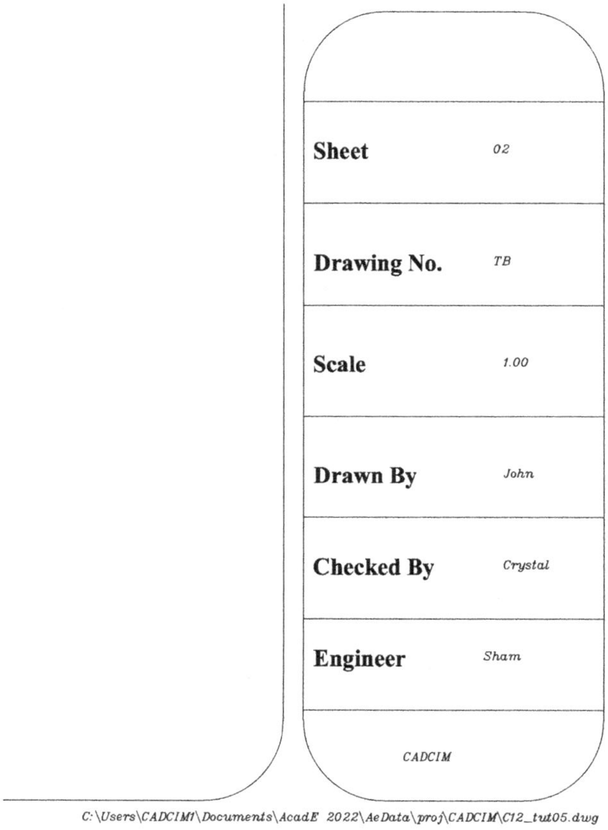

Figure 12-31 The title Block updated

Saving the Drawing File

1. Choose **File > Save** from the menu bar to save the drawing file.

Self-Evaluation Test

Answer the following questions and then compare them to those given at the end of this chapter:

1. Which of the following dialog boxes is displayed when you right-click on an active project and then choose the **Properties** option from the shortcut menu?

   (a) **Drawing Properties** (b) **Project Properties**
   (c) **Properties** (d) None of these

2. The __________ option is used to plot the full drawing set of a project or the selected drawings.

3. The __________ button is used to execute the pending updates on the drawings of an active project.

4. The __________ can be edited by using **Wordpad** and **Notepad**.

5. In the __________ area of the **Project Properties** dialog box, you can specify format for the reference numbers of a ladder.

6. In the **Drawing Properties** dialog box, you can define the settings for a new drawing and project. (T/F)

7. The project files are created and maintained using the **PROJECT MANAGER**. (T/F)

8. The **Title Block Update** option is used to update the attribute values of the title blocks. (T/F)

Review Questions

Answer the following questions:

1. Which of the following reference files, if saved in the project folder, will change the LINEx label description of the **Project Description** dialog box?

   (a) .WDL (b) .LOC
   (c) .WDD (d) .INST

2. Which of the following dialog boxes is displayed when you right-click on an active project and choose the **Title Block Update** option from the shortcut menu displayed?

   (a) **Setup Title Block Update** (b) **Title Block Setup**
   (c) **Update Title Block** (d) None of these

3. Which of the following buttons if chosen in the **Update other drawings?** message box saves the required information in the project task list?

   (a) **OK** (b) **Skip**
   (c) **Task** (d) **All**

4. The **Project Task List** button is used to execute the pending updates on the drawing files of a project. (T/F)

5. The template files have *.dwt* extension. (T/F)

6. You cannot change the settings of a drawing within a project. (T/F)

EXERCISES

Exercise 1

Open the *DEMO05.DWG* drawing file from the **CADCIM** project and save it as *C12_exer01.dwg*. Also, add this drawing file to the **NEW_PROJECT** project. Change the component tag format of the drawing as %N-%F. Also, update it using the **Update/Retag** tool.

(Expected time: 15 min)

Exercise 2

Create a new drawing with the name *C12_exer02.dwg* in the **NEW_PROJECT** project and then change the default settings of the drawing in the **Drawing Properties** dialog box as: **Sheet:** = 1 in the **Sheet Values** area of the **Drawing Settings** tab; **Spacing** = 1; **Width** = 5 in the **Ladder Defaults** area of the **Drawing Format** tab. Also, change the line reference numbers of the ladder in the **Drawing Properties** dialog box and insert a ladder with 5 rungs with 1st reference number as 10, as shown in Figure 12-32. **(Expected time: 15 min)**

Exercise 3

Create a panel template as *Panel_Template.dwt*. Also, insert the following enclosures in it: MANUFACTURER = RITTAL, TYPE = OUTDOOR ENCLOSURES, STYLE = NEMA 3R, and catalog = 9783040. Figure 12-33 shows the *panel_template.dwt* template file.

(Expected time: 20 min)

Hint: Change the scale of the panel in the **Insert Footprint** dialog box to 0.2.

Figure 12-32 *Ladder diagram for Exercise 2*

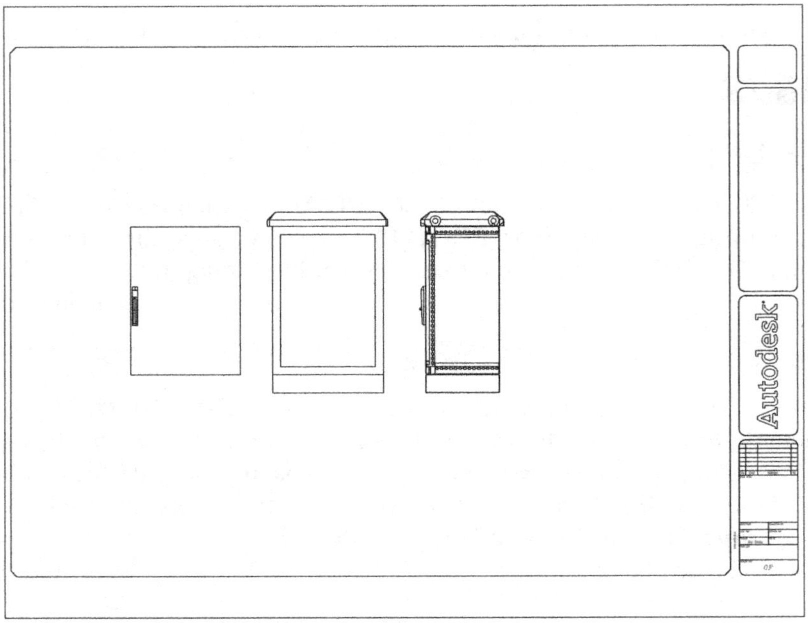

Figure 12-33 *Panel template for Exercise 3*

Answers to Self-Evaluation Test

1. b, **2. Plot Project**, **3. Project Task List**, **4.** reference files, **5. Format Referencing**, **6.** F, **7.** T, **8.** T

Chapter 13

Creating Symbols

Learning Objectives

After completing this chapter, you will be able to:

- *Create symbols*
- *Customize the icon menu*
- *Use the Mark/Verify, Export to Spreadsheet, and Update from Spreadsheet tools*

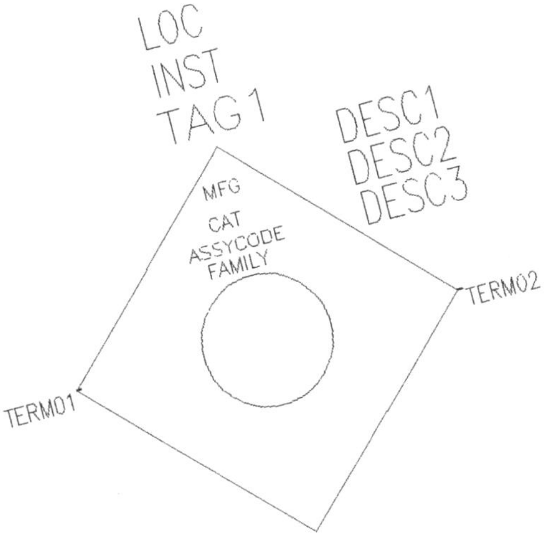

INTRODUCTION

In this chapter, you will learn to create symbols for the components, terminals, panel components, and so on. You will also learn to customize the icon menu. Later in this chapter, you will learn about some miscellaneous tools such as **Mark/Verify**, **Export to Spreadsheet**, and **Update from Spreadsheet**.

CREATING SYMBOLS

Command: AESYMBUILDER

The **Symbol Builder** tool is used to create new symbols such as filters, drives, controllers, and so on, or convert the existing symbols as per your requirement. Using this tool, you can create the symbols easily and quickly. You can also create the electrical symbols using the AutoCAD tools but that takes more time. Also, this tool is used to convert the existing non-AutoCAD Electrical symbols to the AutoCAD Electrical symbols. The symbols that you create using this tool will be compatible with AutoCAD Electrical and they will be displayed in the schematic reports. To create a symbol, choose the **Symbol Builder** tool from the **Symbol Builder** drop-down in the **Other Tools** panel of the **Schematic** tab, as shown in Figure 13-1; the **Select Symbol / Objects** dialog box will be displayed, as shown in Figure 13-2.

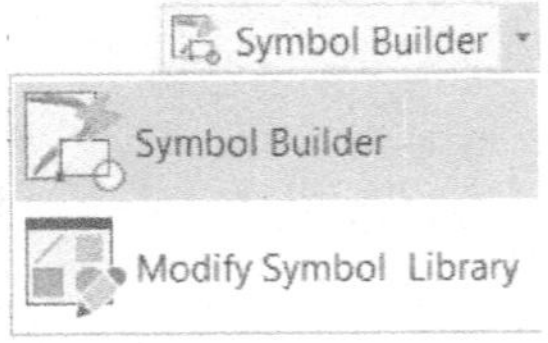

Figure 13-1 *The* ***Symbol Builder*** *drop-down*

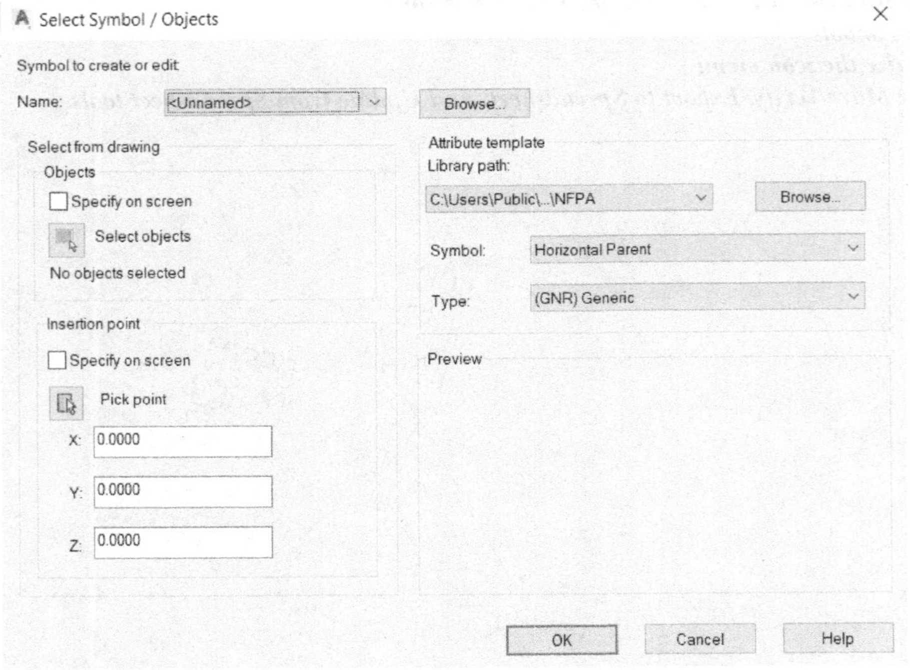

Figure 13-2 *The* ***Select Symbol / Objects*** *dialog box*

Using this dialog box, you can create a new symbol or edit an existing symbol. The **Name** drop-down list consists of the block definitions that are present within the active drawing. You can also specify the block or the drawing by choosing the **Browse** button. The preview of the selected drawing file will be displayed in the **Preview** area and the other options in the **Select from drawing** area of the dialog box will be deactivated. The options in the **Objects** area are used to select the objects that you need to create or edit.

Note

The objects can be attributes, existing blocks, attribute definitions, and any symbol graphics.

The options in the **Insertion Point** area are used to specify the insertion point for the symbol. The options in the **Attribute Template** area are used to specify the library path, symbol category, and type for the attribute template. The **Preview** area is used to display the preview of the block that you have selected from the **Name** drop-down list or the preview of the objects that you have selected from the drawing. After specifying the required options in the **Select Symbol/ Objects** dialog box, choose the **OK** button; the **Symbol Builder** environment and the **Symbol Builder Attribute Editor** palette will be invoked. Next, choose the **Block Editor** tab; the **Block Editor** environment will be invoked, refer to Figure 13-3. The drawing area of the **Block Editor** environment has a dull background which is known as the authoring area. In addition to the authoring area, the **BLOCK AUTHORING PALETTES** is provided in the **Block Editor** environment.

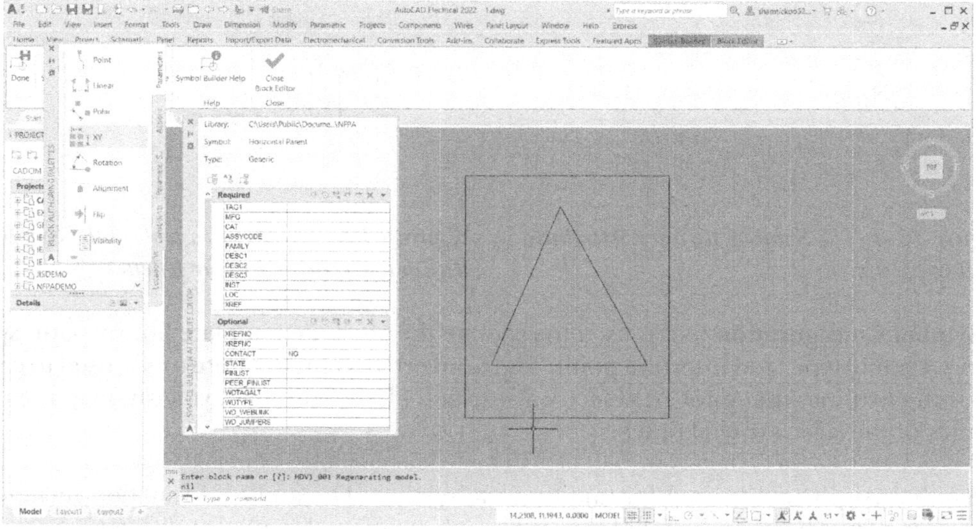

Figure 13-3 *The* ***Block Editor*** *environment*

Note

In order to learn more about the AutoCAD ***Block Editor*** *toolbar and* ***BLOCK AUTHORING Palettes****, refer to* ***AutoCAD 2022: A Problem-Solving Approach, Basic and Intermediate, 27th Edition*** *textbook by Prof. Sham Tickoo.*

The **Symbol Builder Attribute Editor** palette is shown in Figure 13-4. This palette is used to insert, add, modify, remove, and delete attributes. Notice that the library path of the attribute template, symbol category, and type of the attribute template are displayed on the upper left corner of the **Symbol Builder Attribute Editor** palette. Different tools and rollouts in this palette are discussed next.

The **Symbol Builder Attribute Editor** palette consists of tools that are common to any symbol type. These tools are shown in Figure 13-5 and are discussed next.

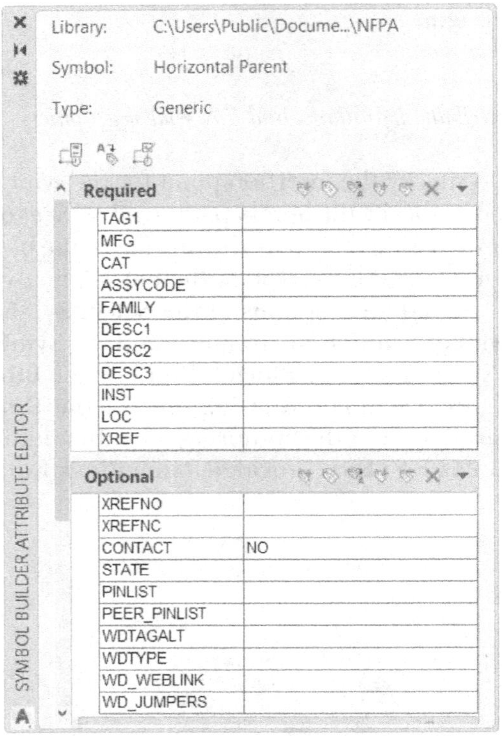

Figure 13-4 The ***Symbol Builder Attribute Editor*** *palette*

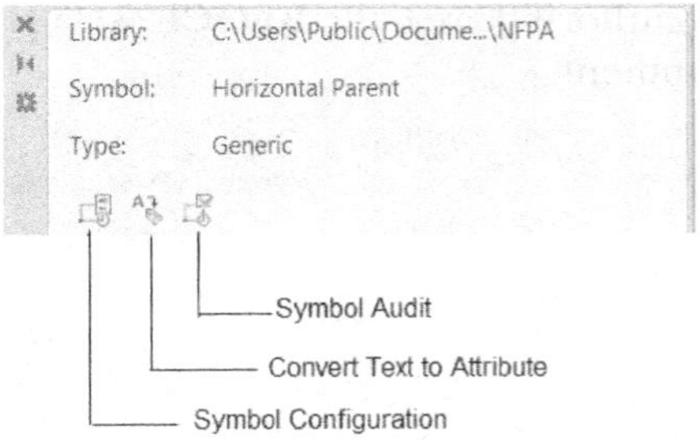

Figure 13-5 The common tools of the ***Symbol Builder Attribute Editor*** *palette*

The **Symbol Configuration** tool is used to redefine the attribute template library path, symbol category, symbol type, and insertion point. The **Convert Text to Attribute** tool is used to convert the existing text into the AutoCAD Electrical attribute. Using this tool, you can map the text to attributes of the selected symbol type.

The **Symbol Audit** tool is used for auditing the symbol. The audit information includes the information of attributes and symbol name. The information of symbol auditing is based on the type of the symbol. Using this tool, you can check the errors found in a symbol.

The **Required** rollout consists of the attributes that are mandatory for a symbol. This rollout consists of the grid area and various tools. The tools and options in the **Required** rollout are shown in Figure 13-6.

The **Optional** rollout consists of the attributes that are not mandatory to be inserted into the symbol. The tools in the **Optional** rollout are similar to that of the **Required** rollout and have been discussed in the previous section.

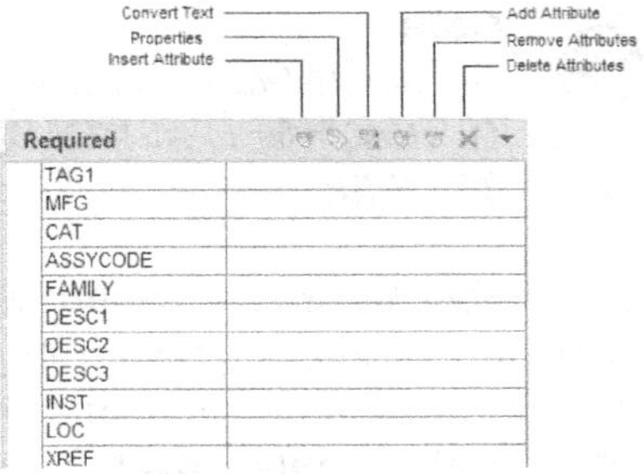

Figure 13-6 The ***Required*** *rollout*

Using the **POS** rollout, you can insert the position attributes into a symbol. You can insert up to 12 position attributes. This rollout consists of the **Add Next** button, which is used to insert the next available attribute in the symbol.

Using the **RATING** rollout, you can insert the rating attributes into a symbol. You can insert up to 12 rating attributes. This rollout consists of the **Add Next** button, which is used to insert the next available attribute in the symbol. The **Wire Connection** rollout is used to select the style and direction of the wire connection attributes. The **Insert Wire Connection** tool is used to insert the selected wire connection to the attribute. The buttons in this rollout are used to add optional wire connection attributes, change attribute properties, move wire connection attribute, and so on. Note that the pins attributes will automatically be inserted in the symbol after the wire connection attributes are added to it.

The **Link Lines** rollout consists of the **Direction** field and the **Insert Link Lines** tool. The options in this rollout are used to select the direction for the link line attributes and to insert the attribute on the symbol.

Once you have added the attributes and completed the symbol graphics using the **Symbol Builder Attribute Editor** palette, choose the **Done** button from the **Edit** panel of the **Symbol Builder** tab; the **Close Block Editor: Save Symbol** dialog box will be displayed.

After specifying the required options in the **Close Block Editor: Save Symbol** dialog box, choose the **OK** button; the **Close Block Editor** message box will be displayed. Choose the **Yes** button in this message box to insert the block or the **No** button to exit the **Block Editor** environment.

CUSTOMIZING THE ICON MENU

Command: AEMENUWIZ

You can customize the icon menu using the **Icon Menu Wizard** tool. The **Icon Menu Wizard** tool is used to modify or add icons of schematic symbol or panel symbol to the icon menu. Using this tool, you can create a new submenu, add icons that will be used for inserting the component or circuits, delete icons, cut, copy, and paste icons. To modify an icon menu, choose the **Icon Menu Wizard** tool from the **Other Tools** panel of the **Schematic** tab; the **Select Menu file** dialog box will be displayed, as shown in Figure 13-7. Using this dialog box, you can edit the default menu files such as *ace_nfpa_menu.dat* for schematic symbols and *ace_panel_menu.dat* for panel symbol.

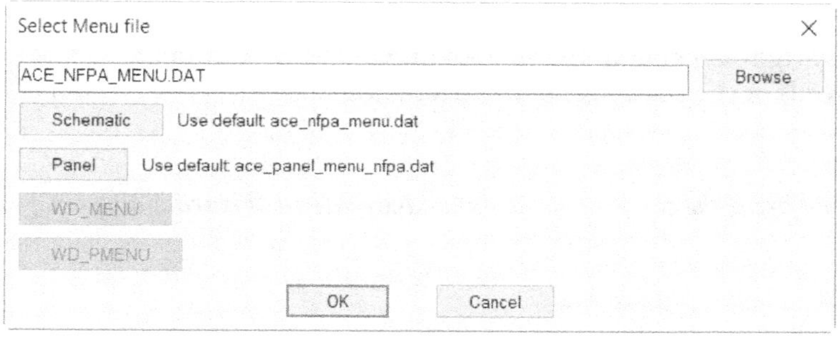

Figure 13-7 *The* ***Select Menu file*** *dialog box*

Enter the name of the icon menu file that you need to edit in the edit box. Alternatively, you can choose the **Browse** button to select the icon menu file. When you choose this button, the **Select ".dat" icon menu file** dialog box will be displayed. Next, select the **.dat* file from this dialog box and choose the **Open** button; the name and path of the icon menu (*.dat*) file will be displayed in the edit box. You can also choose the **Schematic** button to display the default schematic icon menu file in the edit box. If you choose the **Panel** button, the default panel icon menu file will be displayed. By default, the *ACE_NFPA_MENU.DAT* file is displayed in the edit box. Next, choose the **OK** button in the **Select Menu file** dialog box; the **Icon Menu Wizard** dialog box will be displayed, as shown in Figure 13-8. The options in the **Icon Menu Wizard** dialog box are almost similar to those in the **Insert Component** or **Insert Footprint** dialog box, discussed in previous chapters, but the **Icon Menu Wizard** dialog box has an additional drop-down list, **Add**.

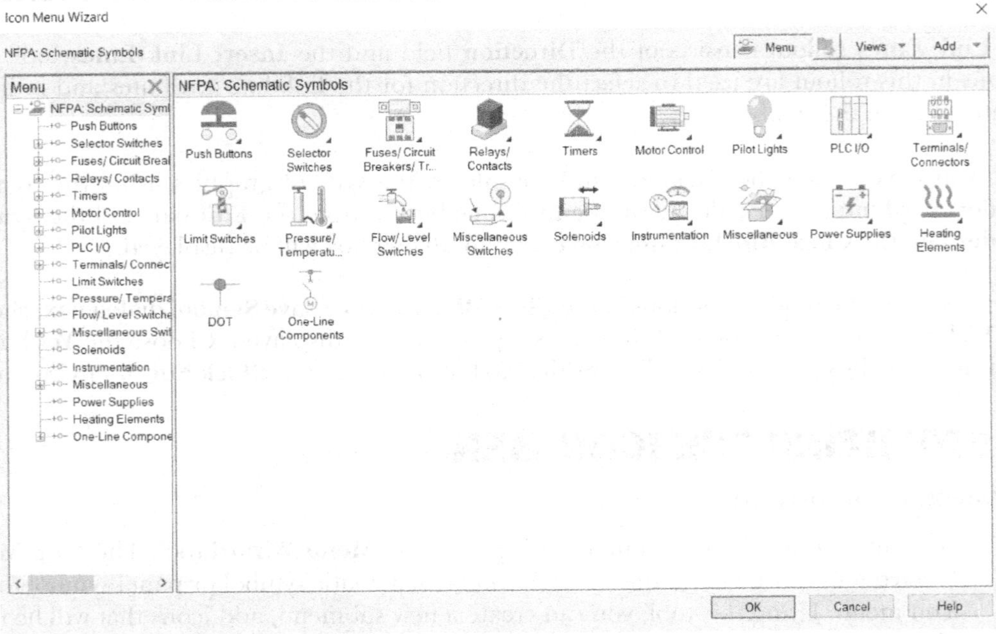

*Figure 13-8 The **Icon Menu Wizard** dialog box*

The options in the **Add** drop-down list are used to add a component, command, new circuit, existing circuit and submenu icon in the menu. Click on the **Add** drop-down list; various options will be displayed, as shown in Figure 13-9. You can also access the options present in this drop-down list by right-clicking in the symbol preview window (where icons are displayed) of the **Icon Menu Wizard** dialog box.

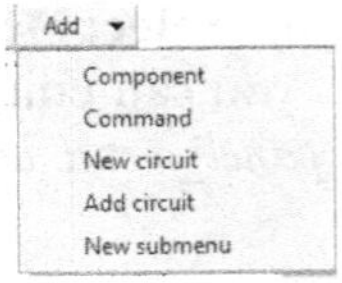

*Figure 13-9 The **Add** drop-down list displayed*

Note

*You cannot insert the components using the **Icon Menu Wizard** dialog box.*

Exporting Data to the Spreadsheet

Command: AEEXPORT2SS

The **To Spreadsheet** tool is used to export the data from the active drawing or project to an external file. This file can be of different formats such as Excel file format (.xls), Access file format (.mdb), Tab - delimited ASCII, and comma-delimited ASCII. Note that the database will be automatically refreshed before it is exported to an output file. To export data, choose the **To Spreadsheet** tool from the **Export** panel of the **Import/Export Data** tab; the **Export to Spreadsheet** dialog box will be displayed, as shown in Figure 13-10.

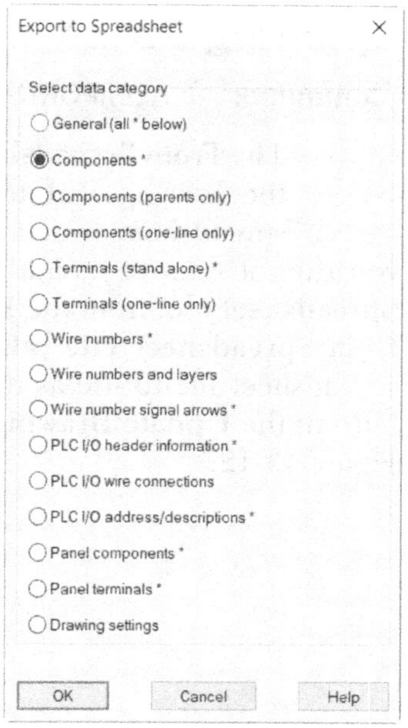

***Figure 13-10** The **Export to Spreadsheet** dialog box*

Select the data category that you need to export from the **Select data category** area of the **Export to Spreadsheet** dialog box. Choose the **OK** button; the corresponding **Data Export** dialog box will be displayed, as shown in Figure 13-11. The options and the name of this dialog box depend on the option that you have selected from the **Export to Spreadsheet** dialog box.

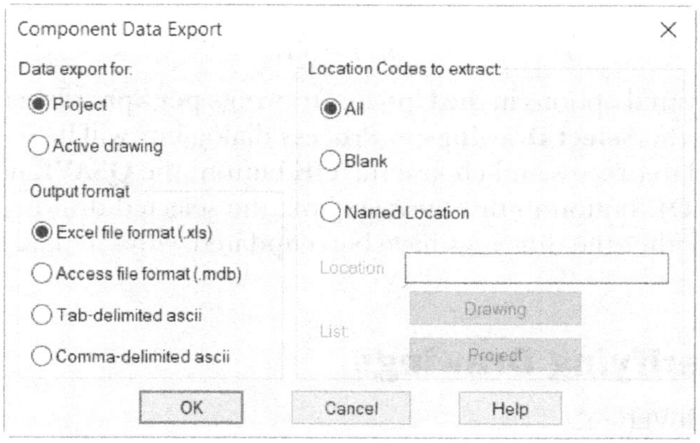

***Figure 13-11** The **Component Data Export** dialog box*

After specifying the options in the **Component Data Export** dialog box, choose the **OK** button; the **Select Drawings to Process** dialog box will be displayed, if the **Project** radio button is selected in the **Data export for** area. Next, select the drawings and choose the **Process** button. The options in this dialog box have already been discussed in the earlier chapters. Choose the **OK** button in the **Select Drawings to Process** dialog box; the **Select file name for Project-wide XLS output** dialog box will be displayed. If the **Active Drawing** radio button is selected in the **Data export for** area, then the **Select file name for drawing's XLS output** dialog box will be displayed. Next, specify the desired location for the exported data file and enter the name for the file in the **File name** edit box. Choose the **Save** button; the file will be saved at the specified location. Now, to view or edit this file, open it in the spreadsheet or database program.

Updating Data from the Spreadsheet

Command: AEIMPORTSS

The **From Spreadsheet** tool is used to import data from an edited file and then modify the drawings within the active project according to information present in the selected spreadsheet or database. Using this tool, you can only update the existing symbols but you cannot remove or add the symbols in the drawings. To update data, choose the **From Spreadsheet** tool from the **Import** panel of the **Import/Export Data** tab; the **Update Drawing from Spreadsheet File (xls, mdb, or csv Format)** dialog box will be displayed. Select the spreadsheet file to use as the source from the **Look in** drop-down list and choose the **Open** button; the **Update Drawings per Spreadsheet Data** dialog box will be displayed, as shown in Figure 13-12.

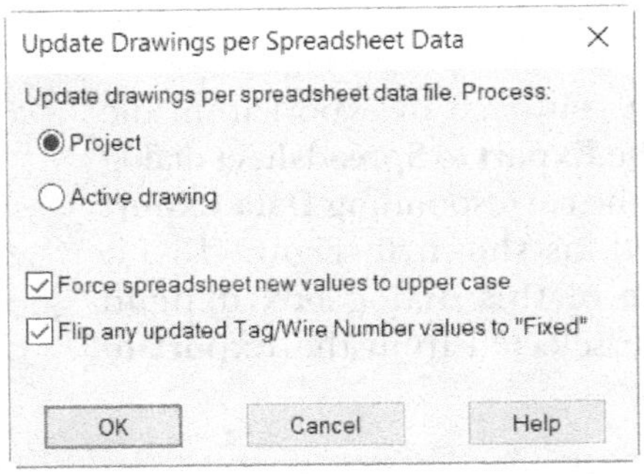

***Figure 13-12** The **Update Drawings per Spreadsheet Data** dialog box*

After specifying the required options in the **Update Drawings per Spreadsheet Data** dialog box, choose the **OK** button; the **Select Drawings to Process** dialog box will be displayed. Select the drawings that you need to process and choose the **OK** button; the **QSAVE** message box will be displayed. Choose the **OK** button in this message box; the selected drawings will be updated. Now, you can check whether the drawings have been updated with the data that you imported from the external file.

Marking and Verifying Drawings

Command: AEMARKVERIFY

The **Mark/Verify DWGs** tool is used to add an invisible mark on the components, wires, wire numbers, and 1st reference of the ladder present into a drawing before sending it to the client for review or to the user to modify a drawing. When the drawings are returned from the client, you can use the **Verify** option to generate a report of the changes. The report consists of a list of added, copied, changed, and deleted components and wire numbers. The changes made in the drawing using AutoCAD LT, AutoCAD, or AutoCAD Electrical will be indicated by the **Mark/Verify DWGs** tool.

Marking Drawings

To mark drawing(s), choose the **Mark/Verify DWGs** tool from the **Project Tools** panel of the **Project** tab; the **Mark and Verify** dialog box will be displayed, as shown in Figure 13-13. In this dialog box, the **Mark/Verify drawing or project** area is used to specify whether to mark or verify the active drawing or the active project. The options in the **What to do** area are used to mark the electrical components, non-AutoCAD Electrical blocks, and lines/wires. These options are also used to verify the changes that have been made in the drawing and remove all AutoCAD Electrical marked data. The **Previous** button is used to display the previous report. The **Active drawing statistics** area displays the details of marked data found on the drawing such as date, time when the drawing was marked, the initials of the person by whom it was marked, and any comments that you have added.

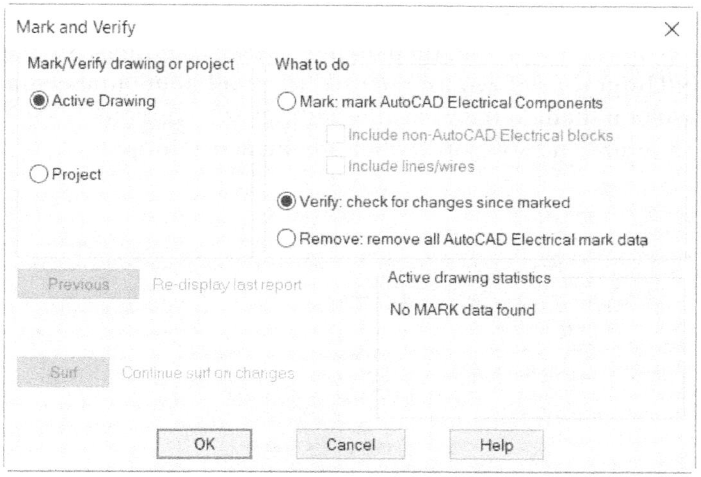

Figure 13-13 *The* ***Mark and Verify*** *dialog box*

Note

The invisible marks are added to the component tags and wire numbers. The appearance or functioning of the drawings will not be affected by these marks but the drawing size may increase by a small amount.

Verifying Drawings

You have learned to add the invisible marks to the drawing(s) before sending it to the client for review or before editing. Now, to verify the changes, which have been made by the client or after editing, choose the **Mark/Verify DWGs** tool from the **Project Tools** panel of the **Project** tab; the **Mark and Verify** dialog box will be displayed. The options in this dialog box have already been discussed in this chapter.

The **Verify: check for changes since marked** and the **Active Drawing** radio buttons are selected by default. Also, note that the marked components and wires status will be displayed in the **Active drawing statistics** area. Choose the **OK** button in the **Mark and Verify** dialog box; the **REPORT: Changes made on this drawing since last Mark command** dialog box will be displayed This dialog box displays the changes that have been made to the drawings that were marked.

Using Project-Wide Utilities

Command: AEUTILITIES

The **Utilities** tool is used to work on wire numbers, component tags, attribute text, wire types, and item numbers of an active project. This tool is also used to define script and run it for an active project. To work on these utilities, choose the **Utilities** tool from the **Project Tools** panel of the **Project** tab; the **Project-Wide Utilities** dialog box will be displayed, as shown in Figure 13-14. The areas and options in this dialog box are discussed next.

The radio buttons in the **Wire Numbers** area are used to work on wire numbers in the active project. The options in the **Signal Arrow Cross-reference text** drop-down list are used to remove or retain all signal arrow cross reference texts in an active project. The options in the **Parent Component Tags: Fix/Unfix** drop-down list are used to set all component tags in an active project to fixed, normal, or retain them to their existing settings. Similarly, options in the **Item Numbers: Fix/Unfix** drop-down list are used to set all item numbers in an active project to fixed, normal, or retain them to their existing settings. The options in the **Change Attribute** area are used to set the size and style of attributes in an active project.

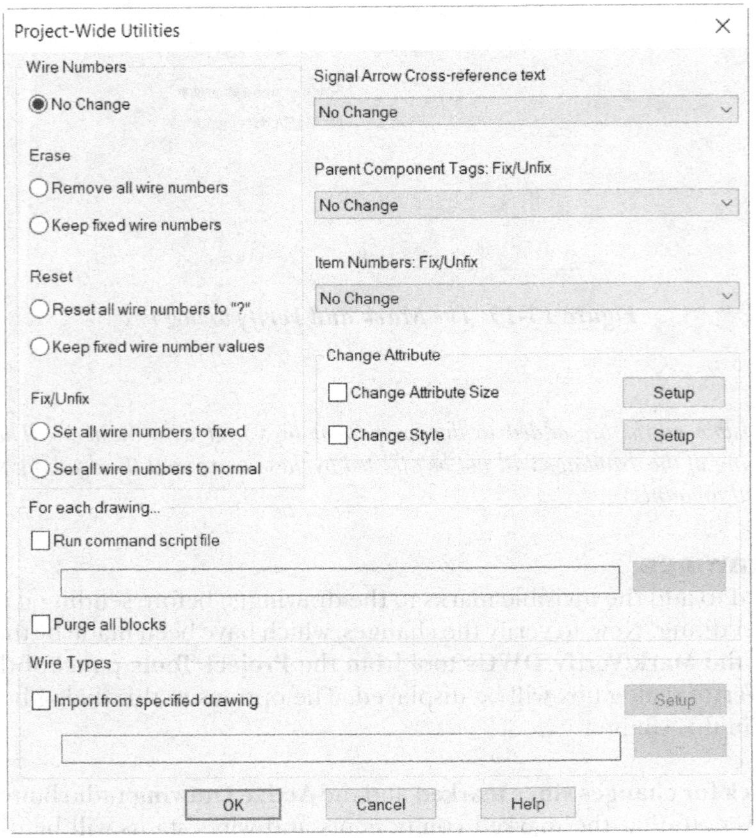

***Figure 13-14** The **Project-Wide Utilities** dialog box*

The **For each drawing** area is used to define script and run it for an active project. The **Wire Types** area is used to import wire types from other drawings. After specifying the desired options in the **Project-Wide Utilities** dialog box, choose the **OK** button; the **Batch Process Drawings** dialog box will be displayed. If you choose the **Project** button in this dialog box and then choose **OK,** the **Select Drawings to Process** dialog box will be displayed. Select the drawings in which you want to make changes and choose **OK**; the selected drawings will be updated. Similarly, if you want to make changes in the active drawing only, select the **Active Drawing** radio button from the **Batch Process Drawings** dialog box and choose **OK**; the changes will be carried out in the active drawing only.

TUTORIALS

Tutorial 1

In this tutorial, you will create a symbol, insert attributes into it, and then save it. Next, you will insert the symbol created into the drawing. **(Expected time: 30 min)**

The following steps are required to complete this tutorial:

a. Create a new drawing.
b. Insert a ladder in the drawing.
c. Create a symbol, insert attributes, and insert wire connection attributes to the symbol.
d. Save and insert the symbol.
e. Save the drawing file.

Creating a New Drawing

1. Create a new drawing *C13_tut01.dwg* in the **CADCIM** project with the *ACAD_ELECTRICAL.dwt* template and move it to the *TUTORIALS* subfolder, as already discussed in the previous chapters.

Inserting a Ladder

1. Choose the **Insert Ladder** tool from **Schematic > Insert Wires/Wire Numbers > Insert Ladder** drop-down; the **Insert Ladder** dialog box is displayed.

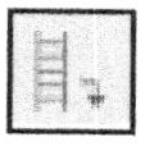

2. Set the following parameters in the **Insert Ladder** dialog box:

Width: **12.000**
1st Reference: **100**
1 Phase: Select this radio button
Spacing: **5.000**
Rungs: **4**
Yes: Select this radio button

Keep the values in the rest of the edit boxes intact.

3. Choose the **OK** button in the **Insert Ladder** dialog box; you are prompted to specify the start position of the first rung. Enter **8,18** at the Command prompt and press ENTER; the ladder is inserted in the drawing.

Creating a Symbol and Inserting Attributes in it

1. In order to create a symbol, first draw a circle in the drawing by choosing the **Center, Radius** tool from the **Circle** drop-down in the **Draw** panel of the **Home** tab or by choosing **Draw > Circle > Center, Radius** from the menu bar; you are prompted to specify the center point for the circle. Enter **7.5,30** at the Command prompt and press ENTER; you are prompted to specify the radius of the circle.

2. Enter **0.5** at the Command prompt and press ENTER; the circle is inserted in the drawing.

3. Choose **Draw > Polygon** from the menu bar; you are prompted to specify the number of sides.

4. Enter **4** at the Command prompt and press ENTER; you are prompted to specify the center of the polygon.

5. Select the center of the circle as the center of the polygon; you are prompted to enter an option.

Note
To snap the center of a circle, right-click on the ***Object Snap*** *button in the Status Bar; a shortcut menu is displayed. Choose the* ***Object Snap Settings*** *options from the shortcut menu; the* ***Drafting Settings*** *dialog box is displayed. Select the* ***Center*** *check box from the* ***Object Snap*** *tab and choose the* ***OK*** *button to save the changes made in this dialog box. Press F3, if the object snap is off.*

6. Enter **I** at the Command prompt and press ENTER; you are prompted to specify the radius of the circle.

7. Enter **6,30** at the Command prompt and press ENTER; the polygon is inserted in the drawing, as shown in Figure 13-15.

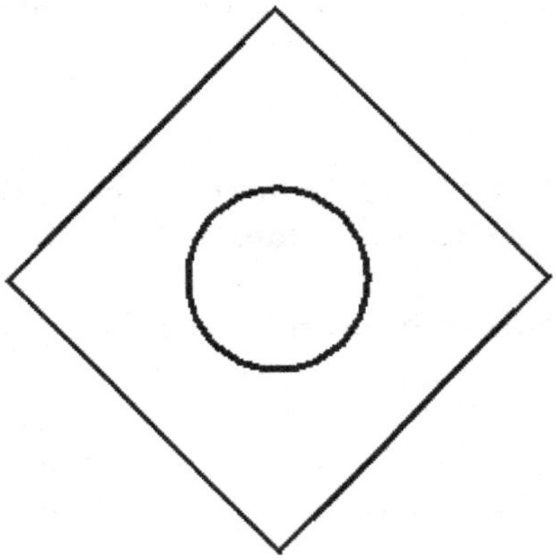

Figure 13-15 *The polygon inserted in the drawing*

8. Choose the **Symbol Builder** tool from **Schematic > Other Tools > Symbol Builder** drop-down; the **Select Symbol / Objects** dialog box is displayed. 

9. By default, the **Unnamed** option is selected in the **Name** drop-down list. Do not change this option as this is used to create the symbol from scratch.

10. Choose the **Select objects** button from the **Select from drawing** area; you are prompted to select the objects. Select the circle and polygon and then press ENTER; the **Select Symbol / Objects** dialog box is displayed again and the preview of the polygon inscribed in circle is displayed in the **Preview** area of the **Select Symbol / Objects** dialog box.

11. In the **Attribute template** area, choose the **Browse** button located next to the **Library path** drop-down list; the **Browse For Folder** dialog box is displayed. In this dialog box, make

sure the path *C:\users\public\ public doc...\NFPA* is displayed in the **Library path** drop-down list. Next, make sure **Horizontal Parent** is selected in the **Symbol** drop-down list, and **GNR (Generic)** is selected in the **Type** drop-down list. Keep rest of the values intact.

12. Choose the **OK** button in the **Select Symbol / Objects** dialog box; the **Block Editor** environment, the **Symbol Builder Attribute Editor** palette, the **Symbol Builder** and **Block Editor** tabs, and the **Block Authoring Palettes - All Palettes** palette are displayed. If the **Block Editor** environment is not invoked, choose the **Block Editor** tab to invoke it.

13. Select the **TAG1** row from the **Required** rollout of the **Symbol Builder Attribute Editor** palette and then choose the **Properties** button; the **Insert / Edit Attributes** dialog box is displayed. Enter **0.25** in the **Height** row of the **Text** area. Next, choose the **OK** button to save the changes made and exit the dialog box. 

14. Choose the **Insert Attribute** button; you are prompted to specify the insertion base point of this tag. Also, notice that TAG1 attribute is attached to the cursor. 

15. Enter **7.5,31.65** at the Command prompt and press ENTER; the TAG1 attribute for the symbol is inserted above the symbol, as shown in Figure 13-16.

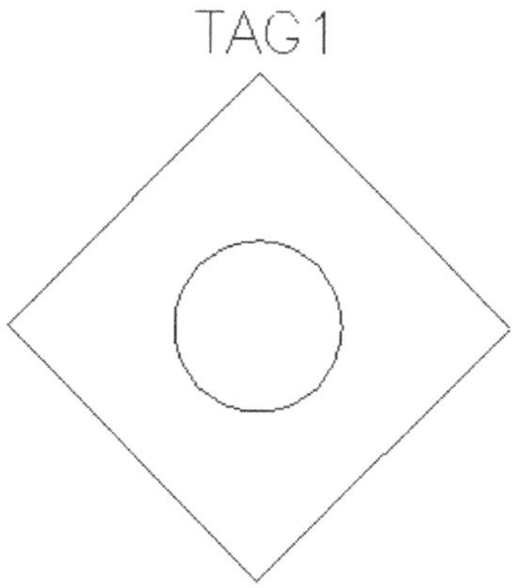

Figure 13-16 *The TAG1 attribute inserted into the symbol*

16. Select the **MFG** row from the **Symbol Builder Attribute Editor** palette and then choose the **Properties** button; the **Insert / Edit Attributes** dialog box is displayed. Enter **0.1** in the **Height** row. Choose the **OK** button to exit the dialog box.

17. Choose the **Insert Attribute** button; you are prompted to specify the insertion base point of the tag.

Note

You can also insert an attribute to the symbol by right-clicking on the attribute and choosing the ***Insert Attribute*** *option from the shortcut menu displayed.*

18. Insert MFG tag in the drawing, refer to Figure 13-17.

19. Change the height of the CAT, ASSYCODE, and FAMILY attributes to **0.1** in the **Insert / Edit Attributes** dialog box as discussed earlier and insert them into the symbol using the **Insert Attribute** button, refer to Figure 13-17. Make sure the **Snap Mode** button is deactivated to insert these attributes at desired places.

20. Similarly, change the height of the DESC1, DESC2, DESC3, INST, and LOC attributes to **0.25** using the **Properties** button in the **Required** rollout and then insert these attributes to the symbol using the **Insert Attribute** button, as shown in Figure 13-18.

21. To insert wire connections into the symbol, scroll down in the **Symbol Builder Attribute Editor** palette and make sure **Left/None** is selected in the **Direction/Style** row of the **Wire Connection** rollout.

22. Choose the **Insert Wire Connection** button from the **Wire Connection** rollout; you are prompted to select Left or (Top/Bottom/Right/rAdial).

23. Enter **6,30** at the Command prompt and press ENTER; the TERM01 attribute is inserted into the symbol and you are prompted to select the location for TERM02. Enter **R** at the Command prompt and press ENTER.

Figure 13-17 Different attributes inserted into the symbol

Figure 13-18 Different attributes inserted into the symbol

24. Enter **9,30** at the Command prompt and press ENTER; the TERM02 is inserted into the symbol. Figure 13-19 shows the TERM01 and TERM02 attributes inserted into the symbol.

25. Press ENTER to exit the command.

26. Select the **TERM01** attribute displayed in the **Pins** rollout and choose the **Properties** button; the **Insert / Edit Attribute** dialog box is displayed.

27. Enter **0.25** in the **Height** row and choose the **OK** button; the height of TERM01 is changed.

28. Similarly, change the height of the TERM02 attribute to **0.25**.

Saving and Inserting the Symbol

1. To save the symbol, choose the **Done** button from the **Edit** panel of the **Symbol Builder** tab; the **Close Block Editor: Save Symbol** dialog box is displayed.

2. Select the **Wblock** radio button from the **Symbol** area, if it is not selected.

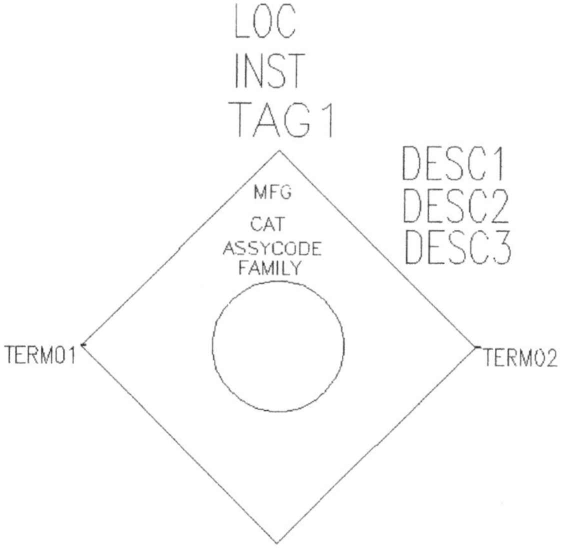

Figure 13-19 *TERM01 and TERM02 attributes inserted into the symbol*

3. Choose the **Pick point** button from the **Base point** area; you are prompted to specify the insertion base point. Select the center of the circle as the base point.

4. Enter **_MY SYMBOL** in the **Unique identifier** edit box and click in the **Symbol name** edit box; the name of the symbol is displayed in the **Symbol name** edit box as HDV1_MY SYMBOL. Keep the rest of the values in this dialog box intact.

5. Choose the **OK** button in the **Close Block Editor: Save Symbol** dialog box; the **Close Block Editor** message box is displayed.

6. Choose the **Yes** button in the **Close Block Editor** message box for inserting the symbol; you are prompted to specify the insertion point for the symbol.

7. Enter **13.5,18** at the Command prompt and press ENTER; the **Insert / Edit Component** dialog box is displayed. Enter **100A** in the edit box of the **Component Tag** area and choose the **OK** button; the symbol is inserted into the ladder, as shown in Figure 13-20. Figure 13-21 shows the zoomed view of the symbol inserted into the ladder.

Saving the Drawing File

1. Choose **File > Save** from the menu bar or choose **Save** from the **Application Menu** to save the drawing file, *C13_tut01.dwg*.

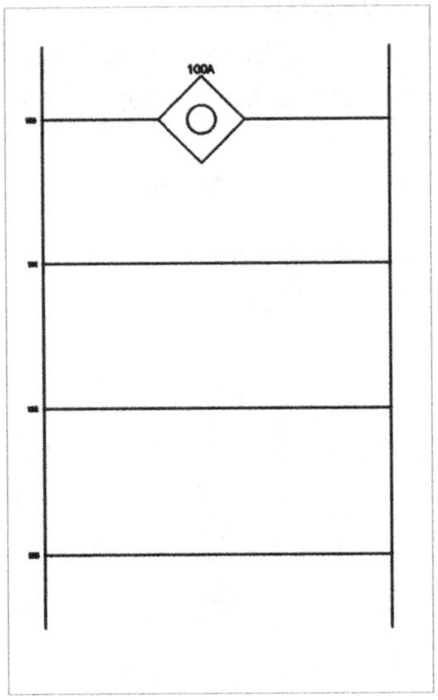

Figure 13-20 *Symbol inserted into the ladder*

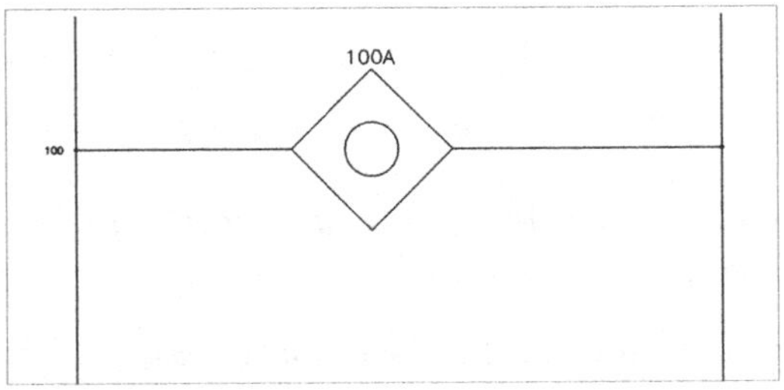

Figure 13-21 *The zoomed view of the symbol inserted into the ladder*

Tutorial 2

In this tutorial, you will add the symbol that you created in Tutorial 1 of this chapter to the **Insert Component** dialog box using the **Icon Menu Wizard** tool. You will then insert the component into the drawing. **(Expected time: 20 min)**

The following steps are required to complete this tutorial:

a. Open, save, and add the drawing to the active project.
b. Add a new icon to the menu.
c. Insert the component.
d. Save the drawing.

Opening, Saving, and Adding the Drawing to the Active Project

1. Open the *C13_tut01.dwg* from the **CADCIM** project and activate the CADCIM project.

2. Save the *C13_tut01.dwg* file with the name *C13_tut02.dwg*. You can also download this file from the CADCIM website. The path of the file is as follows:

   Textbooks > CAD/CAM > AutoCAD Electrical > AutoCAD Electrical 2022: A Tutorial Approach

3. Add the drawing *C13_tut02.dwg* to the **CADCIM** project, as discussed in the previous chapters.

Adding a New Icon to the Menu

1. To add the icon of the new symbol, choose the **Icon Menu Wizard** tool from the **Other Tools** panel of the **Schematic** tab; the **Select Menu file** dialog box is displayed.

2. Choose the **Schematic** button, if *ACE_NFPA_MENU.DAT* is not displayed in the edit box.

3. Next, choose the **OK** button; the **Icon Menu Wizard** dialog box is displayed.

4. Click on the **Add** drop-down list; various options are displayed. Select the **Component** option from the drop-down list; the **Add Icon - Component** dialog box is displayed.

5. Enter **MY SYMBOL** in the **Name** edit box of the **Icon Details** area.

6. Next, choose the **Pick <** button on the right of the **Image file** edit box; you are prompted to select the block.

7. Select the symbol you created in Tutorial 1 of this chapter; the name of block (HDVI_MY SYMBOL) is displayed in the **Image file** edit box.

8. Clear the **Create PNG from current screen image** check box, if it is selected.

9. Next, choose the **Browse** button on the right of the **Block name** edit box; the **Select File** dialog box is displayed. Browse to *"C:\Users\Public\Public Documents\Autodesk\Acade 2022\ Libs\ NFPA"*. Enter **HDV1_MY SYMBOL** in the **File Name** edit box and choose the **Open** button; the path and location of *HDV1_MY SYMBOL.dwg* is displayed in the **Block name** edit box.

10. Choose the **OK** button in the **Add Icon - Component** dialog box; the icon is added to the **Icon Menu Wizard** dialog box, as shown in Figure 13-22.

11. Choose **OK** in this dialog box to save the changes made and exit the dialog box.

Inserting the Component

1. In order to insert the component, choose the **Icon Menu** tool from **Schematic > Insert Components > Icon Menu** drop-down; the **Insert Component** dialog box is displayed.

2. Select **MY SYMBOL** from this dialog box; you are prompted to specify the insertion point.

3. Enter **13,3** at the Command prompt and press ENTER; the **Insert / Edit Component** dialog box is displayed. Make sure **103** is displayed in the edit box of the **Component Tag** area

and choose the **OK** button; the symbol is inserted in the rung 103 of the ladder, as shown in Figure 13-23. Figure 13-24 shows the zoomed view of the symbol inserted into the ladder.

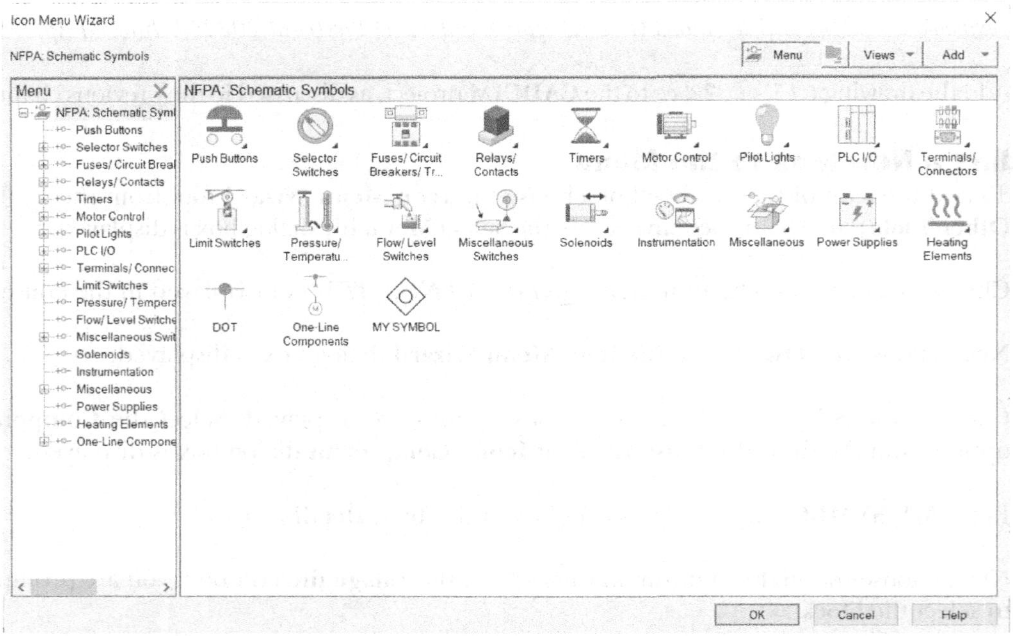

Figure 13-22 *The **Icon Menu Wizard** dialog box showing the icon of the symbol*

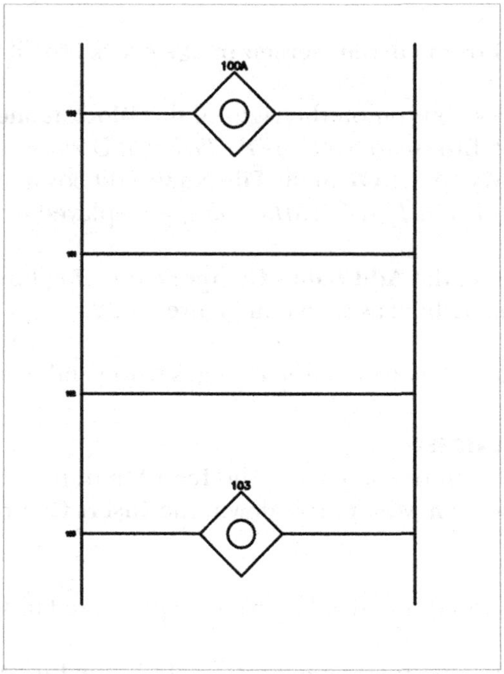

Figure 13-23 *Symbol inserted into the ladder*

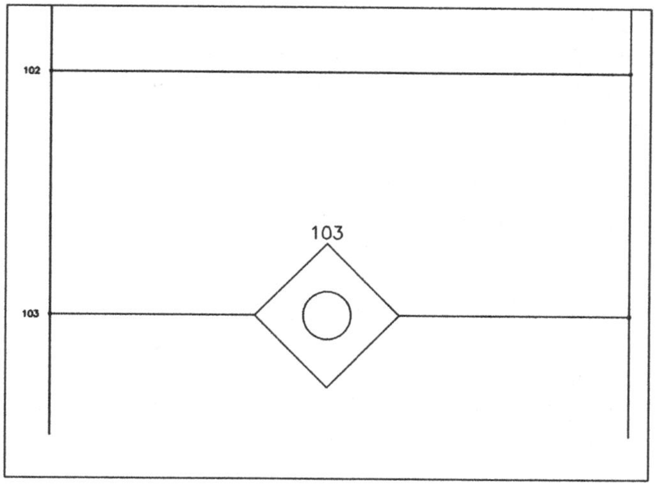

Figure 13-24 The zoomed view of the symbol inserted into the ladder

Saving the Drawing File

1. Choose **Save** from the **Application Menu** or **File > Save** from the menu bar to save the drawing file.

Tutorial 3

In this tutorial, you will export data from a drawing file to an excel sheet, make changes in that sheet, and then import the sheet data to the drawing file. **(Expected time: 15 min)**

The following steps are required to complete this tutorial:

a. Open, save, and add the drawing to the active project.
b. Export the data.
c. Modify the data.
d. Import the data.
e. Save the drawing.

Opening, Saving, and Adding the Drawing to the Active Project

1. Open the *C05_tut01.dwg* from the **CADCIM** project and activate the **CADCIM** project.

2. Save the *C05_tut01.dwg* file with the name *C13_tut03.dwg*. You can also download this file from the CADCIM website. The path of the file is as follows:

   Textbooks > CAD/CAM > AutoCAD Electrical > AutoCAD Electrical 2022: A Tutorial Approach

3. Add the drawing *C13_tut03.dwg* to the **CADCIM** project, as discussed in the previous chapters.

Exporting the Data

1. Choose the **To Spreadsheet** tool from the **Export** panel of the **Import/Export Data** tab; the **Export to Spreadsheet** dialog box is displayed, as shown in Figure 13-25.

2. In this dialog box, make sure the **Components** radio button is selected. Next, choose the **OK** button; the **Component Data Export** dialog box is displayed.

3. In this dialog box, select the **Active Drawing** radio button from the **Data export for** area. Next, select the **Excel file format (.xls)** radio button from the **Output format** area and then choose the **OK** button; the **Select file name for drawing's XLS output** dialog box is displayed.

4. In this dialog box, enter **C13_tut03_components** in the **File name** text box and select **C:\Users\User Name\Documents\AcadE 2022\AeData\Proj\CADCIM** from the **Save in** drop-down list, refer to Figure 13-26. Next, choose the **Save** button; an Excel file with the name *c13_tut03_components* is created at the specified location.

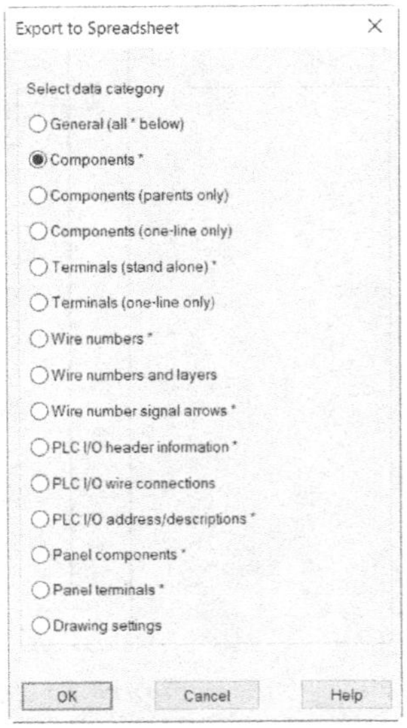

***Figure 13-25** The **Export to Spreadsheet** dialog box*

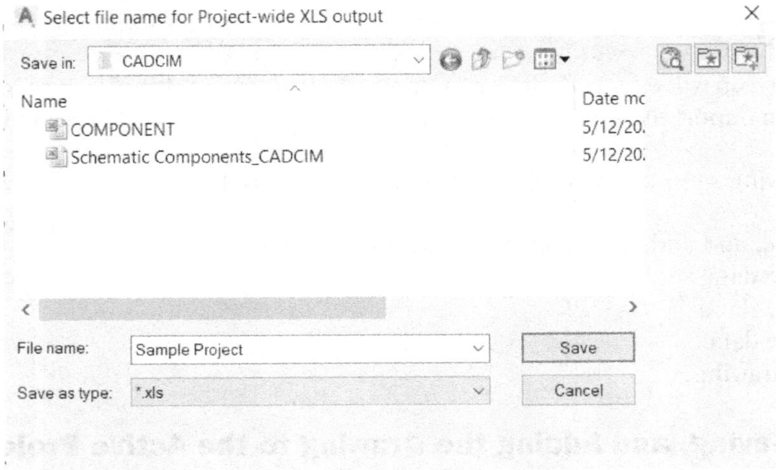

***Figure 13-26** The **Select file name for drawing's XLS output** dialog box*

5. Choose the **To Spreadsheet** tool again from the **Export** panel of the **Import/Export Data** tab; the **Export to Spreadsheet** dialog box is displayed.

6. In this dialog box, select the **Drawing settings** radio button and choose the **OK** button; the **Drawing Settings Data Export** dialog box is displayed. In this dialog box, make sure that the **Project** radio button is selected from the **Data export for** area and the **Excel file format [.xls]** is selected from the **Output format** area. Next, choose the **OK** button; the **Select Drawings to Process** dialog box is displayed.

7. In this dialog box, select all the tutorial files in the **CADCIM** project from the top list (Chapter 2 to Chapter 13) and then choose the **Process** button; all the tutorial files are shifted to the bottom part of the **Select Drawings to Process** dialog box. Next, choose the **OK** button; the **Select file name for Project-wide XLS output** dialog box is displayed.

8. In this dialog box, enter **CADCIM_drawing settings** in the **File name** text box and select the **C:\Users\User Name\Documents\AcadE 2022\AeData\Proj\CADCIM** from the **Save in** drop-down list. Next, choose the **Save** button; an Excel file with the name *CADCIM_drawing settings* is created at the specified location.

Modifying the Data

1. Open the Windows Explorer and browse to the location *C:\ Users\User Name\Documents\AcadE 2022\AeData\Proj\CADCIM*. Next, open the *C13_tut03_components* file from this location.

2. Modify the data in the **DESC1** and **DESC2** columns of this file, refer to Figure 13-27. Next, save and close the file.

	A	B	C	D	E	F	G
1	(PAR1 CH	FAMILY	TAGNAME	DESC1	DESC2	DESC3	(REF)
2	1	PB	PB1	NO	PUSH BUTTON		1
3	1	PB	PB1A	NC	PUSH BUTTON		1
4	1	CR	CR1	CTRL RELAY			1
5	2	CR	CR1	CTRL RELAY	NO		2
6	2	CR	CR1	CTRL RELAY	NC		3
7	1	LT	LT3	GREEN	OFF		3
8	2	CR	CR1	CTRL RELAY	NO		4
9	1	LT	LT4	RED	ON		4
10							
11							
12							

Figure 13-27 *The data changed in the* ***DESC1, DESC2*** *columns*

3. Open the *CADCIM_drawing settings* file from the *C:\ Users\User Name\Documents\AcadE 2022\AeData\Proj\CADCIM* location.

4. Change the data in the **SEC**, **SUBSEC**, and **SHDWGNAM** columns of this file, as shown in Figure 13-28. Next, save and close the file.

Importing the Data

1. Make sure the *C13_tut03* file is open in AutoCAD Electrical. Next, choose the **From Spreadsheet** tool from the **Import** panel of the **Import/Export Data** tab; the **Update Drawing from Spreadsheet File** dialog box is displayed.

2. In this dialog box, browse to the *C:\Users\User Name\Documents\AcadE 2022\AeData\Proj\CADCIM* in the **Look in** drop-down list and then select the *c13_tut03_components* file from the list displayed. Next, choose the **Open** button; the **Update Drawings per Spreadsheet Data** dialog box is displayed.

3. In this dialog box, select the **Active Drawing** radio button from the **Update drawings per spreadsheet data file. Process** area. Next, choose the **OK** button. You will notice that the description of components in the *C13_tut03* file is changed.

4. Choose the **From Spreadsheet** tool again from the **Import** panel of the **Import/Export Data** tab; the **Update Drawing from Spreadsheet File** dialog box is displayed.

5. In this dialog box, browse to the *C:\Users\User Name\Documents\AcadE 2022\AeData\Proj\ CADCIM* in the **Look in** drop-down list and then select the *CADCIM_drawing settings* file from the list displayed. Next, choose the **Open** button; the **Update Drawings per Spreadsheet Data** dialog box is displayed.

6. In this dialog box, select the **Project** radio button from the **Update drawings per spreadsheet data file. Process** area. Next, choose the **OK** button; the **Select Drawings to Process** dialog box is displayed.

	A	B	C	D	E
1	(DWGNAM)	SEC	SUBSEC	SH	SHDWGNAM
2	C02_TUT01	C2	1	01	201
3	C03_TUT01	C3	1	5	301
4	C03_TUT02	C3	2	5	302
5	C03_TUT03	C3	3	5	303
6	C03_TUT04	C3	4	5	304
7	C04_TUT01	C4	1	01	401
8	C04_TUT03	C4	3		403
9	C04_TUT04	C4	4	A	404
10	C05_TUT01	C5	1		501
11	C05_TUT02	C5	2		502
12	C05_TUT03	C5	3		503
13	C05_TUT04	C5	4		504
14	C06_TUT01	C6	1		601
15	C06_TUT03	C6	3		603
16	C06_TUT01_UPDATE	C6	1	01	601
17	C06_TUT03_UPDATE	C6	3	02	603
18	C07_TUT01	C7	1		701
19	C07_TUT02	C7	2		702
20	C07_TUT03	C7	3		703
21	C07_TUT04	C7	4		704
22	C08_TUT01	C8	1		801
23	C08_TUT02	C8	2		802
24	C08_TUT03	C8	3		803
25	C08_TUT04	C8	4		804
26	C09_TUT01	C9	1		901
27	C09_TUT02	C9	2		902
28	C10_TUT01	C10	1		1001
29	C10_TUT2	C10	2		1002
30	C10_TUT03	C10	3	01	1003
31	C10_TUT04	C10	4	02	1004
32	C10_TUT05	C10	5	03	1005
33	C11_TUT01	C11	1		1101
34	C'11_TUT02	C11	2		1102
35	C'11_TUT03	C11	3		1103
36	C12_TUT01	C12	1	01	1201
37	C12_TUT02	C12	2	01	1202
38	C04_TUT02	C4	2	01	402
39	C12_TUT03	C12	3		1203
40	C12_TUT04	C12	4	01	1204
41	C12_TUT05	C12	5	5	1205
42	C13_TUT01	C13	1		1301
43	C13_TUT02	C13	2		1302
44	C13_TUT03	C13	3		1303

***Figure 13-28** The data changed in the **SEC**, **SUBSEC**, and **SHDWGNAM** columns*

7. In this dialog box, select all the tutorial files in the **CADCIM** project from the top list (Chapter 2 to Chapter 13) and then choose the **Process** button; all the tutorial files are shifted to the bottom part of the **Select Drawings to Process** dialog box. Next, choose the **OK** button. If the **QSAVE** message box is displayed, choose the **OK** button in it.

   You will notice that all the tutorial files are opened, changed, and then closed one by one.

   To verify the changes in the drawing settings of the tutorial files, you need to follow the steps given next.

8. Right-click on any of the tutorial files in the **CADCIM** project; the shortcut menu is displayed. Choose **Properties > Drawing Properties** from this shortcut menu, refer to Figure 13-29; the **Drawing Properties** dialog box is displayed.

9. In this dialog box, the values in the **Drawing**, **Section**, **Sub-Section** edit boxes of the **Sheet Values** area are changed, refer to Figure 13-30.

Saving the Drawing File

1. Choose **Save** from the **Application Menu** or **File > Save** from the menu bar to save the drawing file.

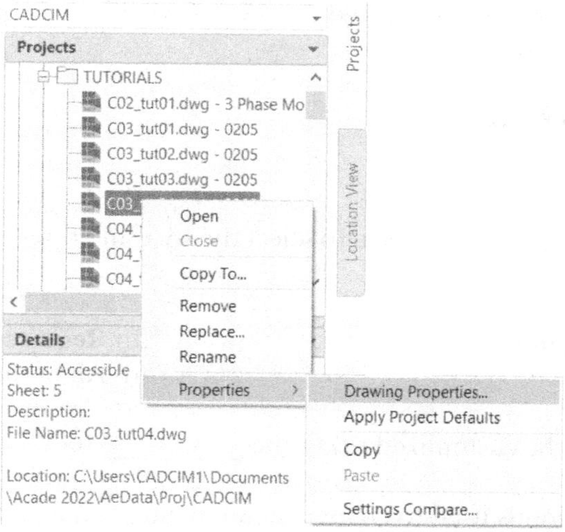

***Figure 13-29** Choosing **Properties > Drawing Properties** from the shortcut menu*

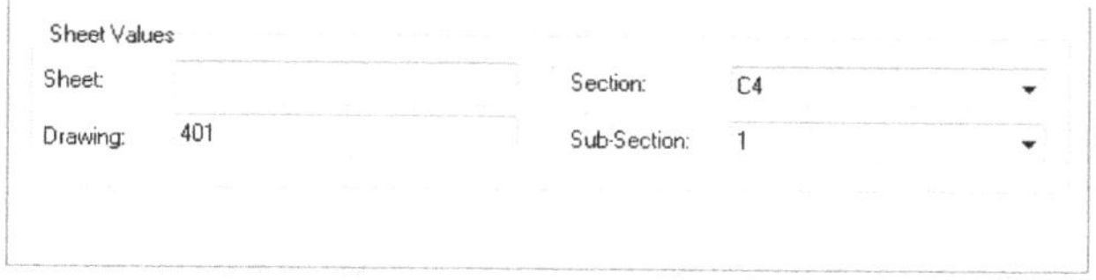

***Figure 13-30** Values changed in the **Drawing, Section, Sub-Section** edit boxes*

Self-Evaluation Test

Answer the following questions and then compare them to those given at the end of this chapter:

1. Which of the following dialog boxes will be displayed if you choose the **Symbol Builder** tool?

   (a) **Symbol Audit** (b) **Symbol Configuration**
   (c) **Select Symbol / Objects** (d) None of these

2. The **To Spreadsheet** tool is used to ______ data from the active drawing or project to an external file.

3. In AutoCAD Electrical, a symbol can be of any size and width. (T/F)

4. The **Icon Menu Wizard** tool is used to add or modify only the schematic symbol libraries. (T/F)

5. The AutoCAD blocks can be converted into AutoCAD Electrical intelligent symbols using the **Symbol Builder** tool. (T/F)

Review Questions

Answer the following questions:

1. Which of the following rollouts is used to select the style and direction of the wire connection attributes?

   (a) **Wire Connection** (b) **Required**
   (c) **Optional** (d) All of these

2. The icon menus can be customized using the __________ tool.

3. The __________ option is used to create a new circuit.

4 The __________ tool is used to import data from an external file to an active drawing or project.

5. The **Mark/Verify** tool is used to add an invisible mark on the components and wire of a drawing before sending it to the client. (T/F)

EXERCISES

Exercise 1

Create a new drawing with the name *C13_exer01.dwg* and insert a single-phase ladder with width = 15, spacing between rungs as 3, and number of rungs = 4. Next, create the symbol and add the attributes to it, as shown in Figure 13-31 and then save it as HDV1_SYMBOL. You will also insert the symbol that you created into the ladder, as shown in Figure 13-32.

(Expected time: 25 min)

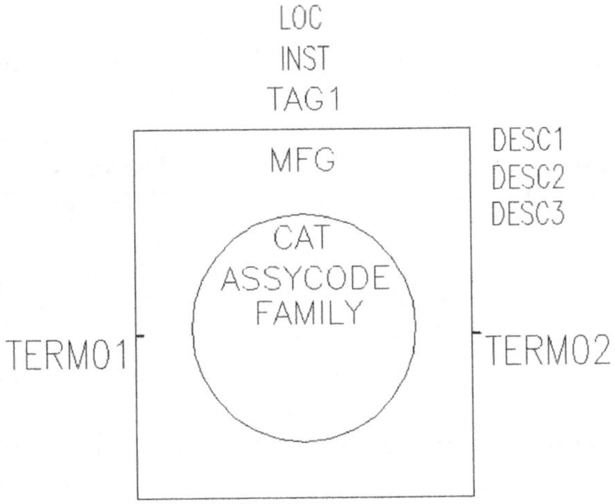

Figure 13-31 *Attributes added to the symbol*

Hint: Change the height of all attributes shown in Figure 13-31 to 0.1 in the **Symbol Builder Attribute Editor** palette. Also, insert TERM01 and TERM02 attributes at the mid-point of rectangle.

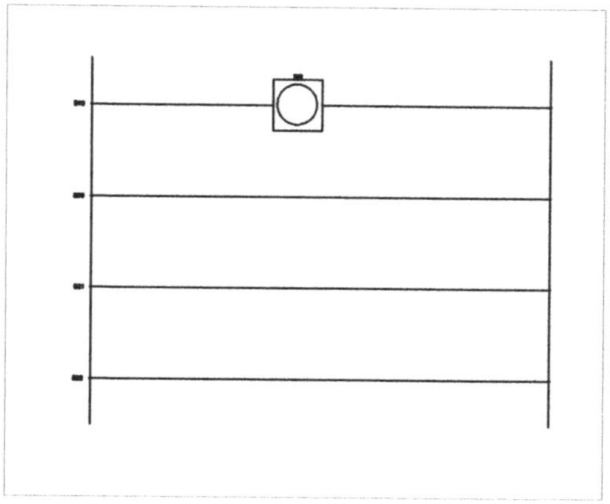

Figure 13-32 *Symbol inserted into the ladder*

Exercise 2

Open the *C13_exer01.dwg* drawing file. Use the **Icon Menu Wizard** tool to create a sub menu in the **Icon Menu Wizard** dialog box, as shown in Figure 13-33, and then add the icon of the symbol that you created in Exercise 1 to the **Miscellaneous** sub menu, as shown in Figure 13-34. Next, insert the symbol into the ladder, as shown in Figure 13-35. **(Expected time: 20 min)**

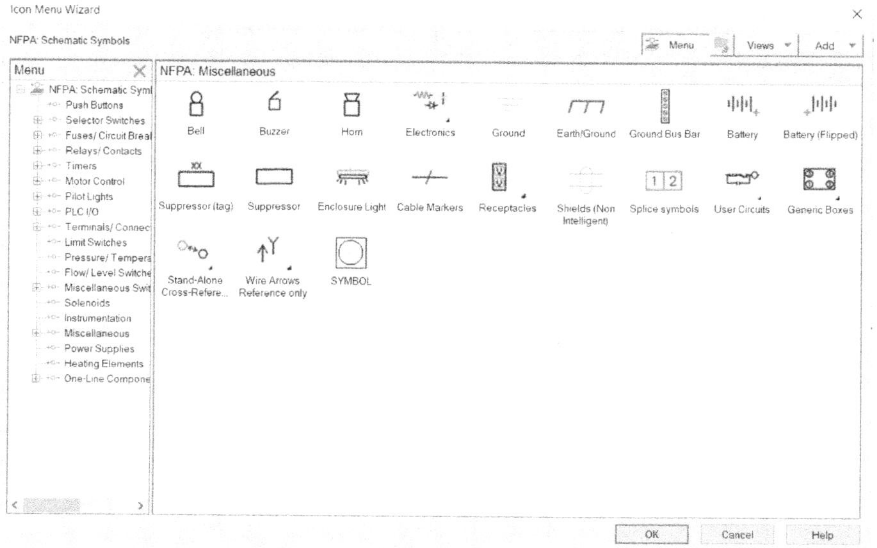

Figure 13-33 *Submenu created in the* ***Icon Menu Wizard*** *dialog box*

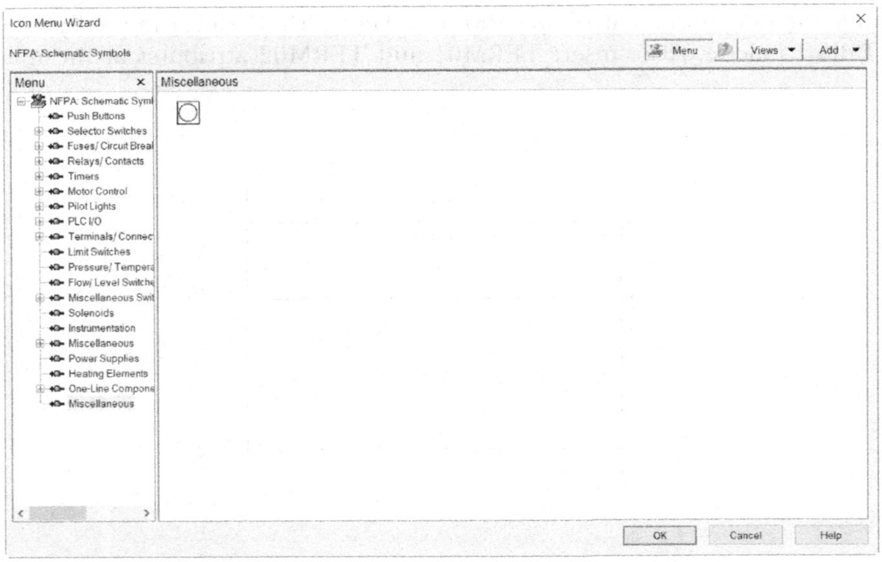

*Figure 13-34 The **Icon Menu Wizard** dialog box displaying the **Miscellaneous** submenu*

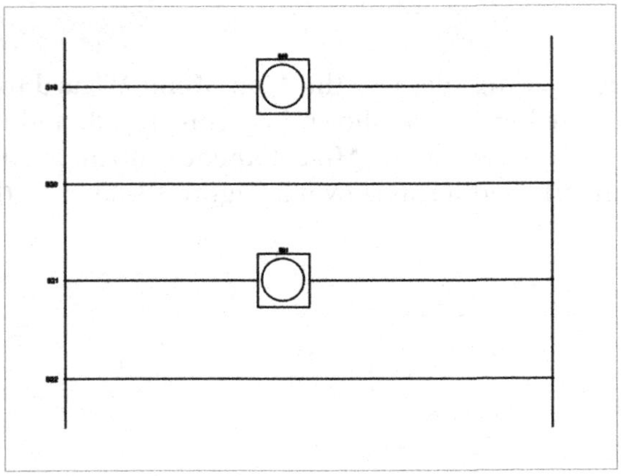

Figure 13-35 Symbol inserted into the ladder

Answers to Self-Evaluation Test

1. c, **2.** export, **3.** T, **4.** F, **5.** T

Student Project

Create a new project with the name **MYPROJECT**. Next, create a new template with the name *myprjtemplate,* refer to Figure SP-1. Using this template, create a new drawing *sprj1.dwg* file in the **MYPROJECT** project. Add description to this project, as shown in Figure SP-2. Insert a three phase ladder with starting line reference number as 101 and then insert user circuit 22 into the drawing at two points, refer to Figure SP-3. Edit components in the drawing to add catalog information and other details to them. Use the attribute editing tools to show, move, and resize the attributes of these components. Use the **Electrical Audit** tool to audit the drawing and correct the errors. Next, create and save format files for component report and bill of material report, refer to Figure SP-4. Now, use the **Automatic Reports** tool to group the format files and generate a cumulative report, refer to Figure SP-5.

Hint: To create *myprjtemplate* template, create a new drawing with *ACAD_Electrical* template. Next, use the **Edit** tool from the **Block** panel of the **Home** tab and change the logo in the block.

Figure SP-1 *The template for* ***MYPROJECT***

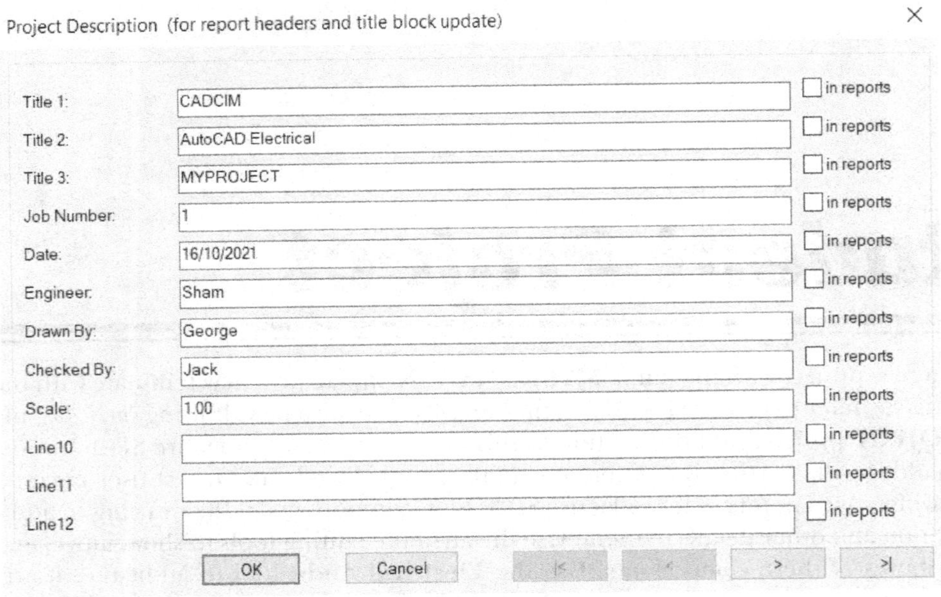

*Figure SP-2 The **Project Description** dialog box*

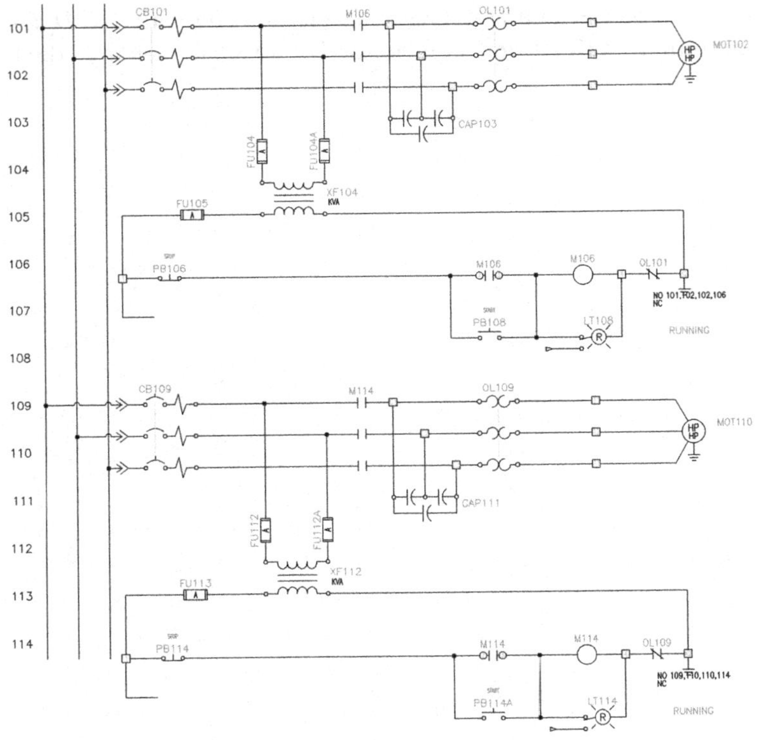

Figure SP-3 User Circuit 22 inserted into the drawing

***Figure SP-4** The **Report Format File Setup** and the **Bill of Material Data Fields to Report** dialog boxes*

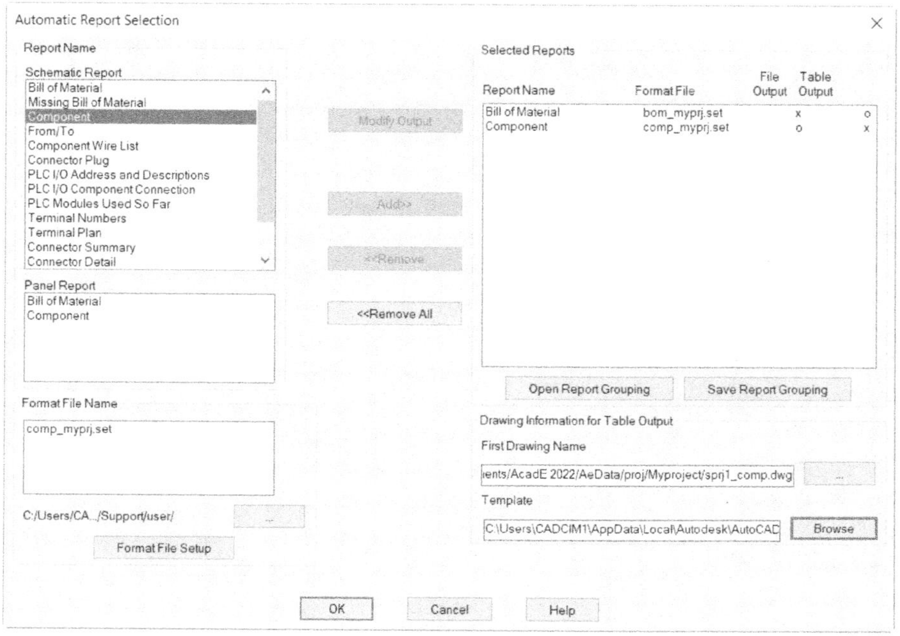

***Figure SP-5** The **Automatic Report Selection** dialog box*

This page is intentionally left blank

Index

Symbols

A

B

C

D

T

U

V

W

X

Z

Other Publications by CADCIM Technologies

The following is the list of some of the publications by CADCIM Technologies. Please visit *www.cadcim.com* for the complete listing.

AutoCAD Electrical Textbooks

- AutoCAD Electrical 2021: A Tutorial Approach, 2nd Edition
- AutoCAD Electrical 2020: A Tutorial Approach
- AutoCAD Electrical 2022 for Electrical Control Designers, 13th Edition
- AutoCAD Electrical 2021 for Electrical Control Designers, 12th Edition
- AutoCAD Electrical 2020 for Electrical Control Designers, 11th Edition
- AutoCAD Electrical 2019 for Electrical Control Designers, 10th Edition

AutoCAD Textbooks

- AutoCAD 2022: A Problem-Solving Approach, Basic and Intermediate, 28th Edition
- AutoCAD 2021: A Problem-Solving Approach, Basic and Intermediate, 27th Edition
- AutoCAD 2020: A Problem-Solving Approach, Basic and Intermediate, 26th Edition
- Advanced AutoCAD 2021: A Problem-Solving Approach, 3D and Advanced, 25th Edition

Autodesk Inventor Textbooks

- Autodesk Inventor Professional 2022 for Designers, 22nd Edition
- Autodesk Inventor Professional 2021 for Designers, 21st Edition

AutoCAD MEP Textbooks

- AutoCAD MEP 2020 for Designers, 5th Edition
- AutoCAD MEP 2018 for Designers, 4th Edition

AutoCAD Plant 3D Textbooks

- AutoCAD Plant 3D 2021 for Designers, 6th Edition
- AutoCAD Plant 3D 2020 for Designers, 5th Edition

Autodesk Fusion 360 Textbook

- Autodesk Fusion 360: A Tutorial Approach, 2nd Edition

Solid Edge Textbooks

- Solid Edge 2021 for Designers, 18th Edition
- Solid Edge 2020 for Designers, 17th Edition

NX Textbooks

- Siemens NX 2021 for Designers,14th Edition
- Siemens NX 2020 for Designers,13th Edition

NX Nastran Textbook

- NX Nastran 9.0 for Designers

SOLIDWORKS Textbooks
- SOLIDWORKS 2021 for Designers, 19th Edition
- SOLIDWORKS 2020 for Designers, 18th Edition

SOLIDWORKS Simulation Textbooks
- SOLIDWORKS Simulation 2018: A Tutorial Approach
- SOLIDWORKS Simulation 2016: A Tutorial Approach

CATIA Textbooks
- CATIA V5-6R2020 for Designers, 18th Edition
- CATIA V5-6R2019 for Designers, 17th Edition

Creo Parametric Textbooks
- Creo Parametric 7.0 for Designers, 7th Edition
- Creo Parametric 6.0 for Designers, 6th Edition

ANSYS Textbooks
- ANSYS Workbench 2019 R2: A Tutorial Approach
- ANSYS Workbench 14.0: A Tutorial Approach

Creo Direct Textbook
- Creo Direct 2.0 and Beyond for Designers

AutoCAD LT Textbooks
- AutoCAD LT 2020 for Designers, 13th Edition
- AutoCAD LT 2017 for Designers, 12th Edition

EdgeCAM Textbooks
- EdgeCAM 11.0 for Manufacturers
- EdgeCAM 10.0 for Manufacturers

Coming Soon from CADCIM Technologies
- Flow Simulation Using SOLIDWORKS 2021
- Finite Element Analysis Using ANSYS Workbench 2021

Made in the USA
Monee, IL
08 November 2026

68028672R00227